AF838921

Streaming Video in the Global South

Streaming Video in the Global South

Edited by
Shakti Jaising and Hadi Gharabaghi

BLOOMSBURY ACADEMIC
NEW YORK • LONDON • OXFORD • NEW DELHI • SYDNEY

BLOOMSBURY ACADEMIC
Bloomsbury Publishing Inc, 1359 Broadway, New York, NY 10018, USA
Bloomsbury Publishing Plc, 50 Bedford Square, London, WC1B 3DP, UK
Bloomsbury Publishing Ireland, 29 Earlsfort Terrace, Dublin 2, D02 AY28, Ireland

BLOOMSBURY, BLOOMSBURY ACADEMIC and the Diana logo are
trademarks of Bloomsbury Publishing Plc

First published in the United States of America 2026

Volume Editor's Part of the Work © Shakti Jaising and Hadi Gharabaghi, 2026
Each chapter © of Contributors, 2026

For legal purposes the Acknowledgments on p. ix constitute an
extension of this copyright page.

Cover design: Eleanor Rose
Cover image: 'Launch Event For Netflix's First Arab Language Original *Jinn*' © Juan
Naharro Gimenez / Getty Images

All rights reserved. No part of this publication may be: i) reproduced or transmitted in any form, electronic or mechanical, including photocopying, recording or by means of any information storage or retrieval system without prior permission in writing from the publishers; or ii) used or reproduced in any way for the training, development or operation of artificial intelligence (AI) technologies, including generative AI technologies. The rights holders expressly reserve this publication from the text and data mining exception as per Article 4(3) of the Digital Single Market Directive (EU) 2019/790.

Bloomsbury Publishing Inc does not have any control over, or responsibility for, any third-party websites referred to or in this book. All internet addresses given in this book were correct at the time of going to press. The author and publisher regret any inconvenience caused if addresses have changed or sites have ceased to exist, but can accept no responsibility for any such changes.

Library of Congress Cataloging-in-Publication Data
Names: Jaising, Shakti, 1976- editor | Gharabaghi, Hadi editor
Title: Streaming video in the Global South / edited by Shakti Jaising and Hadi Gharabaghi.
Description: New York : Bloomsbury Academic, 2026. |
Includes bibliographical references and index.
Identifiers: LCCN 2025029016 | ISBN 9798765135860 hardback |
ISBN 9798765135853 paperback | ISBN 9798765135884 pdf | ISBN 9798765135891 epub
Subjects: LCSH: Streaming video–Developing countries
Classification: LCC HD9697.V543 S77 2026 | DDC 384.55/8–dc23/eng/20250919
LC record available at https://lccn.loc.gov/2025029016

ISBN: HB: 979-8-7651-3586-0
ePDF: 979-8-7651-3588-4
eBook: 979-8-7651-3589-1

Typeset by Integra Software Services Pvt. Ltd.

For product safety related questions contact productsafety@bloomsbury.com.

To find out more about our authors and books visit www.bloomsbury.com
and sign up for our newsletters.

Contents

Illustrations

Figures

Table

Acknowledgments

A project of this nature and scope could not have been realized without the support of our editors at Bloomsbury, chiefly Stephanie Grace-Petinos, whose encouragement and patience with our process has been invaluable. For this we are grateful. Thanks also to the anonymous reviewers for their extremely helpful insights and feedback, and to Sophie Beardsworth, Anne Hunt, and Andrew White for their vital work on production and copyediting.

This volume is the result of the hard work, dedication, and responsiveness of our excellent contributing authors, some of whom joined the project in its final stage and worked within a tight time frame. Our sincere thanks to every one of them. We are also grateful to each other for our roles in this collaboration, and for the opportunity for mutual learning, even with the various professional and labor struggles we faced during the project's course.

Shakti Jaising and
Hadi Gharabaghi, January 2025

Introduction

Shakti Jaising and Hadi Gharabaghi

The rise of subscription video-on-demand (SVOD) services since the late 2000s has intensified the convergence of film and television. A number of excellent studies[1] have tracked the operations of the largest of these services—Netflix, Amazon Prime Video, and Disney+, which continue to grow and compete tightly in an increasingly digitized global mediascape. These studies have shown how the US-born streaming companies' platforms deploy algorithms to personalize content viewers encounter when they log onto nation-specific platform libraries. In some cases, the companies seek out and engage local talent for commissioned productions or "originals," all with the aim of establishing a foothold in diverse markets, and of ultimately ensuring endless growth and long-term profitability. Never before in the history of globalized mass media have capitalist entities from the global North or West operated so systematically and so vastly in other regions of the world.

Netflix is the most profitable[2] of the major streamers, and owing to its size as well as pioneering role in the field, the company has become *the* iconic SVOD service. It is also the most international in its reach. The growing list of "Netflix Nations"[3] means there are only a few territories, such as China, North Korea,

1 Key titles include Chuck Tryon, *On Demand Culture: Digital Delivery and the Future of the Movies* (New Brunswick, NJ: Rutgers University Press, 2013); Amanda D. Lotz, *Portals: A Treatise on Internet-Distributed Television* (Ann Arbor, MI: Maize Books, 2017); Ramon Lobato, *Netflix Nations: The Geography of Digital Distribution* (New York: NYU Press, 2019); Kevin McDonald and Daniel Smith-Rowsey, eds., *The Netflix Effect: Technology and Entertainment in the 21st Century* (New York: Bloomsbury, 2016); Amber M. Buck and Theo Plothe, *Netflix at the Nexus: Content, Practice, and Production in the Age of Streaming Television* (New York: Peter Lang 2019); and Amanda Lotz, *Netflix and Streaming Video: The Business of Subscriber-Funded Video on Demand* (New York: Polity, 2022).

2 Brad Adgate, "Streaming Units Are Beginning to Be Profitable. Here's Why," *Forbes*, October 14, 2024, https://www.forbes.com/sites/bradadgate/2024/10/11/streaming-units-are-starting-beginning-to-be-profitable/.

3 Lobato, *Netflix Nations.*

and Syria, that remain outside its purview. Today, international markets account for almost 70 percent of Netflix's paid subscriber base,[4] and hence for a large share of its overall profits. Netflix's international success has in turn spurred the global expansion of other US streamers.

The international game of streaming giants like Netflix is crucial to understand. At the same time, to study audiovisual streaming's relationship to globalization primarily from the vantage point of the top US corporations is to risk reinforcing these corporations' own self-serving narratives, including their implied technological determinism.[5] Netflix, for instance, has long promoted itself "as a tech company, as opposed to a media company, in order to articulate a brand identity built around disruption… of scheduled television viewing."[6] The company's self-promotion as a disruptive tech company can fuel the impression of Netflix as an autonomous agent, thereby obscuring its location within distinct media cultures—including the one from which it emerges and those into which it installs itself.

This book is not solely about Netflix, but the company figures as an emblematic case for understanding streaming video as (to invoke Raymond Williams) both a "technology and cultural form"[7] that has an increasing impact globally. Because the majority of scholarship on Netflix and other SVOD services has tended to focus on media cultures of the advanced capitalist world or global North, this book chooses to focus instead on the relatively less developed global South, where the vast majority of the world's population lives. By bringing together the categories of "streaming video" and "global South," it contributes to growing interest[8] in mapping the international scope of streaming beyond simply North America and Europe. Individual chapters explore media environments across Asia, Africa, and Latin America; provide close analysis of national and regional

[4] Julia Stoll, "Number of Netflix Paying Streaming Subscribers Worldwide as of 3rd Quarter 2024, by Region," Statista.com, October 18, 2024, https://www.statista.com/statistics/483112/netflix-subscribers/.

[5] For more on the problem with technological determinism, see Raymond Williams, *Television: Technology and Cultural Form* (New York: Routledge, 1974).

[6] Timothy Havens and Ryan Stoldt, "Streaming Channel Brands as Global Meaning Systems," in *From Networks to Netflix: A Guide to Changing Channels*, ed. Derek Johnson, (London: Routledge, 2018), 205.

[7] Williams, *Television.*

[8] Key works that have attended to southern streaming ecologies in some depth include Derek Johnson, ed., *From Networks to Netflix: A Guide to Changing Channels* (London: Routledge, 2018); Lobato, *Netflix Nations*; Stefano Baschiera and Alexander Fisher, eds., *World Cinema on Demand: Global Film Cultures in the Era of Online Distribution* (London: Bloomsbury, 2022); Amanda Lotz and Ramon Lobato, eds., *Streaming Video: Storytelling across Borders* (New York: NYU Press, 2023); Philippe Bouquillion, Christine Ithurbide, and Tristan Mattelart, eds., *Digital Platforms and the Global South: Reconfiguring Power Relations in the Cultural Industries* (New York: Routledge, 2024).

contexts such as Brazil, Egypt, Lebanon, Mexico, the Philippines, and South Asia; and attend both to culturally specific dynamics and to relationships between local and global streamers. By looking at digitized audiovisual culture from the perspective of the global South, the book creates room for greater critical distance from the self-created identities and narratives of the larger transnational streamers like Netflix that have so far tended to emanate from the north.

The book's chapters toggle between textual and political–economic approaches to streaming video, thereby providing interpretations of key texts and genres that circulate via streaming as well as analysis of the infrastructures shaping production, distribution, and reception. The chapters also consider an array of audiovisual formats, including feature-length fiction films, art cinema, documentary, and television series. Taken as a whole, their contributions illuminate how streaming video services operate within diverse cultural contexts, interact with existing forms, and relate to, or transform, established modes of media engagement and habitual ways of seeing.

The North–South Dynamic

In recent years much scholarly work within the humanities and social sciences has turned to analyzing shifting dynamics between the global North and South. Typically, the term "global North" is used to refer to advanced capitalist economies of Western Europe, North America, and East Asia. The "global South," meanwhile, refers mostly to less advanced economies of South and West Asia, Africa, and Latin America, many of which also share a history of resistance against European and US imperialisms. "Global South," which is used more frequently than "global North," gained currency toward the end of the Cold War. Many scholars trace the rise of this term to the publication of the 1980 Brandt Commission report[9] and to the subsequent work of the South Commission, a collective of developing economies that formed to counter the consolidation of economic and political power by G7 nations.[10] Within academia, "global South" started becoming a buzzword in the 1990s, and over the course of the first two decades of the twenty-first century it has increasingly taken the place of "Third

[9] Independent Commission on International Development Issues and Willy Brandt, *North–South: A Program for Survival* (Cambridge, MA: MIT Press, 1980).

[10] For more on the South Commission and the history of North–South relations, see Vijay Prashad, *The Poorer Nations: A Possible History of the Global South* (New York: Verso, 2013).

World" and is often used interchangeably with "non-West" in discourse on postcolonial or developing economies.[11]

The vocabulary of global North and South has been controversial for a few reasons: first, "North" and "South" can suggest homogenous entities, thereby flattening internal differences within these categories. For instance, the term "South" glosses over the differing interests of, on the one hand, dominant nations, like Brazil and India, and on the other hand geopolitically marginalized ones, like Somalia or Bangladesh.[12] Second, a focus on the power differential between "North" and "South" can mask asymmetries and uneven development within both Northern and Southern nations, as well as realities like growing economic inequality and the turn to cultural nationalism that cut across the North–South divide.[13] Third, shifts in geopolitical hierarchy brought about by formations like BRICS, on one hand, and the collapse of prior infrastructures via processes like BREXIT, on the other, demand that we rethink the North–South binary within the context of contemporary globalization.

This volume, in other words, recognizes some of the problems with using the "global South" as a framing category. If it nevertheless chooses to explore the terrain of streaming video by drawing on this category, this is, first, to call attention to media ecologies within developing and emerging economies, which have, until recently, received far less scholarly as well as public attention relative to their counterparts from the advanced capitalist world. Second, a focus on the South allows for insight into what streaming has meant specifically for media ecosystems in the formerly colonized world, where conceptions of "national cinema" and "national television" have long been crucial to imagining national unity and independence from imperial regimes, even if they have also, at times, become entangled within experiments in plutocratic or autocratic dominance.[14] Where the rise of streaming video fits

[11] For more on the rise of "Global South" as a conceptual category, see Arif Dirlik, "Global South: Predicament and Promise," *Global South* 1, no. 1 (2007): 12–23; and Sinah Theres Kloß, "The Global South as Subversive Practice: Challenges and Potentials of a Heuristic Concept," *Global South* 11, no. 2 (2017): 1–17.

[12] For more on this argument, see Prashad, *Poorer Nations*, and Nina Schneider, "Between Promise and Skepticism: The Global South and Our Role as Engaged Intellectuals," *Global South* 11, no. 2 (2017): 18–38.

[13] For this argument, see Shakti Jaising, *Beyond Alterity: Contemporary Indian Fiction and the Neoliberal Script* (Liverpool: Liverpool University Press, 2023).

[14] For more on national cinema and national television in the postcolonial world, see, for instance, Lila Abu-Lughod, *Dramas of Nationhood: The Politics of Television in Egypt* (Chicago: University of Chicago Press, 2005); Shanti Kumar, *Gandhi Meets Primetime: Globalization and Nationalism in Indian Television* (Urbana: University of Illinois Press, 2010); Arvind Rajagopal, ed., *The Indian Public Sphere: Readings in Media History* (New Delhi: Oxford University Press, 2009); and Rudin's chapter in this volume on national cinema in the Philippines.

within histories of media-enabled postcolonial nationalism and how it relates to the prevailing North–South dynamic are key concerns that underlie several of the book's chapters.

But to call attention to the global South as a framing category is not to thereby imply that the North–South dynamic is the only one shaping contemporary media ecosystems. This dynamic is necessary but not sufficient for grasping the complexity of specific national and regional contexts. Nor should attention to the North–South dynamic lead us to necessarily assume a state of absolute dominance by an imperialist North over a vulnerable or resistant South. While, for instance, there is no denying the US-born Netflix's influential role in shaping the global SVOD landscape, it is important to note that such services exist even in nations where Netflix and other US companies have not been able to enter the market: for instance, iQIYI in China, Manbang in North Korea, and Filimo, Namava, and Aparat (an equivalent platform to YouTube) in Iran.

In the case of Iran, services like Namava and Filimo—popular sites with many notable original productions and television series[15]—had already launched by 2016, when Netflix attempted to enter the market.[16] This did not come to fruition, however, as the Iranian government, like the governments of Indonesia and Kenya, objected to Netflix's saturation with "morally inappropriate content."[17] Subsequently, Netflix titles began to become available unofficially within the country, where content piracy is not uncommon (Iran is not a signatory to the Berne Convention of international copyrights laws)[18] and where subscription services are not easily affordable for much of the population. Like Hollywood films and US television titles, Netflix content became accessible in Iran by changing IP addresses and purchasing VPN services. Alternative sites[19] also

[15] Mehdi Semati and Elmira Moosavitabar, "The State and Streaming Services in Iran," in *Triaging the Streaming Wars*, ed. Robert Alan Brookey, Jason Phillips, and Timothy Pollard (New York: Routledge, 2024), 68.

[16] Center for Human Rights in Iran, "Netflix Opening to Iran Ends State Broadcasting's Decades of Monopoly," January 11, 2016, https://iranhumanrights.org/2016/01/netflix-filtered/.

[17] Khabar Gozāri-e Mehr [MehrNews], "[The Reaction of The Office of Audio and Video Regulation Authority to 'Netflix' Offer of Service in Iran]," May 5, 2016, https://www.mehrnews.com/news/3607742/واکنش-مرکز-ملی-فضای-مجازی-به-ارائهسرویس-نت-فلیکس-در-ایران/.; Rhea Yasmine and Alis Haizan, "Indonesia Plans to Impose Censorship Rules on Streaming Services; Analyst Says Move May Turn Viewers to Illegal Sites," August 23, 2023, https://www.channelnewsasia.com/asia/indonesia-streaming-services-ott-platform-netflix-disney-censorship-media-3718571; Sam Kisika, "Netflix Stops Streaming LGBTQ-Specific Movies in Kenya," *Washington Blade*, June 21, 2023, https://www.washingtonblade.com/2023/06/21/netflix-stops-streaming-lgbtq-specific-movies-in-kenya/.

[18] Semati and Moosavitabar, "State and Streaming Services in Iran," 63.

[19] The sites igame.ir, ariapay.me, and sccsell.ir are among a group of sites that offer such services.

stepped in to play a mediating role, by enabling registration for a Netflix account, paid in Iranian currency at a much more affordable rate than what the US streamer might have officially charged. In 2022, the government shut down the mediating sites enabling Netflix content, and the companies owning these sites accused Filimo and Namava of filing complaints with the government.[20] But as popular as Netflix might appear to be among sections of the Iranian population, the company's unofficial presence in a market that is already populated with local streaming companies complicates the archetypical opposition between (Northern) dominance and (Southern) subordination.

Another reason why this opposition might not easily map onto specific streaming landscapes is that even in markets where US streamers do officially prevail, their market presence does not always lead to dominance, as it is contingent on effective alliances with local broadcast or telecommunication companies as well as talent. Furthermore, the power of these transnational streamers, even with effective local alliances, is far from uncontested. To continue with the example of Netflix: in order to enter into other markets in the West Asia and North Africa region, the company had to partner with the Dubai-based satellite TV company Orbit Showtime Network (OSN).[21] And since entering these markets, it has had to compete with an array of streaming services. In crucial ways, local and regional services—like the Arabic-language streamer Shahid or the Spanish-language ViX in Mexico and Latin America—are not simply secondary to Anglo-American companies like Netflix. These local and regional providers often prove more influential in terms of reaching audiences and shaping the kinds of content produced within the linguistic and cultural contexts they serve.

In India, where Netflix launched in 2016, it has had to partner with local telecommunication companies. Recognizing the enormous economic divide in the country, and considering the importance of mobile phones for not only communication but also entertainment purposes, Netflix introduced a cheap, mobile-only subscription plan for smartphones and tablets. At the time of its entry, executives aimed to capture a new market with the potential of providing the "next 100 million" subscribers.[22] However, as of mid-2024 the reported

[20] Semati and Moosavitabar, "State and Streaming Services in Iran," 68.

[21] Netflix, "OSN Signs Middle East's First Partnership Deal with Netflix," February 18, 2028, https://about.netflix.com/en/news/osn-signs-middle-easts-first-partnership-deal-with-netflix.

[22] Nyay Bhushan, "Netflix CEO Reed Hastings on Why the 'Next 100 Million' Subscribers Will Be 'Coming from India,'" *Hollywood Reporter*, February 26, 2018, https://www.hollywoodreporter.com/tv/tv-news/netflix-ceo-reed-hastings-why-next-100-million-subscribers-will-be-coming-india-1088301/.

number of actual subscribers is closer to thirteen million. Netflix is only the number four subscription-based streaming service in India due to its incapacity to attract a broad range of viewers outside of the "premium market," which consists of largely English-speaking, urban, middle- and upper-class audiences. In recent years, the company has expanded its partnerships with telecom and broadband companies, leading to new kinds of prepaid tariff plans that come bundled with streaming video subscription. In addition, it has grown its support for the production and distribution of Indian films, high-budget series, as well as stand-up routines by Indian comedy stars,[23] in an attempt to expand its appeal within South Asia as well as across the diaspora.

Netflix's various struggles across West and South Asia suggest that the company's size and global reach need not necessarily translate into hegemony over global South markets, where subscription-based services tend to be something of a luxury, and where local services may often have an advantage, owing to a mix of political and cultural factors. Further, as the company's attempts at expanding its audience base compel it to support and disseminate (including in the global North) the work of producers and talent from the South, this raises the question of the extent to which the rise of streaming giants like Netflix signals a new wave of Western, media-based cultural domination. Or does this rise in fact complicate what UNESCO (referring to late twentieth-century broadcast and satellite television) described as the "one-way traffic" of media from richer to less developed nations of the world?[24]

The early chapters of this book (Navarro, Vukoder, Rios et al.) engage most directly with questions of "cultural imperialism"[25] and "platform imperialism,"[26] primarily through the example of Netflix and its relationship with, and operations within, the global South. Later chapters attend closely to postcolonial media cultures in contexts like Lebanon, South Asia, and the Philippines (Mouawad and Nassour, Vitali, Rudin) and to emblematic televisual genres from the South, like the Egyptian *musalsalat* and the Mexican *telenovela* (Korneev, Smith). The final chapters investigate the character of global "cinema" in the

23 Gargi Sarkar, "Can Netflix Capitalise on India Boom to Upset JioCinema Plans?," Inc42, August 12, 2024, https://inc42.com/features/how-is-netflix-rewiring-its-india-business/.

24 Hifzi Topuz, "Television's One-Way Traffic," *UNESCO Courier* 4 (1977): 16–17, https://unesdoc.unesco.org/ark:/48223/pf0000046566.

25 For more on cultural imperialism, see, for instance, Tanner Mirrlees, *Global Entertainment Media: Between Cultural Imperialism and Cultural Globalization* (New York: Routledge, 2013).

26 Dal Yong Jin, *Digital Platforms, Imperialism and Political Culture* (London: Routledge, 2015).

age of streaming—through questions of how subtitling engenders translation and mistranslation (Kim) and how US streaming services like Netflix interrupt or reinforce the historical sidelining of films from Africa and Palestine (Fisher, Alkassim).

A recurring issue across many of the chapters pertains to who can afford subscription-based services in the global South and also who is able to write, act, direct, or take on other creative roles within streamers' "original" productions. In other words, as streaming companies compete in offering an endless flow of bingeworthy content, what does this mean for the agency of both audiences and creators, including those from historically marginalized groups? By addressing such questions on the transnational, regional, and local levels, the chapters contribute to a reckoning with "streaming culture,"[27] which might be seen as a particularly illuminating expression of the ethos as well as inequities of contemporary global capitalism.

The Contents of This Volume

This volume coalesces the research of a diverse group of film and media scholars who investigate audiovisual streaming's restructuring of production, distribution, and reception cultures across the global South. The opening chapters prompt reflection on the extent to which the present conjuncture marks a new phase in US media's material and cultural domination over much of the world. Vinicius Navarro weighs in on the North–South dynamic through the category of the "global Netflix documentary," which has been crucial for constructing Netflix's brand identity as a cosmopolitan platform with global reach (across over 190 nations).[28] This category of standardized documentary "travels" the world in search of commodifiable local color, thereby becoming a metonym for Netflix itself, Navarro argues. Like the Discovery Channel documentary of the 1980s and 1990s, the twenty-first-century global Netflix documentary—which includes nature docs like *Tales by Light* (2015) and food documentary series like *Street Food* (2019)—invokes a familiar, explicitly colonial, touristic gaze, so that the global South emerges in it as an exotic site of alterity and cultural

[27] David Arditi, *Streaming Culture: Subscription Platforms and the Unending Consumption of Culture* (Leeds: Emerald, 2021).

[28] World Population Review, "Netflix Users by Country 2024," https://worldpopulationreview.com/country-rankings/netflix-users-by-country.

difference. By analyzing how this colonial imagination domesticates difference, and by pointing to how Netflix's algorithmic curation might limit, rather than open up to, plurality, Navarro pushes against the company's self-promotion as a harbinger of global cultural diversity.

In the subsequent chapter, Bret Vukoder historicizes the North–South dynamic that Navarro unpacks via the global Netflix documentary. Also attending to global cosmopolitanism, Vukoder exposes how this ideology operated in the Cold War era, when it provided a liberal cover for the United States Information Agency (USIA), the superpower's massive transnational media/propaganda operation, whose purpose was to sell US policy and culture by "telling America's story" globally. Like Netflix, the USIA sought to localize its US-aligned audiovisual content by engaging cinematographers, directors, researchers, and other crew members from within the national contexts in which it established production and distribution infrastructures. If the USIA was selling internationally a government-sponsored national "brand," then Netflix now sells profit-driven propaganda on the global stage—making the government media agency a useful, even if unexpected, paradigm for understanding the streaming video corporation. While rejecting a simple one-to-one correspondence between Netflix and the USIA, Vukoder shows why this global agency and its considerable film archives might nonetheless serve as a historiographic tool with which to grasp the systemic logic underlying Netflix's transnational operations.

Together, the first two chapters suggest that the dynamics characteristic of the North's unequal relations with the South live on today. The third chapter further uncovers and explains these dynamics through the example of Brazil, Netflix's second-largest market after the US.[29] If an argument for US media-enabled cultural imperialism is to be made, then Brazil is where it might find compelling support. Analysing Netflix original series between 2016 and 2023, Daniel Rios, Melina Meimaridis, and Daniela Mazur show how, on one hand, the US company's entry into Brazil offers opportunities for new forms of storytelling and new creative voices to be heard, relative to the ossified production culture consolidated under conglomerate Grupo Globo's media monopoly. On the other hand, Netflix's US-centric production norms and standardized series format exercise pressure on Brazilian storytellers, often compelling stories to "feel local" but nonetheless fit within parameters the company deems "universal." Netflix maintains strong oversight over the production of Brazilian shows like the postapocalyptic class

[29] Ibid.

allegory *3%* (2016–20). Storytellers, in turn, feel pressured to narrate series like *Sintonia*, about favela youth, by invoking aesthetic paradigms and models that have already been tried and tested in the global sphere (e.g., those deployed in the international film hit *City of God* [2002]). As a result, when creating content designed for Netflix's platform, local producers often craft narratives reliant on easily legible stereotypes about Brazil and Brazilianness. Beyond this, Netflix engages in practices of "platform imperialism" by ascribing authorship rights to the contracting production company, at the expense of screenwriters and other creators.

The case of Brazil brings into relief a common thread running through many of the chapters: US SVOD services like Netflix are often welcomed in areas of the global South as an alternative to state as well as private-sector media monopolies and oligopolies. At the same time, these services are also seen as normalizing new, restrictive production codes that impinge on creative and cultural autonomy. A standardized production model is often imposed across genres and cultural contexts, leading to the sense that streaming video is yet another industry of mass production. Netflix in particular functions as what one *New Yorker* magazine article described as a "universal power converter, plugging in and adapting successful show formats to different parts of the world," leading to the conception of an "ideal Netflix show," which Bela Bajaria, the company's chief content officer, sees as a "'gourmet cheeseburger,' offering something 'premium and commercial at the same time.'"[30]

If the first three chapters support the notion that US streamers are reproducing in their relationship with the global South power dynamics characteristic of cultural imperialism, then chapter 4 somewhat complicates this notion through its attention to the West Asia and North Africa region, and Lebanon's mediascape in particular. Wissam Mouawad and Pamela Nassour point out that, while Netflix was the first transnational streamer to launch in 2016 (followed by Amazon Prime Video and Disney+), it was the Arabic-language service Shahid that almost a decade earlier led the rise of streaming video in the region. By charting a genealogy of televisual culture in Lebanon, the authors show how the entry into this particular market of key global, regional, and local SVOD services has depended on the nation's liberal telecom laws and relatively free internet access. The current streaming landscape in Lebanon is in constant flux, with local platforms like M.Media ceasing to operate—victims of the fragile political

30 Rachel Syme, "How Much Netflix Can the World Absorb?," *New Yorker*, January 9, 2023.

environment—and others like Netflix and OSN+ entering into collaboration. But platforms like Shahid that cater to the taste and reception habits of many, if not all, populations in the region have a higher chance of attracting a consistently large share of the audience base. Shahid also benefits from the preestablished network of its parent company—the Saudi, state-owned media conglomerate MBC—and from attracting peak viewership around Ramadan, when families across many Arab countries gather to watch television serials. Netflix has tried to join in this tradition, by, for instance, releasing its series *Jinn* (2019), a Jordanian fantasy drama, during Ramadan. However, this series fell short of meeting audience expectations, and in some contexts also proved controversial for its narrative choices. On the whole, Lebanon in particular, and the West Asia and North Africa markets in general, present a markedly different view relative to Brazil of how the North–South dynamic plays out in the arena of streaming video.

The following chapters focus on trends and on-the-ground realities of audiovisual streaming within as well as across vast regions of the global South. Valentina Vitali points to a variety of streaming services, or OTT (over-the-top) platforms, emerging in South Asia, including large players like Zee5 and smaller ones like OHO Gujarati, Samuh Bhutan, and Cinema Ceylon. By opening up new avenues for audiovisual distribution within the region, these platforms, Vitali argues, are allowing for fresh interaction between national cultures that have long been divided by intraregional hierarchies and conflicts. Within this climate, platforms like Netflix, Amazon, and Zee5 also introduce an expanded range of opportunities for women filmmakers and for women-centered content—all of which potentially lends legitimacy to these platforms, while simultaneously mediating a new kind of feminist politics in South Asia. Drawing on feminist philosopher Nancy Fraser's distinction between "affirmative" and "transformative" politics, Vitali analyzes the aesthetic strategies of a set of women-centered OTT-produced and -distributed South Asian series. Ultimately, she uses the series *Churails* (2020) and *Qatil Haseenaon ke Naam* (2021)—both released via the Indian streamer Zee5—to respectively exemplify the difference between representations that simply open up space for "affirmative" identification and those that also generate "transformative" responses to gender injustice.

Paul Julian Smith's analysis in the following chapter also attends to women-centered content—in this case from Mexico. Smith shows how the centrality of female protagonists has long been a feature of Mexican television drama, especially of telenovelas produced by legacy media. The merger in recent years of

Mexican conglomerate Televisa with US Spanish-language network Univision, and the emergence out of the merged TelevisaUnivision of streaming service ViX, has led to an infusion of capital and an expanded audience base, enabling Mexican creators to innovate with both form and content. Within this context, the telenovela has spawned new subgenres (with education on social issues built into the storytelling), but it has also left its imprint on other genres popularized by streaming, including the political thriller and the bio series, which were once male focused. Smith's analysis of the Mexican mediascape raises the possibility of two-way, rather than one-way, traffic between North and South—by suggesting that it may not just be Northern streamers like Netflix who affect media production cultures in places like Mexico but also Mexican genres like the telenovela that might be influencing aesthetic transformations in genres with global currency. Furthermore, as against the tendency to focus primarily on the agency of streaming video services, who are often seen as revolutionizing legacy TV, Smith points to legacy media's enduring role in the present—and to how large sections of the population in Mexico, including those who cannot afford access to subscription-based platforms, remain loyal to the set rhythms and scheduled programming of broadcast television.

Like Smith, Egor Korneev suggests not absolute rupture but rather continuity between linear and digital television cultures. Korneev's chapter engages the history of state-produced national television in Egypt and shows how this national television's classic product from the 1980s and 1990s—the *musalsalat*, akin to the Latin American telenovela—is kept alive via YouTube. The Egyptian public service broadcaster's uploading of *musalsalat* from an earlier era onto its YouTube channel is enabled by YouTube being an advertising-supported, free, content-sharing streaming as well as social media platform. Even with the entry into the Egyptian market of SVOD services like Netflix and Shahid, many viewers desire to rewatch classic *musalsalat*. Attending to viewer comments on the state media's YouTube channel, Korneev concludes that audiovisual streaming (including both SVODs and YouTube) has not led to a radical break with the legacies of state-controlled Egyptian television but rather has reshaped an earlier species of televisual content into an object of nostalgia. This nostalgia is partly for a past that is perceived to have allowed for greater social cohesion, as well as for a mode of collective spectatorship (associated with the now largely defunct state-funded television) that contrasts with the individualized modes of viewing normalized in the streaming era.

If Korneev's chapter looks at the fate of national television in the era of streaming, then Daniel Rudin turns to that of "national cinema," in this case of the Philippines. Rudin argues that the contradictions that have from the beginning informed the concept of national cinema come to the surface with the rise of audiovisual streaming. Key to the chapter's argument is Filipino filmmaker Mikhail Red's award-winning psychological horror film *Deleter* (2022), which thematizes the trauma experienced by online content moderators in the Philippines who work for large social media platforms. The film is significant for its commentary on the Philippines' place within contemporary global capitalism, but it highlights as well director Mikhail Red's self-conscious aesthetic and distribution strategies, which rely simultaneously on the infrastructures of national cinema in the Philippines (which guarantee a sizable domestic audience) and of transnational streaming services like Amazon Prime Video (which make the film available to the nation's diaspora and beyond). Rudin argues that Red's openness to transnational capital undercuts what national cinema was designed to protect—that is, the economic autonomy of Filipino cinema. But the filmmaker's approach to film financing and distribution also ends up exposing problems inherent in the discourse of "national cinema," which emerges out of an elite-led postcolonial nationalism that begins as economic protectionism and over time deteriorates into narrow cultural nationalism.

Rudin's discussion of the relationship between streaming and national cinema leads into Hiju Kim's analysis of subtitled cinema in the age of streaming. As its chief case study, the chapter offers a close analysis of the critically acclaimed, Mubi-distributed *Decision to Leave* (2022) by South Korean director Park Chan-wook. The film is an emblematic example of contemporary Korean cinema that travels internationally via streaming services, but it is also a self-conscious commentary on the process of subtitling and its attendant translations and mistranslations. By moving between Korean, written and spoken Mandarin Chinese, and English, the film illuminates and expresses awareness of a core quality of subtitled cinema—that is, how this cinema creates, as Kim puts it, a multilingual, immersive experience for viewers, while sacrificing ambiguity and cultural complexity. Translation emerges in *Decision to Leave* as not simply an agent of clarity; rather, through the various mistranslations that are part of the plot, the film raises questions about the nature of mediated communication. But alongside this self-conscious storytelling, the film's translation into multiple languages for streaming audiences adds another extratextual layer to its commentary. Ultimately, by focusing on subtitling as a crucial process catalyzing the "Korean Wave" of popular culture, and by highlighting the roles of professional as well as fan subtitlers, Kim sheds light on a process that is rarely attended to but that is integral to the transmission of audiovisual content across national and linguistic boundaries.

Following Kim's analysis of subtitling practices in streaming-distributed cinema, Alexander Fisher's chapter considers the role of streaming video in opening up new production and distribution opportunities for hitherto marginalized cinemas from Africa. While global streaming video platforms like Netflix, Amazon Prime Video, and Apple TV+ have provided access to releases from relatively underdeveloped African film industries, they have also—by grouping together diverse African cinemas through category tags like "Made in Africa" and "African Movies"—contributed to flattening out differences between various cinemas emerging out of the African continent. Drawing on the example of Netflix UK, Fisher highlights a tendency, owing to the streamer's preference algorithm, to conflate various African cinemas. This at once reinforces and undermines the traditional othering of African cinemas, and impacts as well on how African filmmakers approach their work. Fisher's close analysis of three Anglophone African crime films that feature on Netflix UK—*Nairobi Half Life* (dir. David Gitonga, Kenya, 2012), *The Girl in the Yellow Jumper* (dir. Loukman Ali, Uganda, 2020), and *Gold Coast Lounge* (dir. Pascal Aka, Ghana, 2020)—points to their shared postmodern self-awareness as African films that rely on Hollywood genre conventions. Through their example he argues that the "transcultural" (blurring distinctions between global and local) and "transmedial" (emanating from the convergence of film, video, and new media) are distinctive features of the "viewing environment" structured by streaming-video-distributed cinema.

The final chapter goes beyond the major streamers to consider a universe of underground streaming video distribution, which can circulate rarely seen films from vanished archives. To illuminate this underground universe, Samirah Alkassim uses as case studies three films about the Palestinian struggle that were made in 1970s Lebanon by women filmmakers and that have been unavailable until their recent circulation via informal, streaming-enabled networks. The films—*Les Femmes Palestiniennes* by Jocelyn Saab (1974), *Atfal Walakin* by Khadijeh Habashneh (1979), and *Khotwa Khotwa* by Randa Chahal (1979)—deploy "archive effects" to engage audiences with disappeared or shadow archives and to speak of the history of Palestinian resistance in ways that, Alkassim argues, remain uncannily relevant within the present context of war and crisis in the Middle East. But the films' obscurity outside of the informal networks in which they circulate also raises questions about what is left out of the content libraries of leading streaming video platforms.

The informally circulating archival films Alkassim discusses present an interesting point of juxtaposition with Netflix's library of "Palestinian Stories," which, while initially promising access to content from a historically marginalized location, was almost completely—and controversially—deleted in 2024.[31] This exposes as well the ephemerality of streaming libraries. As one media scholar puts it, "Streaming is never a fixed site, artifact, or set of practices. It requires ongoing attention."[32] Indeed, despite enormous investment of capital, the formal distribution circuit created by streaming platforms offers little assurance of continued access to films or series. Viewers are at the mercy of streaming companies, who may withdraw content at will or with the ending of licences.

The ever-shifting nature of streaming video is borne out also by the rise and fall of particular platforms (like M.Media in Lebanon or Watch It in Egypt) and by the recent turn to advertising revenue by corporate SVOD services. Within this context, the advertising-supported YouTube, once seen as categorically distinct from the SVOD sphere, is increasingly emerging as Netflix's top competitor in many global South contexts, such as India, where its free access brings to it a much larger audience base as well as talent pool.[33] Netflix and other streaming video services have been known to invest in creative talent emanating from the more diverse and less affluent youth culture that free social media platforms such as YouTube and also Facebook foster. These developments suggest that, in order to stay viable in global South markets, subscription-based streaming video services like Netflix might in the near future find themselves needing to increasingly feed off, and also directly compete with, such freely accessible platforms—for locally resonant talent and wide-reaching content.

While ultimately the purpose of this volume is not to settle on a single conclusion regarding the current global media configuration, and while the

[31] Tyler Nesler, "Netflix Urged to Reinstate Deleted Palestinian Films," *Streaming Media*, October 28, 2024, https://www.streamingmedia.com/Articles/News/Online-Video-News/Netflix-Urged-to-Reinstate-Deleted-Palestinian-Films-166601.aspx.

[32] T. L. Taylor, "Foreword: Broadcasting Our Lives, Broadcasting Ourselves," in *Real Life in Real Time: Live Streaming Culture*, ed. Johanna Brewer, Bo Ruberg, Amanda L. L. Cullen, and Christopher Persaud (Cambridge, MA: MIT Press, 2023), xiv.

[33] For more on the relationship between social media platforms and streaming video services in India, see Smith Mehta, *The New Screen Ecology in India: Digital Transformation of Media* (London: BFI, 2023).

complexity of streaming video landscapes in the global South cannot be simply subsumed under the framework of cultural imperialism (however valid this framework remains for understanding particular cases and contexts), what is undeniable is that the South as a whole is going to continue to play an increasingly important role in providing markets for all sorts of streaming services. Data from 2023 suggests that the cumulative GDP of global South nations exceeded that of advanced capitalist nations for the first time in centuries.[34] The precise implications of such developments cannot be assumed beforehand, but they underscore the need to attend to the shifting dynamics of globalization and their impacts on media economies and cultures beyond the North.

Bibliography

Abu-Lughod, Lila. *Dramas of Nationhood: The Politics of Television in Egypt*. Chicago: University of Chicago Press, 2005.

Adgate, Brad. "Streaming Units Are Beginning to Be Profitable. Here's Why." *Forbes*, October 14, 2024. https://www.forbes.com/sites/bradadgate/2024/10/11/streaming-units-are-starting-beginning-to-be-profitable/.

Arditi, David. *Streaming Culture: Subscription Platforms and the Unending Consumption of Culture*. Leeds: Emerald, 2021.

Baschiera, Stefano, and Alexander Fisher, eds. *World Cinema on Demand: Global Film Cultures in the Era of Online Distribution*. London: Bloomsbury, 2022.

Bhushan, Nyay. "Netflix CEO Reed Hastings on Why the 'Next 100 Million' Subscribers Will Be 'Coming from India.'" *Hollywood Reporter*, February 26, 2018. https://www.hollywoodreporter.com/tv/tv-news/netflix-ceo-reed-hastings-why-next-100-million-subscribers-will-be-coming-india-1088301/.

Bouquillion, Philippe, Christine Ithurbide, and Tristan Mattelart, eds. *Digital Platforms and the Global South: Reconfiguring Power Relations in the Cultural Industries*. New York: Routledge, 2024.

Buck, Amber M., and Theo Plothe. *Netflix at the Nexus: Content, Practice, and Production in the Age of Streaming Television*. New York: Peter Lang, 2019.

Center for Human Rights in Iran. "Netflix Opening to Iran Ends State Broadcasting's Decades of Monopoly." January 11, 2016. https://iranhumanrights.org/2016/01/netflix-filtered/.

[34] Vijay Prashad, "The Emergence of a New Non-alignment: The Twenty-Fourth Newsletter," Tricontinental Institute for Social Research, June 15, 2023, https://thetricontinental.org/newsletterissue/new-non-alignment/.

Dirlik, Arif. "Global South: Predicament and Promise." *Global South* 1, no. 1 (2007): 12–23.

Havens, Timothy, and Ryan Stoldt. "Streaming Channel Brands as Global Meaning Systems." In *From Networks to Netflix: A Guide to Changing Channels*, edited by Derek Johnson, 321–31. London: Routledge, 2018.

Independent Commission on International Development Issues and Willy Brandt. *North–South: A Program for Survival*. Cambridge, MA: MIT Press, 1980.

Jaising, Shakti. *Beyond Alterity: Contemporary Indian Fiction and the Neoliberal Script*. Liverpool: Liverpool University Press, 2023.

Jin, Dal Yong. *Digital Platforms, Imperialism and Political Culture*. London: Routledge, 2015.

Johnson, Derek, ed. *From Networks to Netflix: A Guide to Changing Channels*. London: Routledge, 2018.

Khabar Gozāri-e Mehr [MehrNews]. "[The Reaction of the Office of Audio and Video Regulation Authority to 'Netflix' offer of service in Iran]." May 5, 2016. https://www.mehrnews.com/news/3607742/واکنش-مرکز-ملی-فضای-مجازی-به-ارائه‌سرویس-نت-فلیکس--در-ایران

Kisika, Sam. "Netflix Stops Streaming LGBTQ-Specific Movies in Kenya." *Washington Blade*, June 21, 2023. https://www.washingtonblade.com/2023/06/21/netflix-stops-streaming-lgbtq-specific-movies-in-kenya/.

Kloß, Sinah Theres. "The Global South as Subversive Practice: Challenges and Potentials of a Heuristic Concept." *Global South* 11, no. 2 (2017): 1–17.

Kumar, Shanti. *Gandhi Meets Primetime: Globalization and Nationalism in Indian Television*. Urbana: University of Illinois Press, 2010.

Lobato, Ramon. *Netflix Nations: The Geography of Digital Distribution*. New York: NYU Press, 2019.

Lotz, Amanda D. *Portals: A Treatise on Internet-Distributed Television*. Ann Arbor, MI: Maize Books, 2017.

Lotz, Amanda D. *Netflix and Streaming Video: The Business of Subscriber-Funded Video on Demand*. New York: Polity, 2022.

Lotz, Amanda, and Ramon Lobato, eds. *Streaming Video: Storytelling across Borders*. New York: NYU Press, 2023.

McDonald, Kevin, and Daniel Smith-Rowsey, eds. *The Netflix Effect: Technology and Entertainment in the 21st Century*. New York: Bloomsbury, 2016.

Mehta, Smith. *The New Screen Ecology in India: Digital Transformation of Media*. London: BFI, 2023.

Mirrlees, Tanner. *Global Entertainment Media: Between Cultural Imperialism and Cultural Globalization*. New York: Routledge, 2013.

Nesler, Tyler. "Netflix Urged to Reinstate Deleted Palestinian Films." *Streaming Media*, October 28, 2024. https://www.streamingmedia.com/Articles/News/Online-Video-News/Netflix-Urged-to-Reinstate-Deleted-Palestinian-Films-166601.aspx.

Netflix. "OSN Signs Middle East's First Partnership Deal with Netflix." February 18, 2018. https://about.netflix.com/en/news/osn-signs-middle-easts-first-partnership-deal-with-netflix.

Prashad, Vijay. "The Emergence of a New Non-alignment: The Twenty-Fourth Newsletter." Tricontinental Institute for Social Research, June 15, 2023. https://thetricontinental.org/newsletterissue/new-non-alignment/.

Prashad, Vijay. *The Poorer Nations: A Possible History of the Global South.* New York: Verso, 2013.

Rajagopal, Arvind, ed. *The Indian Public Sphere: Readings in Media History.* New Delhi: Oxford University Press, 2009.

Sarkar, Gargi. "Can Netflix Capitalise on India Boom to Upset JioCinema Plans?" Inc42, August 12, 2024. https://inc42.com/features/how-is-netflix-rewiring-its-india-business/.

Schneider, Nina. "Between Promise and Skepticism: The Global South and Our Role as Engaged Intellectuals." *Global South* 11, no. 2 (2017): 18–38.

Semati, Mehdi, and Elmira Moosavitabar. "The State and Streaming Services in Iran." In *Triaging the Streaming Wars*, edited by Robert Alan Brookey, Jason Phillips, and Timothy Pollard, 22–41. New York: Routledge, 2024.

Stoll, Julia. "Number of Netflix Paying Streaming Subscribers Worldwide as of 3rd Quarter 2024, by Region." Statista.com, October 18, 2024. https://www.statista.com/statistics/483112/netflix-subscribers/.

Syme, Rachel. "How Much Netflix Can the World Absorb?" *New Yorker*, January 9, 2023.

Taylor, T. L. "Foreword: Broadcasting Our Lives, Broadcasting Ourselves." In *Real Life in Real Time: Live Streaming Culture*, edited by Johanna Brewer, Bo Ruberg, Amanda L. L. Cullen, and Christopher Persaud, ix–xv. Cambridge, MA: MIT Press, 2023.

Topuz, Hifzi. "Television's One-Way Traffic." *UNESCO Courier* 4 (1977): 16–17. https://unesdoc.unesco.org/ark:/48223/pf0000046566.

Tryon, Chuck. *On-Demand Culture: Digital Delivery and the Future of the Movies.* New Brunswick, NJ: Rutgers University Press, 2013.

Williams, Raymond. *Television: Technology and Cultural Form.* New York: Routledge, 1974.

World Population Review. "Netflix Users by Country 2024." https://worldpopulationreview.com/country-rankings/netflix-users-by-country.

Yasmine, Rhea, and Alis Haizan. "Indonesia Plans to Impose Censorship Rules on Streaming Services; Analyst Says Move May Turn Viewers to Illegal Sites." August 23, 2023. https://www.channelnewsasia.com/asia/indonesia-streaming-services-ott-platform-netflix-disney-censorship-media-3718571.

1

The Global Netflix Documentary: Local Difference, Branding, and the Global South

Vinicius Navarro

The global South has been key to Netflix's worldwide expansion. In 2011, Latin America became one of the first markets explored by the company, following its initial push to reach audiences outside the United States. India was added in 2016, the same year the service became available in most countries in Asia and Africa. This expansion came on the heels of Netflix's transformation from a DVD rental service to a subscription video-on-demand platform, a move that sealed its status as an internet company. The second decade of the twenty-first century also saw Netflix grow from distributor to producer of original media content. In addition to reaching audiences throughout the global South, Netflix now produces local-language programs in countries such as Brazil, India, and Mexico, to name a few. Already an important market for Netflix, the global South became thus a valuable feature in the company's global brand.

This chapter looks at the global South by focusing on one specific genre: the global Netflix documentary. It sees this type of documentary as a metonymic extension of the company itself, treating the documentaries as emblematic of Netflix's global brand. Netflix's interest in documentary precedes, of course, its rise as a media studio and its growth as a global internet company. As Joshua Glick notes, documentary offerings have figured prominently on the company's website since its launch in 1998, helping distinguish Netflix from the large video rental companies that existed then, most of which treated documentary as a specialized niche relegated to the periphery of their media inventories.[1]

[1] Among the existing rental companies of the time, Glick mentions Blockbuster (which "stressed big budget new releases"), Hollywood Video (where mainstream American cinema stood alongside "a stronger selection of foreign films and cult classics"), Movie Gallery (known for its "adult section"), and Suncoast Motion Picture Company (which "specialized in the for-sale market of underground films and collector editions of studio features"). Joshua Glick, "Platform Politics: Netflix, the Media Industries, and the Value of Reality," *World Records* 5 (2021): 61.

The beginning of the streaming service in 2007 and the company's expansion into the production of original content amplified this interest in documentary, creating a closer connection between documentary media and the Netflix brand. Documentaries bring distinction to Netflix's content library, contributing "quality entertainment" to the service's website.[2] The global Netflix documentary, in its turn, resonates with the kind of cosmopolitanism that characterizes the company's brand image.

No single feature suffices to define the global Netflix documentary. Just as Netflix documentaries broadly speaking encompass a wide variety of themes and genres—from biographies of celebrities to nature films, from true crime to social-problem documentaries—the term "global documentary" suggests a multifaceted and at times slippery category.[3] At the level of content, global Netflix documentaries "travel" the world with an appetite for stories centered on both human and nonhuman subjects, allegedly revisiting a familiar narrative that implicates local specificity and global connectivity. Nature documentaries, for example, cover geographically diverse corners of the globe, while favoring a planetary, supranational perspective on the world (as is the case with the ambitious series *Our Living World*). Documentaries that address social problems focus on local stories in order to appeal to our sense of shared humanity (a trope explored in the third season of *Tales by Light*, a series I discuss below). And those that celebrate global diversity turn to culturally specific experiences but end up revealing suggestive parallels between different cultures (a common feature in travel documentaries centered on food culture, for example).

From a formal standpoint, these documentaries include a variety of approaches, sometimes evoking conventions reminiscent of travel photography and *National Geographic*, at other times incorporating features familiar from the more recent history of reality television. Despite the wide range of genres and topics involved, though, some attributes remain constant from one documentary to another. Like most projects connected to the company's name, these documentaries are expected to contain entertainment value, allegedly bridging the gap between a tradition of independent-minded documentary

2 Sudeep Sharma, "Netflix and the Documentary Boom," in *The Netflix Effect: Technology and Entertainment in the 21st Century*, ed. Kevin McDonald and Daniel Smith-Rowsey (New York: Bloomsbury, 2016), 183–97.

3 Netflix provides its own meticulous and often unorthodox list of categories. A documentary that could be simply placed under a broad genre such as nature or wildlife is also described as a family documentary and tagged as inspiring or captivating. For a more orthodox list of categories, see Glick, "Platform Politics," 64.

filmmaking and the commercial exigencies of the platform. Polished aesthetics and high production values further amplify this popular appeal, as does the emphasis on conventional narratives centered on individual characters. All this points to a certain "quality standard," in which differences of content and style—as well as diversity of culture, ethnicity, nation, and landscape—are outweighed by the production values adopted by the company.

For the purposes of this discussion, what I call the global Netflix documentary should also conform to more precise criteria. I am interested in projects that involve not one but multiple locations and, in some cases, movement across national, geographic, or cultural borders. Nationally or locally specific projects per se may not fit these criteria, as they could fail to represent at the textual level the ideal of globality and the sense of cosmopolitanism that are integral to Netflix's brand.[4] While scanning the website for global documentaries, I have also chosen to privilege those that are explicitly associated with the company's name, projects that come with the "Netflix Original" label attached to them. Although I am aware that Netflix's investment in these projects may vary significantly from one documentary to another, I tend to regard the label Netflix Original as a useful indicator of how the documentaries are expected to represent the Netflix brand.

This chapter contributes to two of the debates included in this book: the analysis of content produced for the streaming platform and the discussion of a global-local dynamics based on the presumption of universal storytelling. I also consider the technological affordances of the platform and the significance of the Netflix brand as a sort of agent shaping the encounter between Netflix users and the global South. The latter often appears less as a geopolitical reality than an array of subjects and experiences in the service of the company's brand image—a desirable sign of alterity and, at the same time, an emblem of Netflix's global reach. Underlying my analysis of the global Netflix documentary is also a more general critique of Netflix's refashioning of documentary media as a standardized product that meets the needs of a global video-on-demand platform, the latest chapter in a history that has reshaped television documentary after the 1980s.

[4] The association between cosmopolitanism and the company's brand has been discussed by several scholars. See, for example, Axelle Asmar, Tim Raats, and Leo Van Audenhove, "Streaming Difference(s): Netflix and the Branding of Diversity," *Critical Studies in Television* 18, no. 1 (2023): 24–40; see also Evan Elkins, "Streaming Diplomacy: Netflix's Domestic Politics and Foreign Policy," in *The Routledge Handbook of Digital Media and Globalization*, ed. Dal Yong Jin (New York: Routledge, 2021), 150–7.

Difference and the Global Netflix Documentary

It is hardly an exaggeration to say that for a large number of streaming media consumers Netflix has redefined the conventions and expectations attributed to documentary media. For those with little exposure to a tradition dating back to the silent cinema era, Netflix may in fact provide a yardstick with which to approach documentary media in general, thus reducing to a standardized, narrative-oriented product a multifaceted practice with a history that spans over a hundred years. Alexandra Juhasz and Alisa Lebow could very well be describing the Netflix documentary when they criticize the dominance of narrative conventions in contemporary documentary media, even if they do not refer specifically to the company. Storytelling, they note, "is not the most or only effective form for documentary, as affecting as it can be. *Not everything should be molded into a story*" (emphasis in the original).[5] To be sure, this interest in narrative is not new to the history of documentary, but its current prevalence—particularly in US documentary—tends to constrain the practice of documentary making and impoverish the experiences available to documentary audiences. Similarly, prescribing to documentary media a standard largely based on entertainment value risks distorting the ethos that has traditionally guided documentary practices, however many and heterogeneous those practices may have been.[6]

Just as it reduces documentary media to an easily consumable product—a product subject to the exigencies of commercial distribution—Netflix also squeezes global diversity into neatly wrapped packages. There is a correlation here between predictability of experience and standardization of form: the latter provides a reliable container to the former. The documentaries, in other words, render conveniently digestible what might otherwise appear foreign or unusual.

The rationale replicates the logic of tourism, according to which difference and alterity must also be turned into consumer items. Tourism favors the picturesque over the mundane, the exotic over the ordinary—all from the safety of a preplanned, relatively predictable ritual. "The tourist moves from a familiar place to a far place and then returns to the familiar place."[7] Tourism can

5 Alexandra Juhasz and Alisa Lebow, "Beyond Story: An Online, Community-Based Manifesto," *World Records* 2 (2018): Article 3, https://worldrecordsjournal.org/beyond-story-an-online-community-based-manifesto/.

6 For a general account of ethics in documentary, see Bill Nichols, *Representing Reality: Issues and Concepts in Documentary* (Bloomington: Indiana University Press, 1991).

7 John Urry and Jonas Larsen, *The Tourist Gaze 3.0* (London: Sage, 2011), 12.

also reproduce existing forms of privilege and social hierarchies, both within a particular society and across national borders. The rich are more likely to travel than the poor; the global North is more likely to cast its tourist gaze on the global South than the other way around. Given the intimate relationship between tourism and photographic media, these disparities can easily be extended to the ways the world is pictured by leisure travelers and reproduced by various tourism industries.

The global Netflix documentary evokes the casual curiosity of a tourist while promising a safe "return" to a familiar place. The sights and experiences on the screen give the impression of a heterogenous and variegated geography, of different social realities and culturally distinct situations. Yet it all ends up being diluted by Netflix's entertainment brand and its predictable and ultimately ethnocentric perspective on the world—the safe return to a familiar place—which not coincidently uphold traditional hierarchies between North and South, "center" and "periphery," the documentary makers and their subjects. That this return to a familiar place is a deliberate gesture—a strategic move within the culture of Netflix—became apparent in the company's 2020 "One Story Away" promotional video, which was part of a large global advertising campaign. Just under two minutes long, the video is essentially a collage of images from Netflix's films and programs (mainly fictional), stitched together by a voice-over narration that addresses the viewer as a close interlocutor. (Ava DuVernay does the voice-over narration in the American version of the video.) "There's a lot you may not know," the narrator reminds us. "But that's exactly what makes a story worth watching. Because in the end, we're only one story away." In the vast world of Netflix, we learn from the video, there is no risk of getting lost, as long as one sticks to the company's narrative-oriented, formulaic type of entertainment, of which documentaries are an integral part.[8]

A long list of cultural references renders the documentaries' global imagery reassuringly familiar, occasionally aligning colonial history and consumer culture. Oftentimes the global Netflix documentary recalls what Martin Roberts refers to as "coffee-table globalism," a term he applies to late twentieth-century films such as *Powaqqatsi* (dir. Godfrey Reggio, 1988) and *Baraka* (dir. Ron Fricke, 1992) but which may also describe older cultural developments, "extending from the founding of the National Geographic Society in 1888, through Edward Steichen's *Family of Man* photography exhibition and book of the 1950s, to the

[8] Netflix YouTube Channel, "One Story Away," accessed November 20, 2023, https://www.youtube.com/watch?v=IqkVUfYMZWM.

contemporary global mythologies of the Discovery Channel."[9] (Of these, the Discovery Channel turns out to be a particularly useful reference, as I discuss below.) Underlying each of these developments—and connecting them all—is a "Eurocentric mythology of the world" conveniently wrapped in a self-serving liberal humanism that glosses over deep global divides.[10] This form of globalism subsumes cultural and racial differences under a vague, albeit fitting, notion of shared humanity—a gesture that ultimately does little to confront social disparities or dismantle cultural hierarchies.

The global Netflix documentary updates to the age of streaming this supercilious attitude toward cultural differences. However, the self-serving humanism that Roberts sees in the "coffee-table globalism" of yesteryear seems less credible now, or less earnest than it once was. In its place stands a different kind of common denominator, less an abstract ideal of humanity than a media connector that promotes its own brand of globalism, namely, Netflix itself.

Consider a documentary like *Tales by Light* (2015), a multi-episode project shot on every continent, with a particular emphasis on locations in the global South. Directed by the Australian cinematographer Abraham Joffe and produced by Canon Australia, the series scans the world for picturesque sites, non-Western peoples, adventure, and wildlife.[11] Apart from a condescending admiration for non-Western cultures and a ready-to-wear reverence for the natural environment, there is little that connects one episode to another. Each one looks at different subjects and explores new locations. What really keeps the series together, what gives coherence to the project, is a narrative gimmick. The episodes follow one or more Western photographers (mostly White photographers) in their journeys of discovery, a trope reminiscent of colonial cultural imageries—and colonial travel writing in particular. In season 2, for example, we follow two renowned wildlife photographers based in Kenya, watch their interactions with a group of Maasai people, and witness their encounters with "big cats." We also see an underwater photographer as he travels to western Brazil in search of a green anaconda, the heaviest snake on the planet, then heads north to the Caribbean, where he dives among sharks. Lastly, we watch a war photographer as he sets

9 Martin Roberts, "*Baraka*: World Cinema and the Global Culture Industry," *Cinema Journal* 37, no. 3 (1998): 67.

10 Ibid., 66.

11 Interestingly, before becoming available on Netflix in 2016, the first season of *Tales by Light* screened on the National Geographic Channel subscription network in Australia and New Zealand, a useful reminder of the genealogy of the global Netflix documentary. Canon, "Tales by Light," accessed January 7, 2024, https://www.canon.com.au/about-canon/news-and-press-releases/tales-by-light.

out searching for "dignified" forms of death, a project that takes him to India and Namibia. The series has more to show, but this cursory description should suffice to evoke a familiar West-centered representational tradition.[12] *Tales by Light* aligns the perspectives of the photographers featured in the project with the worldview offered by the documentary series, which in turn represents the kind of globalism embodied by the Netflix brand.

One obvious charge against *Tales by Light* is that it renders interchangeable the various locations featured in the project, reducing the presumed diversity of the world to a condescending fascination with the natural, "the primitive." The actual places and experiences documented matter less than the gaze cast on them. Outdated as it seems, this attitude vis-à-vis global diversity gains new relevance in the context of today's streaming media services, where interchangeability of location becomes an asset well suited to the exigencies of a global video-on-demand platform. More than representing a backward-looking cultural practice, the presumption of interchangeability serves a particular regime of media circulation. It fulfills the expectation that any experience may be turned into a consumer item, regardless of its specific character or provenance. If *Tales by Light* befits a platform like Netflix, it is precisely because it conflates global diversity and consumer choice, rendering virtually irrelevant the gap between the actual experiences and their availability on the website. Diversity becomes inseparable from diversion, the dominant idiom on the platform.[13]

In this, too, Netflix gestures toward an earlier era, revisiting and updating significant media developments of the late twentieth century. This time, instead of looking at the outdated Cold War-era globalism of the "Family of Man" tradition, what we watch is a successful neoliberal alliance between documentary and entertainment culture.

[12] Season 3 of *Tales by Light* adds a new layer to the series' West-centric perspective on the global South. The first two episodes feature a White photographer from New Zealand working in partnership with UNICEF to document child labor and harsh living conditions in Bangladesh. Here the nostalgic colonial gaze that prevails in the series is replaced by a White savior trope, and the colonial "other" seen in earlier episodes becomes a victim to be rescued by a well-intentioned emissary from the global North. The project of documenting child labor—which the photographer featured in those two episodes shares, of course, with the documentary maker—recalls what Brian Winston once described as documentary's "tradition of the victim." See Winston, "The Tradition of the Victim in Griersonian Documentary," in *New Challenges for Documentary*, ed. Alan Rosenthal (Berkeley: University of California Press, 1988), 269–87. For a harsh critique of this kind of humanitarian approach to the documentary "other," see Pooja Rangan, *Immediations: The Humanitarian Impulse in Documentary* (Durham, NC: Duke University Press, 2017).

[13] I am borrowing the term "diversion" from John Corner's analysis of reality television and documentary media, to which I return shortly. See Corner, "Performing the Real: Documentary Diversions," *Television & New Media* 3, no. 3 (2002): 255–69.

From Documentary to Factual Entertainment—and Back Again

This interest in exploiting the entertainment value of nonfiction media precedes the advent of Netflix's streaming service by at least a couple of decades. Two developments in particular deserve attention here. The first comes from the history of documentary proper, more specifically the history of TV documentary in the United States; the other involves the refashioning of factual media into a lucrative form of entertainment. The last decades of the twentieth century saw a renewed interest in documentary within commercial television, thanks in part to the expansion of the cable TV industry in the United States. Although theatrical documentary also showed clear commercial aspirations at the turn of the century, reaching unprecedented box office success in the United States in the early 2000s, the history of television offers an earlier and, in some ways, closer precedent to the popularization of documentary media on Netflix, an American company with global ambitions whose investment in the genre suggestively recalls the success of the Discovery Channel a few decades earlier.[14] The second development I discuss here is the expansion of factual entertainment on television at the end of the twentieth century, the clearest example of which may be found in reality television. Both developments contributed to a rapidly changing television landscape—in the United States and beyond. And each provides a path to understanding Netflix's approach to documentary media as a form of diversion that befits the company's packaging of consumer-friendly globalism.

This history matters for a couple of reasons. First, it shows that Netflix's emphasis on documentary as a form of diversion constitutes, in part, an extension of existing trends in the American television industry, the impact of which was not confined to US markets. Additionally, the reference to factual entertainment situates the Netflix documentary within a narrative that exceeds the history of documentary per se, one in which the notion of factuality became far more pliable than it had been for most of the twentieth century.

[14] Netflix's connection with theatrical documentary remains significant, as its association with prestigious documentary filmmakers such as Ava DuVernay and Errol Morris easily shows. For a brief account of the commercial success of American documentary in the early twenty-first century, see Paul Arthur, "Extreme Makeover: The Changing Face of Documentary" and Patricia Aufderheide "The Changing Documentary Marketplace," *Cineaste* 30, no. 3 (Summer 2005): 18–23 and 24–8, respectively.

What happened to American television documentary in the 1980s is part of the broader history of postnetwork TV in the United States. The growth of cable television at the time allowed programmers to address distinct demographics and attend to niche TV markets, a feature that differentiated the commercial appeal of cable TV from the targeting of mass audiences that characterized the network era. On the one hand, the vastly increased number of channels available resulted in larger diversity of programs, which at times also translated into diversity of social representation; on the other hand, targeting dispersed yet specific audiences became a common industry practice in the cable era. Documentary might have looked like a weak contender in this new television environment, considering its declining importance for American television during the previous decades.[15] Yet, as it turned out, cable revitalized the relationship between commercial television and documentary, although not without redefining some of the expectations attached to the latter. Documentary took on a new importance on American TV, and entertainment value, once a contentious issue between journalists and TV producers, became a pivotal element in this resurgence of television documentary.[16]

No media company represented this renewed interest in documentary like the Discovery Channel. As Cynthia Chris writes, Discovery "[devoted] virtually all of its programming to a genre that had nearly disappeared from for-profit television: the documentary."[17] Starting modestly in 1985, by the turn of the century Discovery "had become the cornerstone of a media conglomerate in control of four basic cable networks."[18] The emphasis on nonfiction might have sufficed to explain the connection between the Netflix documentary and American cable television from the 1980s. But there are more specific parallels that deserve attention too.

[15] The early 1960s are usually described as the Golden Age of television documentary in the United States. Interest in documentary among the commercial networks declined after that and did not resurge until the 1980s. In the intervening years, documentary programming migrated to public television. See Cynthia Chris, "All Documentary, All the Time? Discovery Communications Inc. and Trends in Cable Television," *Television & New Media* 3, no. 1 (2002): 7–28; Michael Curtin, "'Packaging Reality': The Influence of Fictional Forms on the Early Development of Television Documentary," *Journalism Monographs* 137 (1993): 1–37.

[16] According to Michael Curtin, producers tried to introduce Hollywood's narrative conventions to American television documentary in the late 1950s and early 1960s but were often in conflict with the networks' correspondents, who favored a clearer separation between entertainment and fact-based, issue-oriented media. Curtin, "'Packaging Reality,'" 15.

[17] Chris, "All Documentary, All the Time?," 8.

[18] Ibid., 7.

Discovery was a global media company with a recognizable brand. As Chris notes, both its global ambitions and the emphasis on branding contributed decisively to the company's successful expansion.[19] Still other parallels become apparent if we look at the types of documentary associated with the Discovery brand: travel and exploration, history and nature films, to name a few obvious genres that would be favored by Netflix years later. Perhaps most significantly, much like Netflix, Discovery incorporated entertainment value into its brand, thus reviving an interest in narrative conventions that had surfaced in American television documentary a few decades earlier.[20] Besides obvious parallels in aesthetics, genre conventions, and subject matter, Discovery's approach to documentary thus foreshadows Netflix's own refashioning of nonfiction media into a commercial product. More than simply expand the category of documentary media, what both Discovery and Netflix proposed was a transformation of it, a change in institutional practice that brought documentary making ever closer to media genres that have historically stood apart from it.

Alongside these changes in television documentary, the 1990s saw a phenomenal growth in factual entertainment, much of it under the label "reality TV," a somewhat elastic yet extremely suggestive category. As it became a popular cultural phenomenon, starting in the late 1980s, reality TV quickly transformed factual television programming. Indeed, it transformed the very concept of factuality. Although reality television never claimed for itself the mission—or the status—of documentary, its presence on TV programming schedules worldwide had a noticeable impact on documentary media, whether we are considering production methods, aesthetic and narrative conventions, or audience expectations. Looking at British nonfiction television at the turn of the century, for example, John Corner refers to a "radical dispersal of documentary energies across the schedules," a phenomenon to which he applies the term "postdocumentary."[21] What is apparent here is the proximity between documentary and other forms of factual programming that might have little to do with it, shows like *Big Brother*, for example, which Corner specifically discusses in his article.

This association between documentary and reality TV did not, of course, start in the 1990s with the global rise of factual entertainment. Anna McCarthy

[19] Ibid., 8–9.
[20] See Curtin, "'Packaging Reality.'"
[21] Corner, "Performing the Real," 263.

traces this crossover back to the 1950s, when hidden-camera shows in the United States combined documentary interest and entertainment value.[22] A more specific example would surface a couple of decades later, with the 1973 broadcasting on US public television of the documentary series *An American Family*, which has been described both as direct cinema and as "an early form of reality TV."[23] The last years of the twentieth century, however, offered a new environment for nonfiction media, making it more difficult to insulate documentary from entertainment. It was not just a matter of recognizing the "documentary roots" of reality TV but also of acknowledging the impact of the latter on documentary culture. As Corner puts it, "it is hard to see how *any* form of television documentary program can remain completely unaffected by the new ecology of the factual" (emphasis in the original).[24]

That these shifts were global in scope makes them all the more relevant. Reality TV programs "traveled" easily across national borders because they were designed to be culturally adaptable. Rather than rely on broad generic formulas—the standard practice in entertainment media—the programs were often based on templates, or formats in the media industry's jargon, that were specific enough to be sold to television industries worldwide, yet also flexible enough to be "filled" with specific local content.[25] Take *Big Brother*, for example, the program discussed by Corner. Created by Endemol, a company originally based in the Netherlands, *Big Brother* became a global sensation as it was adapted by television industries around the world, with each new iteration of the program featuring a local cast and incorporating local cultural elements. To the popularity of factual entertainment, TV producers thus added the appeal of local culture. A Netflix reality show like *Iron Chef: Quest for an Iron Legend* (2022), for instance, which has iterations shot in Mexico and Brazil, is a direct

22 Anna McCarthy, "'Stanley Milgram, Allen Funt, and Me': Postwar Social Science and the 'First Wave' of Reality TV," in *Reality TV: Remaking Television Culture*, ed. Susan Murray and Laurie Ouellette (New York: New York University Press, 2004), 19–39.

23 Susan Murray, "'I Think We Need a New Name for It': The Meeting of Documentary and Reality TV," in *Reality TV: Remaking Television Culture*, ed. Susan Murray and Laurie Ouellette (New York: New York University Press, 2004), 41.

24 Corner, "Performing the Real," 265.

25 In this context, a TV format is a set of rules and guidelines that serves as the basis for the production of programs such as *Dancing with the Stars*, *American Idol*, and *America's Got Talent*, among many others. The format is not the program itself, but it constitutes copyrighted material and can be licensed to television industries worldwide as long as the TV producer pays the owner of the format a licensing fee. See Silvio Waisbord, "McTV: Understanding the Global Popularity of Television Formats," *Television & New Media* 5, no. 4 (2004): 359–83. See also Albert Moran, *Copycat Television: Globalisation, Program Formats and Cultural Identity* (Luton, UK: University of Luton Press, 1998).

descendant of formatted reality TV shows from earlier decades. For the sake of comparison, we can think of a Netflix documentary series like *Chef's Table* (2015–19), which also includes episodes shot in Mexico and Brazil, among other countries, and recalls some of the features of those reality shows—most noticeably the emphasis on individual characters and the interest in culinary and cultural diversity—even though in this case the approach to local content departs from the protocols of formatted reality television.

Against this backdrop, the global Netflix documentary appears not as an anomaly but as a paradigmatic example of a particular attitude toward fact-based media that gained prominence in the last three decades. It constitutes an amplification of earlier media developments. But the platform also places documentary media within a novel media ecology, one whose complexity could not have been fully understood in the last decades of the twentieth century. The question to address now is what is new about the forms of documentary experience available on Netflix. How does the platform help shape the experience afforded by the global Netflix documentary?

The Global South as Curated Experience

For practical purposes, the global Netflix documentary does not exist on its own; it appears alongside other titles on the Netflix website. Or, rather, it is placed there according to a specific curatorial logic. Netflix functions as a sort of media programmer: it selects, classifies, arranges, and displays the materials available in its library so as to shape or facilitate a particular kind of experience. More than the viewing of individual texts, that experience presumes engagement with the site and its recommendation system. It is inseparable from the kind of media curation provided by the platform.

This curatorial process is anything but neutral. Like the documentaries themselves, it has the potential to both define and limit our perspective on the global South, even though the selection and arrangement of the library materials may not come across as a deliberate media intervention. Instead, the curatorial process remains somewhat mysterious, hidden from view, although often perceived as a desirable service to the user. Netflix's recommendation system is unusually meticulous. It breaks a large category like "documentary" into highly specific genres, based on criteria that elude more orthodox systems of categorization. These small categories are then matched with the users' viewing habits, creating the impression that the service offers personalized content to the Netflix subscriber.[26] For example, *Tales by Light*, the aforementioned documentary on Western photographers and their journeys to "remote" parts of the world, is described on the Netflix website as "Family Watch Together TV" and "Food and Travel TV"—as well as "Australian" and "Inspiring." Based on my viewing of this documentary and on data collected from other users, Netflix recommended that I also watch *Night on Earth* (Plimsoll Productions, 2020) and *David Attenborough: A Life on Our Planet* (Silverback Films and WWF, 2020), in addition to other nature-related and travel-oriented documentaries, several of which evoked a familiar *National Geographic*-type of imagery, or reduced the world's natural and cultural diversity to the exoticism associated with a tourist excursion. What might have looked like a casual, value-free experience—my browsing of the Netflix catalogue—turned out to be a calculated process with the potential to reinforce a certain outlook on the world, in this case a particular view of the global South, as well as a specific perspective on globality itself.

Although it seems to enrich or facilitate the viewing experience, Netflix's algorithmic curation in fact works against the impression of openness suggested by the company's website, as Neta Alexander argues. Writing in the mid-2010s about Netflix's recommendation system, Alexander laments what she sees as the narrowing of cultural options that results from algorithmic curation, the seeming abundance of items in the Netflix content library notwithstanding.[27] In her analysis, she questions several assumptions commonly associated with the video-on-demand service, including the belief in user agency and the promise of unlimited choice. The latter turns out to be deceptive not just because library titles are often removed from the site but also, and perhaps more importantly, because the presumption of free and unrestricted choice contradicts the business logic behind the recommendation system. That system is premised on the notion

[26] According to the company, recommendation algorithms "provide our members with personalized suggestions to reduce the amount of time and frustration to find something great content [*sic*] to watch.... We do this by using the data about what content our members watch and enjoy along with how they interact with our service." Netflix also uses collaborative filtering in order to make specific recommendations to individual users. Netflix Research, "Recommendations," accessed February 6, 2024, https://research.netflix.com/research-area/recommendations. See also Neta Alexander, "Catered to Your Future Self: Netflix's 'Predictive Personalization' and the Mathematization of Taste," in *The Netflix Effect: Technology and Entertainment in the 21st Century*, ed. Kevin McDonald and Daniel Smith-Rowsey (New York: Bloomsbury, 2016), 103–26.

[27] Alexander, "Catered to Your Future Self." Alexander's article focuses specifically on Netflix's Cinematch recommendation system, which the company launched in 2000. Over time, Netflix added numerous features to its system and tried to improve its capacity to predict user interest. On its "Research" web page, the company states, "We continue to improve our algorithms and look for new areas we can personalize." Netflix Research, "Recommendations."

that erratic user behavior can be turned into relatively predictable patterns.[28] (Because I watched *Tales by Light*, Netflix assumes that I may also be interested in this or that documentary.) "Contingency," "serendipity," and "potentiality"—in fact all that which may upset the pattern—constitutes a sort of "noise."[29]

The target of Alexander's critique is Netflix's role as arbiter of taste, but the argument can be easily extrapolated to include the service's curation of world cultures and global events. There, too, algorithmic curation tends to work against openness and contingency. Cultural and natural diversity must conform to a programming logic based on entertainment value, global differences succumb to the allure of the familiar, and the retrograde perspective on the global South shared by several documentaries is amplified by the recommendation algorithms.

Another way Netflix tricks us into believing in openness and plurality, while actually favoring the centripetal pull of the platform, is by investing in serial narratives. As a structuring logic shared by numerous Netflix shows, serialization operates most noticeably at the textual level. Yet it is hard to think of it without also considering the properties of the platform and the viewing experience afforded by it. In some ways, Netflix's emphasis on serialization can be read as a response to changing media environments and new viewing conditions. Serial narratives encourage loyalty to specific shows, when distracted and "migratory" behavior might have been the default disposition among contemporary media users. Additionally, the instant availability of an entire series has the potential to substantially lengthen the viewing experience. More important in the present context, serialization nurtures the illusion of abundance and diversity that bolsters the platform's alleged globalism. It should come as no surprise, then, that the serial format seems well suited to projects that purport to be *about* the world, documentaries that feature various landscapes or showcase different cultures.

What the serial global documentary also offers is more of the same. If serialization seems to accommodate diversity of content, it also submits what might otherwise seem like disparate realities or incongruous situations to a relatively stable design. All the traveling in *Tales by Light*, to return to a familiar example, results in a somewhat predictable narrative: a photographer takes the viewer on a journey to explore an unfamiliar, often picturesque, corner of the world. The journey functions as an ordering device, a means to render legible

[28] Alexander, "Catered to Your Future Self."
[29] Ibid., 113.

what is presumably foreign to the viewer. *Magical Andes* (Trailer Films, 2019), another *National Geographic*-type documentary that was recommended based on my viewing history, also submits differences of culture and landscape to a narrative pattern in which every site, subject, or experience is expected to look equally digestible and similarly picturesque. The documentary features a variety of locations on South America's Andes Mountains, each paired with "characters" who serve as representatives of a particular locality, culture, or community. As expected, the repetitive pattern built into the structure of the documentary cements a certain perspective on the world—and on the global South more specifically. In each episode, local characters pay tribute to the natural environment and affirm their attachment to the region. In lieu of plurality, addition produces reinforcement, with one story blending into the other as the viewer progresses through the series. What prevails—to use another tourism-related analogy—is a sort of postcard aesthetics, whereby the brevity of the encounters with local subjects ends up flattening all sense of difference, leaving only a lustrous surface, worthy of admiration yet entirely forgettable. Serialization thus allows for the kind of diversity that adorns the global Netflix documentary but ends up producing a distant, generic, vaguely amusing distraction.

The Netflix Brand as Cultural Mediator

Underlying Netflix's "domestication" of the global South is the company's famed brand. Branding is key to the life of a subscription video-on-demand service like Netflix, which operates in an increasingly competitive media landscape and reaches nearly every country on the planet, even if access to it remains uneven throughout the world. Like other companies, Netflix uses its brand to communicate a certain set of values, or, better yet, to share those values with its subscribers and potential customers. Branding presumes some level of "interactivity… between the consumer and producer," as Catherine Johnson points out.[30] It functions as a mediator between a company or organization and its audience, user, or client. The relationship that forms between one and the other, however, is far from equal. "Rather, branding attempts to shape, control and/or manage the values attributed to products and, through this, the uses to

[30] Catherine Johnson, *Branding Television* (New York: Routledge, 2012), 3–4.

which the product is put."[31] In other words, branding involves some authority over the user or consumer. And insofar as it aims at creating a strong sense of identity, it presumes authority over the product or service being offered as well.

Netflix's brand is defined as much by the "quality" entertainment offered by the platform as by the cosmopolitan values associated with it. As Evan Elkins notes, "the company stresses cultural diversity and cosmopolitanism as central components of its brand image."[32] A seemingly paradoxical scenario emerges here, wherein what might seem contrary to the building of a self-contained identity—the appeal of culturally diverse experiences—becomes in fact one of the building blocks of the company's brand image. The paradox, of course, is only apparent since Netflix incorporates diversity itself into its brand, a carefully calculated gesture that strengthens the company's cultural power, as a recent study on Netflix's press releases has shown.[33] From this perspective, documentaries on the global South become a particularly valuable asset. On one hand, the company's brand image serves as a means to mediate the user's viewing experience. (It is through the lens of cosmopolitanism that we look at the global South on Netflix.) On the other, the documentaries become themselves a means to uphold that brand image, a way to illustrate, echo, and amplify the company's corporate and cultural identity.

Documentaries in general lend prestige to Netflix's brand image, even if the sobriety normally attributed to the genre has now become largely nominal.[34] The peripatetic character of the global documentary adds an extra—and much desirable—layer to this relationship with the company's brand. In theory, the documentaries embody the cultural and geographic diversity Netflix purports to celebrate. The question becomes, then, how much of this interest in diversity escapes "the centripetal force of the corporate brand."[35] A series like *Magical Andes*, for example, was shot in several Andean countries, featuring local characters, local stories, and local landscapes. At the same time, the

[31] Ibid., 4.

[32] Elkins, "Streaming Diplomacy," 151.

[33] Asmar, Raats, and Van Audenhove, "Streaming Difference(s)," 30. The article breaks Netflix's branding of diversity into several categories: "racial and ethnic diversity," "gender diversity," "sexual diversity," "age diversity," and "linguistic diversity."

[34] Sharma, "Netflix and the Documentary Boom," 186.

[35] I borrow this phrase from Robert Stam. Commenting on television and internet ads for transnational corporations, Stam notes that they "offer a transnationalist utopia... of a world without center or borders," yet their "wildly dispersive centrifugal energies are ultimately disciplined, in most cases, by the centripetal force of the corporate brand." The comment could easily apply to Netflix's own branding of global diversity. Robert Stam, *World Literature, Transnational Cinema, and Global Media: Towards a Transartistic Commons* (New York: Routledge, 2019), 134.

documentary envelops racial, cultural, and natural differences in themes that transcend local specificity, such as labor, entrepreneurship, love of the land, and care for the natural environment, as if to secure the universal appeal of each story. The unfamiliar becomes relatable, and the local is rendered universal—or universally recognizable. The result is a recalibration of the relationship between local specificity and global reach, whereby a potentially multicultural perspective on the world is obfuscated by the company's bland cosmopolitanism.

For all the emphasis on locally produced content, Netflix's brand thus overrides the significance of localization, the process by which a particular program comes to reflect the needs and interests of local audiences. Take a series like *Street Food* (Boardwalk Pictures), for example, a US-produced documentary from 2019 shot in Asia, Latin America, and the United States. As is often the case, the series relies on a relatively predictable formula. Each episode focuses on a specific city and features a character whose life provides material for the main narrative. In the season devoted to Latin America, that character is usually a woman who owns a food business. The season combines biographical material with an alleged interest in the cultures featured in its six episodes, taking the viewer from Buenos Aires to Salvador, from Oaxaca to Lima, and from Bogota to La Paz. The reiterative nature of the formula recalls the program templates used in formatted reality television shows, but now the potential for transcultural adaptation is largely gone. Instead, the episodes in *Street Food* seem generally self-contained, each existing as a relatively autonomous unit within the documentary series. Whereas the TV format "changes hands" when it is locally adapted, the documentary series remains mostly under the same creative vision, with a few local crew members added in each new location. And although the series represents food cultures from different countries, it does not specifically address their respective local audiences, privileging instead the sensibility of a nondescript, purportedly cosmopolitan user.

Like the platform's recommendation algorithms, the Netflix brand adds a new dimension to the ethnocentric perspective that prevails in the global documentary. The brand extends the cultural production of meaning to the process of media circulation, in this case a process informed by a seemingly open but ultimately narrowly defined identity. This process involves new routes through which media materials circulate. In theory, the global Netflix documentary could connect different audiences in the global South, "delivering" the South to the South and thus circumventing a top-down, North–South flow of information. As a mediator in a potential South–South encounter, however, the Netflix brand reorients this process, subjecting it to the company's own form of globalism. The South thus reaches the South through the agency of the Netflix brand. The documentaries are rebranded as a form of diversion, and the South, always a convenient emblem of alterity, is assimilated into the company's seemingly liberal yet generally ethnocentric cosmopolitanism, leaving largely intact the distribution of cultural power between North and South.

Bibliography

Alexander, Neta. "Catered to Your Future Self: Netflix's 'Predictive Personalization' and the Mathematization of Taste." In *The Netflix Effect: Technology and Entertainment in the 21st Century*, edited by Kevin McDonald and Daniel Smith-Rowsey, 81–100. New York: Bloomsbury, 2016.

Arthur, Paul. "Extreme Makeover: The Changing Face of Documentary." *Cineaste* 30, no. 3 (2005): 18–23.

Asmar, Axelle, Tim Raats, and Leo Van Audenhove. "Streaming Difference(s): Netflix and the Branding of Diversity." *Critical Studies in Television* 18, no. 1 (2023): 24–40.

Aufderheide, Patricia. "The Changing Documentary Marketplace." *Cineaste* 30, no. 3 (2005): 24–8.

Canon. "Tales by Light." Accessed January 7, 2024. https://www.canon.com.au/about-canon/news-and-press-releases/tales-by-light.

Chris, Cynthia. "All Documentary, All the Time? Discovery Communications Inc. and Trends in Cable Television." *Television & New Media* 3, no. 1 (2002): 7–28.

Corner, John. "Performing the Real: Documentary Diversions." *Television & New Media* 3, no. 3 (2002): 255–69.

Curtin, Michael. "'Packaging Reality': The Influence of Fictional Forms on the Early Development of Television Documentary." *Journalism Monographs* 137 (1993): 1–37.

Elkins, Evan. "Streaming Diplomacy: Netflix's Domestic Politics and Foreign Policy." In *The Routledge Handbook of Digital Media and Globalization*, edited by Dal Yong Jin, 150–7. New York: Routledge, 2021.

Glick, Joshua. "Platform Politics: Netflix, the Media Industries, and the Value of Reality." *World Records* 5 (2021): 59–76.

Johnson, Catherine. *Branding Television*. New York: Routledge, 2012.

Juhasz, Alexandra, and Alisa Lebow. "Beyond Story: An Online, Community-Based Manifesto." *World Records* 2 (2018): Article 3. https://worldrecordsjournal.org/beyond-story-an-online-community-based-manifesto/.

McCarthy, Anna. "'Stanley Milgram, Allen Funt, and Me': Postwar Social Science and the 'First Wave' of Reality TV." In *Reality TV: Remaking Television Culture*, edited by Susan Murray and Laurie Ouellette, 23–43. New York: New York University Press, 2004.

Moran, Albert. *Copycat Television: Globalisation, Program Formats and Cultural Identity*. Luton, UK: University of Luton Press, 1998.

Murray, Susan. "'I Think We Need a New Name for It': The Meeting of Documentary and Reality TV." In *Reality TV: Remaking Television Culture*, edited by Susan Murray and Laurie Ouellette, 65–81. New York: New York University Press, 2004.

Netflix Research. "Recommendations." Accessed February 6, 2024. https://research.netflix.com/research-area/recommendations.

Netflix YouTube Channel. "One Story Away." Accessed November 20, 2023. https://www.youtube.com/watch?v=IqkVUfYMZWM.

Nichols, Bill. *Representing Reality: Issues and Concepts in Documentary*. Bloomington: Indiana University Press, 1991.

Rangan, Pooja. *Immediations: The Humanitarian Impulse in Documentary*. Durham, NC: Duke University Press, 2017.

Roberts, Martin. "*Baraka*: World Cinema and the Global Culture Industry." *Cinema Journal* 37, no. 3 (1998): 62–82.

Sharma, Sudeep. "Netflix and the Documentary Boom." In *The Netflix Effect: Technology and Entertainment in the 21st Century*, edited by Kevin McDonald and Daniel Smith-Rowsey, 143–54. New York: Bloomsbury, 2016.

Stam, Robert. *World Literature, Transnational Cinema, and Global Media: Towards a Transartistic Commons*. New York: Routledge, 2019.

Urry, John, and Jonas Larsen. *The Tourist Gaze 3.0*. London: Sage, 2011.

Waisbord, Silvio. "McTV: Understanding the Global Popularity of Television Formats." *Television & New Media* 5, no. 4 (2004): 359–83.

Winston, Brian. "The Tradition of the Victim in Griersonian Documentary." In *New Challenges for Documentary*, edited by Alan Rosenthal, 269–87. Berkeley: University of California Press, 1988.

2

A Cold War Netflix: The US Information Agency as a Key Historiographic Model for the Transnational Streamer

Bret Vukoder

In 2016, Netflix realized its #NetflixEverywhere plan to go "fully global," making the platform "unblocked and accessible (almost) everywhere," reaching 190 countries with programming in thirty languages.[1] Many in the press, media industries, and public marked the expansion as a seminal moment—suggesting the transnational potential of what the new streaming era could bring to production and distribution, while also generating concern over a powerful amplification of American cultural imperialism through popular entertainment. Indeed, Netflix's plan seems to have come to fruition over the past eight years, with the company presently claiming that 70 percent of its members are outside of the United States.[2] The ambivalence and uncertainty over the launch of a global Netflix was in part a function of its presumed novelty. As scholars such as Jean Chalaby, Amanda Lotz, and Ramon Lobato have highlighted, we can perhaps look to early satellite television or multinational networks, such as CNN and MTV, as historiographic predecessors to Netflix's international reach and ambitions. Yet nothing seems to have achieved the global scale of Netflix, with thousands of hours of programming algorithmically curated to the individual and regions in which it is streamed.

Well before Netflix went "fully global," however, the United States government managed a massive and labyrinthine transnational media office of its own—the United States Information Agency (USIA), which sought to "tell

[1] Netflix, "About Netflix," 2024, https://about.netflix.com/en.
[2] Netflix, "Top Investor Questions," 2024, https://ir.netflix.net/ir-overview/top-investor-questions/default.aspx.

America's story" in order to instantiate and sell American policy and culture across the world. Between 1953 and 1999 the USIA, using a variety of delivery technologies, produced and distributed roughly 20,000 moving image titles in 150 countries and sixty languages. The agency—while fundamentally an apparatus of the United States' neoimperialism during the Cold War—circulated a surprisingly diverse breadth of content. This diversity was in part a function of its multinational labor structure, in which artists and technicians outside of the United States were employed or contracted to make films for the agency. Many titles unsurprisingly towed the line of American interests, but several films express distinct cultural and aesthetic idioms of the people and areas in which they were made. In these cases, USIA films sometimes reveal a dialectic with their institutional context, challenging conventional readings of government film and Cold War hegemonies. The archive's diversity is also a product of the multiple formats the USIA employed. Although the majority of the extant USIA archive is documentary, it additionally holds newsreels, cartoons, avant-garde work, soap operas, and even the occasional feature-length narrative film. As with films belonging to Netflix's library, motion pictures that are part of the USIA archive were designed to reach a varying scope of audiences, with some aimed for global release and others catered to specific countries or regions. And like Netflix's newer outreach toward international productions, USIA leveraged the voices and labor of people outside of the United States to further particularize the appeal of their output.

In this chapter, my intention is not to categorically conflate Netflix's current output and operations with those of the USIA's, nor to propose a one-for-one correspondence between these transitional media entities. The USIA and Netflix are in fact quite different, especially in regard to the content type each has produced and licensed. Rather, I aim to position the open and accessible USIA archive[3] as a key historiographic tool by which to frame and investigate the proprietarily hidden *systems* through which Netflix presently conducts its business. Lobato, Scarlata, and Cunningham characterize Netflix, like other global SVODs, as a "'transnationalized American offering with an international-flavored seasoning'[4]... [that] benefit[s] from enormous scale, resources, and

[3] For more details concerning the state of access for the USIA archive, see H. Gharabaghi and Bret Vukoder, "The Motion Pictures of the United States Information Agency: Studying a Global Film and Television Operation," *Journal of e-Media Studies* 6, no. 1 (2022).

[4] L. Albornoz and M. T. García Leiva, "Netflix Originals in Spain: Challenging Diversity," *Critical Studies in Television* 37, no. 1 (2022): 64; qouted in R. Lobato, A. Scarlata, and S. Cunningham, "Conceptualizing the National and the Global in SVOD Original Production," in *Streaming Video: Storytelling across Borders*, ed. A. Lotz and R. Lobato (New York: NYU Press, 2023), 40.

network effects,"[5] but is subject to the legal, economic, and *cultural* dynamics of various international markets. While Netflix's actions and operations are indeed consistent with those of a multinational corporation, they nonetheless also bring to mind aspects of government media. In some recent literature concerning Netflix,[6] scholars have even begun using a lexicon of governmentality to describe the streamer[7] but have stopped short—beyond brief mentions of lobbying efforts and targeted hiring—of engaging the histories, archives, and content of government media.

The USIA's strategies, on the other hand, were certainly in conversation with the paradigm of the multinational corporation that also began to emerge during the Cold War. Its thousands of international employees, transnational flows of media, global infrastructural investments, and outwardly cosmopolitan ethos distinguish the USIA from many of its precedents in government media and propaganda. Inversely to Netflix drawing language and leadership from the world of government, the USIA has its own legacy of hiring from private corporate media, like Edward Murrow, and employing consumerist jargon in its goal to "sell" and "convince" persons of how they could identify themselves in the American brand. In their respective transnational operations, the USIA and Netflix therefore each exhibit a similarly strategic and cultural liminality, neither fully corporate nor governmental.

My goals in this chapter are twofold. First, I will introduce and contextualize several components of USIA motion picture operations and output, highlighting select titles and archival materials from the agency's paper trail. In mapping both discursive and material elements of the USIA bureaucracy, I will suggest how they may represent salient paradigms for understanding the Netflix ecosystem. Second, I will delineate how the transnational designs of both the USIA's and Netflix's production, distribution, and exhibition protocols effectively define and entrench their respective brands across national contexts. Through breadth, saturation, and local involvement within their media output, platforms, and infrastructures, they build a stability that facilitates, respectively, political and market hegemonies on a global stage.

5 Lobato, Scarlata, and Cunningham, "Conceptualizing the National and the Global," 40.

6 Such as Evan Elkins, "Streaming Diplomacy: Netflix's Domestic Politics and Foreign Policy," in *The Routledge Handbook of Digital Media and Globalization*, ed. D. Y. Jin (New York: Routledge, 2021), 150, and Axelle Asmar, Tim Raats, and Leo Van Audenhove, "Streaming Difference(s): Netflix and the Branding of Diversity," *Critical Studies in Television* 18, no. 2 (2022): 24–40.

7 Though Amanda Lotz's term "SVOD" (subscriber-funded video-on-demand service) is perhaps the most precise in describing Netflix, I will be primarily using the more colloquial and less clunky term "streamer."

The Transnational Aspirations and Cosmopolitan Discourse of USIA Motion Picture Operation

Before introducing specific details pertaining to the USIA motion picture bureaucracy, I will provide a snapshot of the agency's institutional history and the discourse surrounding its culture and purported mission, thereby identifying threads that are relevant to unpacking Netflix's current transnational dominance. The USIA, with its multinational presence and reach, was both a reflection of and an active force in shaping what we have come to define as the Cold War. Conventional and reductive historiographies of the period fixate on the bipolar ideological and military struggles between the two superpowers, the United States and Soviet Union. While these two nations certainly embodied the dominant political and cultural forces, the Cold War has a much more complex, diverse, and sprawling understory that goes beyond these homogenizing forces. If anything, the Cold War fragmented the traditional Westphalian sense of international relations[8] that characterized the peaks of the colonial era. Although the USIA was an arm of a system that often sought to reinstall a US-centric imperial world order, its complicated legacy instead shows a multitude of ways that it sometimes fostered an increasingly diverse and decentered world order.

In what follows, I will position the internet age—a period in which Netflix and other US-based streamers have blossomed—as not a distinct but rather comparable historical moment to that of the Cold War. I argue the last thirty years are an extension or a new branch of the sensibilities and dynamics that took root during that era. Relatively speaking, the early internet was a Wild West of content and platforms, but by the early 2000s we began witnessing a concentration of power in the handful of transnational, public-facing, but America-based tech corporations we all know today—like Meta, Apple, Alphabet, and Netflix. Instead of representing a political system, these corporations represent the more amorphous logic of capital—an offshoot of Western ideologies, now almost exclusively defined by a goal of endless growth—that aims to install itself among an impossibly diverse, global consumer base.

The United States after the Second World War faced a similar challenge, as it had to adapt its methods of securing economic and cultural hegemony over a newly interconnected but pluralistic world. After an unprecedently brutal

8 "Westphalian" generally posits a noninterventionist approach to international relations, respecting the sovereignty of nations with well-defined borders. Historically, however, Western nations applied this philosophy to only themselves and not nations within the global South.

war, cultural outlets and economic relationships offered a "softer" means of preserving the United States' power and influence. The majority of American leaders and lawmakers in the late 1940s were hence willing to support and fund initiatives like the Marshall Plan, the Point Four Program, and earlier, less centralized iterations of the USIA.[9] Between 1945 and 1953, the United States established hundreds of US Information Service (USIS) posts and libraries throughout the world and began to invest in media infrastructure, especially in strategically important countries like Japan, Iran, and Korea.[10] The Smith-Mundt Act of 1948, which established the loose foundation of what became the USIA in 1953, vaguely codified these varied initiatives into law. The bill called for "an information service to disseminate abroad information about the United States, its people, and policies" in order to "increase mutual understanding."[11]

Dwight Eisenhower, after intimately experiencing the horrors and costs of the "hot" conflict of the Second World War, campaigned on the potential of the Smith-Mundt Act in 1952, calling information infrastructure an instrument of "tremendous importance" in the new Cold War.[12] Once he arrived in office, he followed up on his campaign promise by reorganizing the diffuse US public diplomacy, propaganda, and psychological warfare offices. For the more covert, less savory, and often reprehensible forms of persuasion, Eisenhower leaned on the young CIA. But for more public-facing expressions of American interests, policies, and culture, Eisenhower created the centralized and independent USIA.

Broadly, the agency was tasked with "telling America's story," but in practice, its global infrastructure and technological tools provided means for telling stories that could validate, resist, or stand independently of the American perspective in countless cultural and aesthetic idioms. In a manner that diverged from traditional propaganda offices, the agency was to "emphasize the *community of interest* that exists among freedom loving peoples" in its goal to "imaginatively [depict] the correlation between US policies and the legitimate aspirations of

[9] Gharabaghi and Vukoder, "Motion Pictures of the United States Information Agency."

[10] For examples of this scholarship, see H. Gharabaghi, "The Syracuse Mission to Iran during the 1950s and the Rise of Documentary Diplomacy," *Journal of Cinema and Media Studies* 60, no. 4 (2021): 9–36; Y. Tsuchiya, *Occupying Eyes, Occupying Voices: CIE/USIS Films and VOA Radio* (Tokyo: University of Tokyo Press, 2012); S. Park-Primiano, "The Tentacular Reach of USIA/S in Postwar South Korea," *Journal of e-Media Studies* 6, no. 1 (2022).

[11] Govinfo.gov, "H.R. 3342—80th Congress (1947–1949), United States Information and Educational Exchange Act of 1948," January 27, 1948, https://www.govinfo.gov/content/pkg/COMPS-1091/pdf/COMPS-1091.pdf.

[12] K. Osgood, *Total Cold War: Eisenhower's Secret Propaganda Battle at Home and Abroad* (Lawrence: University Press of Kansas, 2006), 48.

other peoples of the world."[13] In other words, the agency was ostensibly meant to work alongside the national and local communities with whom they were communicating—expressing varied national interests through both straight, more informational outlets, like the newsreel, and creative means, such as experimental films. In practical terms, this collaboration with local actors across national contexts informed how the USIA designed their transnational bureaucracy. Building from the infrastructure of earlier American information agencies, the agency set up around the world hundreds of libraries, information centers, binational centers, and mobile units to circulate both US-made and locally produced books, radio programs, magazines, performance, fine art, and, of course, film and television. The agency's public affairs officers (PAOs) lived within the roughly 150 countries in which they had presence, often collaborating with a primarily local staff. In fact, more than half of the USIA's total creative and administrative labor was non-American.[14] While their radio wing, the Voice of America, was perhaps the most prolific and well known of their media, motion pictures were thought by many to be the agency's most effective outlet.

Fascinatingly, the language within meetings concerning motion picture infrastructure before the USIA's launch reveals the practical and philosophical shape to the agency's motion picture operations that aimed to leverage local perspectives toward American interests. Planners, while often diminishing the involvement of local people, understood how crucial their involvement and idiomatic appeals were for selling and ensconcing US interests and infrastructure within several countries. In a 1951 meeting between the International Motion Picture Division (IMPD) in the Department of State and several major figures from Hollywood (such as Frank Capra) concerning film circulation abroad, the attendees asked the following question: "Does the United States government… get our money's worth out of this program by… having the picture for the Philippines produced in the Philippines, certain pictures for the Iranians produced in Iran, certain pictures for the Indians produced in India?" To this, Walter Wittich, a former film professor, responded affirmatively, claiming, "This particular version of life in the Philippines is true… and accurate." And Herbert Edwards, head of the IMPD, put it in more concrete terms: "The Filipino Committee that is made of nine very knowledgeable people in the Philippines… advise on everything we do…. They develop the themes…. We own it. We

[13] N. J. Cull, *The Cold War and the United States Information Agency: American Propaganda and Public Diplomacy, 1944–1989* (Oxford: Oxford University Press, 2009), 102.

[14] R. Elder, *The Information Machine: The United States Information Agency and American Foreign Policy* (Syracuse: Syracuse University Press, 1967), 10.

finance it." The message, he added, will not "boomerang against you" because it is a "message from a Filipino to his people."[15] In 1953, the year the USIA was officially launched, Edwards's language regarding local production broadened in order to capture the essence of the USIA's ostensible ethos in motion picture collaborations with local people and governments. Speaking at a congressional hearing to Senator Fulbright, who asked Edwards about the nature of foreign nationals' authorship of the films, Edwards responded, "The ones that are produced abroad are invariably produced by local private concerns, so that the finished product will be completely in the visual idiom of the country.... These people have been able to see themselves on screen."[16]

The global scope of the USIA bureaucracy often clashed with its incredibly local dynamics—sometimes resulting in what could be labeled a dialectic, in which America's interests, reputation, and identity were negotiated within a fluid transnational context. Like with other independent agencies, the director of the USIA was nominated by the president and approved by the Senate. For better or worse, the agency policies, culture, and media output that a given director pushed typically reflected the president whom he was representing. The USIA would often play a major role in coercively imposing presidential will abroad, like in the case of Johnson's and Nixon's catastrophic actions in Vietnam. Alternatively, despite its neoimperialist underpinnings, many of the USIA employees dispersed throughout the world believed and practiced dialogism—the more idealized version of the agency mission. Former agency director George V. Allen referred to this mission as the "two-way street." Discrediting the notion that the USIA was "selling" America, Allen believed the agency should "build international understanding," even if that did not mean acceptance. Furthermore, Allen posited, "we cannot hide our faults any more than we should our virtues,"[17] thereby exhibiting a willingness to acknowledge the worst of America.[18] President Carter took this notion even further, referring

15 US Department of State, "Transcript of Proceedings, Film Advisory Committee, Monday, September 24, 1951," Washington, DC, 1951, C-32, P 218, RG 306, Records of the United States Information Agency, National Archives and Records Administration, College Park.

16 United States Senate, "Overseas Information Programs of the United States: Hearings before a Subcommittee of the Committee on Foreign Relations," 82nd Congress, Second Session, November 20–21, 1952, 212, https://www.google.com/books/edition/Overseas_Information_Programs_of_the_U_S/pdVEAQAAMAAJ?hl=en&gbpv=0.

17 G. V. Allen, "Information Is a Two-Way Street," *Challenge* 7, no. 5 (1959): 52.

18 The USIA's "willingness" to show the "worst" of America was partly an acceptance of the America that was inescapably presented through its popular culture apparatuses—namely, the Hollywood film. Throughout its history, USIA had a complicated relationship with Hollywood, often drawing from its talent pool and relying on its output, while also aiming to keep their distance and show a more "real" version of America not rendered through the Hollywood lens. For more, see Gharabaghi and Vukoder, "Motion Pictures of the United States Information Agency."

to the "two-way street" as a "reverse" or "second mandate" in which the agency would not only "tell the world about our society and policies" but also "tell ourselves about the world so as to enrich our own culture *as well as* give us the understanding to deal effectively with problems among nations."[19]

In terms of USIA media, this dialectic between the global and local translated to a complicated and diverse corpus, building a spectrum of titles ranging from insidious disinformation benefiting a hegemon to visionary, unique, and profoundly vernacular art. Whichever shape the agency took, however, "understanding" and *knowing* was prioritized. And media emerging from the agency was mined for data. The Cold War forged a world that is inescapably global but also boundlessly plural, making the ability to predict, stabilize, and grow—a logic of both nation-states and capital—increasingly difficult. Hence the *hows*, *whys*, and *to what extents* of people's engagement of cultural objects became an incredibly valued insight, whether for government agencies or businesses selling people entertainment, education, or escape.

In this sense, the comparison between the USIA and Netflix is not superficial. Both organizations' actions, beneath their ocean of content, are fundamentally premised on the circulation of cultural objects as a means of sustaining influence—whether political or technological—and growth. Furthermore, both operations are alike in terms of their transnational reach, their navigation of varying local and legal dynamics within different countries, and their interconnected multimedia productions (a newer pursuit for Netflix).[20] Post-2016, Netflix has even developed a lexicon that is strongly reminiscent of the USIA's in its espousing of a universal ideal through a cosmopolitan lens and in its employing of a language of governmentality to describe its methods and plans.

A New Discursive Cosmopolitanism: Netflix's Mission to "Entertain the World"

Extending from its terse mission statement to "entertain the world," Netflix justifies its "global footprint" through the belief that "great stories transcend borders," despite their nation of origin. In what Michael Wayne characterizes

[19] Cull, *Cold War and the United States Information Agency*, 370.

[20] Beyond theatrical releases of their in-house films, Netflix has begun a large expansion into video games. See, for example, J. Peters, "All the News about Netflix's Gaming Efforts," *Verge*, December 2024, https://www.theverge.com/22772589/netflix-video-games-app-news-updates.

as a "muddiness" to "Netflix's discursive reliance on ambiguous notions of local authenticity,"[21] the company leverages the *idea* of local productions to bolster a global brand. Wayne points to how Netflix's head of global TV, Bela Bajaria, emphasizes the company's relationship with "extraordinarily empowered local decision makers" who "make the decisions in their own time zone in their own country and in their own language," which results in the export of "local authentic stories and shows [to] everywhere around the world."[22] Nearly seventy years later, Bajara's sentiment echoes the argument Edwards made to Congress for the value of localized USIS productions. What Edwards refers to as the "visual idiom of the country," Bajara calls "local authentic stories."

This ideal of a harmonious, transnational exchange of moving-image culture raises questions about its practical viability. Moreover, how does Netflix delineate and sell the feasibility of its ambitious vision? To answer these questions, scholars have started to look more deeply at fields like international relations, public diplomacy, and media history to explain Netflix's corporate language, strategy, and identity. Despite the advertised cosmopolitan promise of Netflix's presence in an interconnected world, Ramon Lobato claims that for many countries, "the specter of cultural trespassing"[23] nevertheless lingers. Lobato adds that, while the television industry today is "characterized by dense, overlapping palimpsests of technologies, markets, and viewing habits,"[24] we need to recognize how the "history of broadcast television is closely tied to the history of the nation-state,"[25] which still informs the landscape in which companies like Netflix conduct their business. As large and influential as some US-based streamers are, they cannot just open up shop within other countries on a whim. And relative to selling a material consumer product—such as an iPhone—the dissemination of and investment in *culture* carries a slew of complicated and idiosyncratic variables that warrant a more delicate approach on the part of international streamers.

The transnational strategy central to Netflix's model and mission, therefore, mirrors that of the diplomatic and aid wings of a nation-state seeking to instantiate its influence in the world. This strategy calls for a synthesis of cutting-edge technology and on-the-ground human relationships. In Netflix's recent efforts

[21] M. L. Wayne, "Place in Netflix Original Police Drama: Local Signifiers and Global Audiences," in *Streaming Video: Storytelling across Borders*, ed. A. Lotz and R. Lobato (New York: NYU Press, 2023), 58.

[22] Ibid., 54.

[23] R. Lobato, *Netflix Nations: The Geography of Digital Distribution* (New York: NYU Press, 2019), 64.

[24] Ibid., 52.

[25] Ibid., 50.

to shape transnational policies by "engaging directly in domestic and global struggles over ideological and technological power," they are, as Evan Elkins has aptly claimed, "a political organization" practicing what he calls "streaming diplomacy."[26] Indeed, the company's post-2016 expansion hearkens back to the first two decades of the Cold War and the race to expand the range and impact of media—such as radio and broadcast television—as a diplomatic *and* political tool. In 1956, for example, USIA Deputy Director Abbot Washburn couched the agency's liberalist, cosmopolitan vision of "universal peace, freedom, and progress" for "all peoples throughout the world" in discourse about the agency's technological and media capacities shared in the context of "person-to-person contact." In what may have previously been seen as an "insurmountable" challenge, according to Washburn, the agency could boast, "modern technology [that] has provided the tools with which to do the job… there is no area so remote, no people so isolated, but that the contact can be made."[27] In other words, for the USIA, the media was the diplomacy, and the "person-to-person contact" was the means by which it was delivered.

In carrying out its "streaming diplomacy," Netflix embraces and capitalizes on its reputation as the vanguard of streaming technology and design. Even before the launch of its streaming services in 2007, though, Netflix had already begun to lay the foundation of a brand premised on choice and speed. Beginning in 1997 and stretching all the way into 2023, Netflix provided consumers with access to one of the deepest DVD and Blu-ray libraries in the world, directly mailing customers movie and television titles of all types in the now iconic red envelopes. As Lobato notes, the hyperbolic "disrupter" status with which we associate their streaming service was a label the company forged in its analog era, calling itself the "world's largest online DVD rental service" back in 2002.[28] Yet it was the now infamous "Netflix Prize," a public competition the company announced in 2006 to develop the best search algorithm, that loudly *announced* their commitment to technological innovation. Occurring over three years, the contest sparked amazing innovations in recommendation algorithms, but the company ended up making its own system behind a proprietary wall. The contest, it appeared, was just as much about performing innovation and "disruption" as it was about

[26] Elkins, "Streaming Diplomacy," 150.
[27] A. Washburn, "U.S. Information Agency," *World Affairs* 119, no. 4 (1956): 102.
[28] Lobato, *Netflix Nations*, 20.

actually using it.[29] In more recent statements to investors, the company still vaguely promises "freedom and flexibility," as well as "rapid innovation,"[30] such as ultra-4K resolution, for its consumers. However, the company also publishes its own *Tech Blog*, in which it frequently releases esoteric but legible content about the nuts and bolts beneath the hood of its platform.[31] These rhetorical demonstrations of its technological capacities attest to the competitive edge of its platform.

As with the USIA, Netflix's branding of its technological innovativeness has recently been complemented by rather traditional notions of diplomacy. As a signifier of this traditionalism, the company hired President's Obama's former national security advisor and ambassador to the UN, Susan Rice, for its board of directors in 2018.[32] While details regarding its policy lobbying and international outreach throughout the world are difficult to pin down, some of the language within its investor statements implies a diplomatic sensibility similar to the USIA's underlying these operations. Noting how Netflix is now "virtually available everywhere," the company "understand[s]… that there are cultural differences and some variances in content tastes around the world" and that these "consumers' and governments' expectations will rise." But as it continues to expand internationally, Netflix's promised "approach," much like the USIA's promise of a "two-way street,"[33] is "to listen, learn, and improve rapidly."[34]

Over the last two decades, Netflix has cultivated a transnational brand that feels familiar, usable, and diverse, with a seeming inscrutability to its superior technology and library. By powerfully promoting its advanced media technology through diplomatic strategies, the company has laid the groundwork for what Dal Yong Jin calls "platform imperialism." Jin identifies this phenomenon as an "asymmetrical relationship of interdependence in platform technologies and political culture between the West, primarily the US, and many developing countries."[35] In the age of streaming, "Netflix" has taken on a near synecdochical

[29] B. Hallinan and T. Striphas,"Recommended for You: The Netflix Prize and the Production of Algorithmic Culture," *New Media & Society* 18, no. 1 (2016): 117–37.

[30] Netflix, "Netflix's View: Streaming Entertainment Is Replacing Linear TV," https://ir.netflix.net/ir-overview/long-term-view/default.aspx.

[31] Netflix, *Netflix Tech Blog*, https://netflixtechblog.com.

[32] Elkins, "Streaming Diplomacy," 154.

[33] Allen, "Information Is a Two-Way Street," 52.

[34] Netflix, "Netflix's View."

[35] D. Y. Jin, *Digital Platforms, Imperialism, and Political Culture* (New York: Routledge, 2015), 12.

status in the world of nonlinear television, and through language and action, the US-born company has played a major role in mythologizing itself as such.

An Ongoing Legacy of Transnational Production, Distribution, and Exhibition

In media history, there are few precedents to Netflix's transnational footprint in terms of the scope of its creative and administrative labor, the variety of in-language content produced throughout the world, and the extent of the legal and technological obstacles the company continuously manages. The closest historical parallels include midcentury Hollywood's "runaway productions," which moved away from the studios in Los Angeles and looked abroad for new shooting locations, cheaper talent and labor, and favorable tax situations. In some cases, companies went as far as to build satellite studios, staffed primarily by non-American employees.[36] Similarly, with the emergence of cable and satellite television, seminal networks like CNN and MTV internationalized themselves, hiring correspondents and crews in several locations outside of the United States, even developing regional-specific channels.[37] But of these examples, the USIA, because of its idealistic mission to transcend borders, coupled with the scope of its multinational staff and network of local productions, provides perhaps the most productive historical lens through which to make sense of Netflix today.

In this section, I will move beyond a discourse analysis of Netflix's and USIA's rhetorical posturing and highlight some concrete details regarding the USIA's production, distribution, and exhibition bureaucracies and the film products they created. Rather than claiming this story, writ large, serves as a straight analogue for Netflix today, I want to position the USIA's transnational motion-picture operations as a historiographic model that helps to highlight overlooked features of Netflix's operations today. The most relevant details for understanding the streaming era pertain to the USIA's media bureaucracy, and historical instances when local labor, interests, and idioms are at tension with top-down policy goals are especially revealing.

[36] D. Steinhart, *Runaway Hollywood: Internationalizing Postwar Production and Location Shooting* (Oakland: University of California Press, 2019).

[37] Lobato, *Netflix Nations*, 104.

Yet the proficiency of the agency's globally branched bureaucracy—when things run as planned and projects are executed successfully—can be just as telling.

The agency ran at scale for forty-three-plus years with thousands of employees and hundreds of productions annually, made and distributed across 150 countries. The scope and diversity of labor, technology, and equipment allowed for a type of reach not even achievable to studio-era Hollywood. In fact, the USIA described its Motion Picture Service in 1965 as "the most far-reaching film distribution network in existence… seen by an estimated 750 million people annually in theaters, television, and through nontheatrical showings arranged by USIS."[38] It is a reach, however, that has likely been exceeded by Netflix today. And exploring the persistence of this Cold War "information machine," as Robert Elder called it,[39] can perhaps illuminate the machinery behind the continued market dominance of Netflix, which similarly leverages its transnational operations to achieve massive size and reach and to develop at once diverse and familiar offerings.

Countless graphics the USIA produced over the years chart a clean and clear picture of the hierarchy to their bureaucracy, with the sitting president always at the top and PAOs[40] at the bottom, overseeing their local post and employees. But the reality of their operations—especially in the arena of motion-pictures production—presents a far more dynamic and interesting history. Many American filmmakers and administrators were also part of this bureaucracy. They worked domestically to produce titles and to secure rights for others, eventually circulating them within the agency's vast distribution and exhibition networks. At the same time, the USIA relied heavily on all types of labor from outside the United States, in fact employing more non-Americans than Americans.[41] In the context of motion pictures, locally hired people performed a variety of jobs—serving as projectionists, researchers, cinematographers, editors, or even directors, among many other roles. And while many transnational or local productions were intended for wider distribution, just as many were intended

[38] United States Information Agency, "Facts about USIA," 1965, 5, https://www.google.com/books/edition/Facts_about_USIA_Audience_the_World/pO5GAQAAIAAJ?hl=en&gbpv=0.

[39] Elder, *Information Machine.*

[40] To further clarify, PAOs were USIA employees who managed USIS posts and lived with the community in which they worked.

[41] Elder, *Information Machine*, 10.

for and made exclusively in the language of audiences of the country in which they were made.[42]

The extent to which the USIA relied on foreign production units varied from year to year, becoming less common by the 1980s, with the agency making television programs more frequently in-house at their Washington, DC, studio. During the first three decades of the Cold War, however, more than half of the agency-made films[43] were produced abroad, at USIS posts or by contracted foreign production companies. The USIA, as previously noted, recognized the value of films being made in the cadence of the cultures into which they were disseminated. But they also saw these transnational productions—requiring extensive cooperation—as good opportunities for person-to-person diplomacy. Though the effecting of US policy was typically the goal of a given motion picture—and the productions were accordingly overseen by PAOs—some films would not address policy at all and simply be catered to the specific tastes or interests of an area. As Elder notes, the hope with such films was to "open the door for other USIA products."[44] At larger posts, like some in Brazil, in which local motion picture production was common, the agency would leave film officers to their own devices, doing little "backstopping" as the community team created a consistent output of titles. By the 1970s, the USIA had produced a film or show in over seventy-five different countries.[45]

To get a picture of the medley of people involved in foreign productions, and also of the plurality of contexts in which USIA motion pictures were exhibited abroad, I will spotlight a few examples. In 1959, Silvio Peluffo (Figure 2.1), who would sometimes take photographs for the USIS in Tripoli, decided to direct a documentary for the USIA "in his spare time." His film, *Ghirza*, which "received rave notices in Tripoli newspaper,"[46] explores the Tripoli Castle and Roman ruins

[42] A wider distribution could involve upward of a hundred countries in some cases, but interesting, per a domestic distribution ban in the Smith-Mundt Act of 1948, these globally circulated titles (and any USIA-associated title) could *not* be seen within the United States (with a few rare exceptions). The idea of a universalized America in USIA films, then, was shaped and experienced only by those *outside* of America. This universalized version, perhaps, is like the more borderless, corporate vision exhibited in Netflix originals today. For more details, see Gharabaghi and Vukoder, "The Motion Pictures of the United States Information Agency."

[43] United States Information Agency, "4th Report to Congress, January 1–June 30, 1955," 8, https://www.google.com/books/edition/Report_to_Congress/yoTPDTsbXhgC?hl=en.

[44] Elder, *Information Machine*, 244.

[45] Ibid., 248.

[46] United States Information Agency, *USIA Correspondent* 1, no. 6 (May 1959): 8, RG 306, p. 191, Records of the United States Information Agency, National Archives and Records Administration, College Park.

Figure 2.1 The director Peluffo, a local contributor to USIS Tripoli, smiles as he holds his camera. *Ghirza*, directed by Silvio Peluffo © Liberty Productions 1959. *USIA Correspondent*, 1 (6): 8, RG 306, p. 191, Records of the United States Information Agency, National Archives and Records Administration, College Park.

in the area. Meanwhile, in Pusan, Hong Wan-Pyo, a USIS employee who was acknowledged for his "noteworthy performance" in a semiannual agency report, set up "USIS clubs" for which he would arrange screenings, sometimes from one of the agency's mobile vans equipped for film projection.[47] In Tegucigalpa,

[47] United States Information Agency, "7th Report to Congress, July 1–December 31, 1956," 30, https://www.google.com/books/edition/Report_to_Congress/yoTPDTsbXhgC?hl=en.

Marcial Sevilla, a USIS cinematographer, was part of a team making a film about a delegation of American doctors visiting Honduras.[48] And in the Philippines, former president Ferdinand E. Marcos himself served as a "guest narrator" in 1967 for the film *Civic Action at Home and Abroad*.[49]

The motion-picture labor done for the USIA abroad is, in and of itself, worthy of historical investigation. Yet such specific cases are perhaps best employed historiographically when measured against the larger, purportedly cosmopolitan mission of the USIA and when the films themselves are analyzed relative to their policy contexts. What does it mean for a multinational team to participate in the "telling" of "America's story" or in the narration of American culture, policies, or ideologies? Alternatively, what does it mean to narrate one's own story within this American bureaucratic machine? In many cases, films are aligned with US interests and policies and typically executed in a by-the-books documentary mode. But in others, even some that are not particularly obvious, we can perceive a film—and the cultures it represents—to be in tension with the system in which it was created.

One such example is director Chun Sun Myung's 1965 *Island Doctor*, produced in South Korea (Figure 2.2).[50] The film gives a documentary account of a Korean physician, Timothy Rhee, who moves to the remote island Ulleung-do in the Sea of Japan to "invest his humanity" as the small community's "first doctor." To underscore a message that champions the introduction of Western medicine in a place that has been otherwise unexposed to it, the film plays images of the old and new off one another, suggesting the traditional and modern can be reconciled. The documentary is bookended by images of a weathered 2,000-year-old tree sitting atop a snowy cliff—a signifier of the island's timelessness and distinct character. Having withstood harsh weather for two millennia, the tree suggests that so too can the people of the island weather new conditions and ideas. Rhee, a man of the world (and a stand-in for UN-supported medical aid), disparages the "ignorance" of other, more traditional methods, such as an "ineffective

48 United States Information Agency, *USIA World* 9, no. 9 (May 1976): 8, 306-USW, Box 25, Records of the United States Information Agency, National Archives and Records Administration, College Park.

49 United States Information Agency, *USIA World* 1, no. 6 (November 1967): 5, 306-USW, Box 10, Records of the United States Information Agency, National Archives and Records Administration, College Park.

50 Chun Sun Myung, dir., *Island Doctor*, film, 1965, United States Information Agency, 306.3318, Records of the United States Information Agency, National Archives and Records Administration, College Park.

Figure 2.2 Clockwise from left: the ancient tree whose image bookends the film; Dr. Timothy Rhee and his team are greeted as they arrive at Ulleung-do's port; and Dr. Rhee and team treat a hesitant patient. *Island Doctor*, directed by Chun Sun Myung © Liberty Productions 1965. 306.3318, Records of the United States Information Agency, National Archives and Records Administration, College Park.

medicine pot." Yet, reflecting on his training, he knows he has to meet the island community where they are at. Rhee's realization is captured in a scene in which he gazes up at photographs of his idols on his bookshelf—Mahatma Ghandi, Father Damien of Molokai, and Albert Schweitzer, who he studied alongside earlier in his life. Though his first patient, as the film explains, refuses to thank him, Rhee recounts Schweitzer's words of advice: "Anyone who proposes to do good must not expect people to roll stones out of his way."

Although *Island Doctor* plays as a quiet, modest, though beautiful film, it is nevertheless emblematic of the USIA's (and United States') sense of itself in the Cold War world. The multiple infrastructures that the USIA and other US agencies brought were designed to instantiate a reconciliation of new material realities with those things that were elemental to an area's distinct culture. And Rhee, a Western-trained but Korean physician, is employed as a symbol of that potential reconciliation. Beyond the film text itself, some of *Island Doctor*'s production details reinforce this notion, tying the policy messaging to the *localness* of the film itself. In his history of the USIS in South Korea, Han Sang Kim defines two eras of USIS-Korea documentaries during the earlier decades of the Cold War. Before 1958, he claims, most of these titles were an extension of the USIS itself, labeled as "educational films" and leaving the local filmmakers' names off the credits. However, in the late 1950s, the documentaries were increasingly categorized as "cultural films." In titles like *Island Doctor*, the Korean directors' names were now attached, and (suggesting artistic vision behind the film)[51] the documentaries were often entered into film festivals.[52] But Kim calls it an "ironic fate"[53] for many of these Korean directors. The newfound agency, artistry, and attribution behind the documentaries was still part and parcel to the United States' neoimperialist institution. In *Island Doctor*, the new medical infrastructure, while certainly beneficial, could not be fully separated from the source from which it came.

Island Doctor, while filled with striking images of a distinct landscape and community, is actually quite typical for a USIA production. It is a one-reel documentary, one of three formats that were the agency's bread and butter. The most frequently seen within the archive, the one- or two-reel documentary would cover a variety of subjects, from the story of how waste management systems work[54] in the United States to the American tour of the Lebanese pop star Fayrouz.[55] Alongside this mode, newsreels, especially in the first two

[51] H. S. Kim, "Uneven Screens, Contested Identities: USIS, Cultural Films, and the National Imaginary in South Korea, 1945–1972," PhD diss., Seoul National University, 2013.

[52] *Island Doctor* was indeed entered into the 12th Asian Film Festival in 1965. Ibid., 232.

[53] Ibid., 235.

[54] US Department of Health Education and Welfare, *The Stuff We Throw Away*, film, 1970, United States Information Agency, 306.10876, Records of the United States Information Agency, National Archives and Records Administration, College Park.

[55] Morten Parker, dir., *Fayrouz*, film, 1972, United States Information Agency, 306.5084, Records of the United States Information Agency, National Archives and Records Administration, College Park.

decades of the Cold War, also constituted a major output for the agency and were the most likely to have been produced on a consistent basis in other countries by a non-American filmmaking team. Interestingly, USIA newsreels often aggregated news from America with stories specific to their own country's or region's politics, industries, and cultures. Coverage of Eisenhower's State of the Union address, for example, could be placed alongside footage from a community arts festival occurring in the newsreel's home country. To date, we have identified nearly twenty distinct series emerging from several regions—such as 306-CNR (Chile Newsreels), 306-IN (News of Iran), and 306-PT (Pakistan Today). In addition to one-reel documentaries and the newsreel, the "dialogue" was a hugely prolific format. In the 1980s this format's prominence coincided with the USIA's investment in televisual and satellite technologies. Programs from record groups like 306-WNET (Worldnet Today) and 306-ED (Electronic Dialogue) generally featured long, specialized discussions between domestic or international journalists and American politicians or experts within a given field.

Although most titles in the archive fall under one of these three umbrellas, the agency frequently dipped its toes in other formats. Like Netflix today, the goal was to strike a balance of consistency and diversity in subject for audiences. And also like Netflix today, saturation—always offering *something* to watch and a *place and means* to watch—was crucial to the USIA's branding efforts. Demonstrating willingness to support alternative formats and artists, the agency would occasionally sponsor or distribute the work of avant-garde filmmakers, circulating films by artists like Stan Brakhage[56] and directly funding projects such as Curtis Harrington's *The Four Elements*.[57] The agency also deployed animated shorts, such as the ham-fisted *Escuela de Asesinos* (School for Assassins),[58] and puppet art, like the 306-HAJ (Tales of the Hoja Nasreddin) series. With a similar reticence to Netflix's theatrical release strategies for in-house productions, the USIA would even help fund and shoot feature-length narrative films. These

[56] For example, Curtis Harrington, dir., *Prelude: Dog Star Man*, film, 1961, United States Information Agency, 306.8489, Records of the United States Information Agency, National Archives and Records Administration, College Park.

[57] Chun Sun Myung, *The Four Elements*, film, 1965, United States Information Agency, 306.4585, Records of the United States Information Agency, National Archives and Records Administration, College Park.

[58] United States Information Agency, *School of Assassins* (*Escuela de Assesinos*), film, 1967, 306.2997, Records of the United States Information Agency, National Archives and Records Administration, College Park.

titles, often leaving the USIA name unattached, would have limited release scopes within mainstream cinemas, showing alongside popular films produced by the domestic culture industries. One noteworthy example is the 1959 film *Give Us This Day* (or *Ama Namin* in Tagalog), which appears to be a feature-length documentary (with staged scenes) about an inspiring teacher. According to the *USIA Correspondent*, an internal agency periodical, "The picture was made by the Lebran Studio of Manila under supervision of USIS Manila and IMS [the Motion Picture Service]. Such well-known Filipino actors and actresses as Manuel Conde, Charito de Leon, and Simoun Almario play in it."[59] In a similar vein, the agency even expanded into a telenovela in Latin America, producing 26 episodes of *Nuestro Barrio* in the 1960s for the audiences in the region.[60]

The twenty thousand titles of USIA and the thousands that Netflix has produced or licensed mean nothing if they are not accessible for viewing. Fittingly, both organizations have a history of investing in transnational distribution infrastructure. One could argue the resistance Netflix sometimes faces today when it expands is in part a function of the pervasiveness of American media during the Cold War—a legacy in which the USIA played an important part. Although Hollywood had a global footprint since the 1910s, USIA media, with its heavy investment in distribution, reached places the studios never could. In several cases, though these stories often very overstated, agency films were people's first exposure to motion pictures.[61] Within earlier iterations of the agency, and into the 1960s, the USIA relied on "mobile film units," which in essence were off-road-capable vehicles (vans, boats, horse-drawn carriages, among other means) that could double as portable movie theaters. In agency literature, these vehicles were mythologized as being capable of "awakening a spirit,"[62] "calm[ing] a belligerent group,"[63] or even motivating "30 campesinos

[59] United States Information Agency, *USIA Correspondent* 1, no. 6 (May 1959): 3, RG 306, p. 191, Records of the United States Information Agency, National Archives and Records Administration, College Park.

[60] Elder, *Information Machine*, 249; and United States Information Agency, *Nuestro Barrio*, television series, 1962–8, 306-NB, Records of the United States Information Agency, National Archives and Records Administration, College Park.

[61] See W. Dizard, *The Strategy of Truth: The Story of the U.S. Information Service* (Washington, DC: Public Affairs Press, 1961), 99; and H. Gharabaghi, "American Mice Grow Big! The Syracuse Audiovisual Mission in Iran and the Rise of Documentary Diplomacy," PhD diss., New York University, 2018.

[62] United States Information Agency, *USIA Correspondent* 6, no. 1 (January 1964): 12, 306-USC, Box 1, Records of the United States Information Agency, National Archives and Records Administration, College Park.

[63] United States Information Agency, *USIA Correspondent* 6, no. 12 (December 1964): 11, 306-USC, Box 3, Records of the United States Information Agency, National Archives and Records Administration, College Park.

from the backcountry of El Salvador [to] voluntarily [give] their time to build a makeshift road through the jungle underbrush so a USIS mobile film unit could travel into the remote area."[64] With television, these mobile units became less necessary, but the romanticization of "reach" persisted. For example, Alvin Snyder, the agency's head of motion pictures during Reagan's presidency, waxes poetic about the agency's early investment in satellite television: "We could turn [satellite] on virtually within minutes of when we wanted it. Because a satellite signal beam, like a giant spotlight from space, can cover enormous areas, we would be able to fit both Europe and Africa within the same cone of coverage."[65]

With an ability to reach nearly everywhere outside of the Soviet Union's sphere of influence, USIA motion pictures capitalized on all types of exhibition spaces and technologies, playing in commercial theaters, USIS libraries and centers, embassies, over television, and in several other types of nontheatrical spaces. In their ethnographic study of Cold War Iceland, which was open to both US and USSR media, Hafsteinsson and Grétarsdóttir describe the lengths offices like the USIA were willing to go in order to show their films. Anticipating the reach of streaming, USIA films played in "well-equipped cinemas" but also in "private homes, community halls, prisons, church associations, agricultural associations, youth clubs, and schools."[66] The aesthetics and technologies of exhibition, as well, were often central to their motion picture mission. Across Turkey in 1960, a Turkish State Railways car was transformed into a portable theater, making its way through nine towns in the country and drawing as many as 1,800 people to watch a film projected by a railroad employee.[67] In the early 1960s, the agency (with a team of local carpenters) constructed temporary Circarama theaters in several parts of South Asia.[68] During its visit to Calcutta (Kolkata), the theater drew 1,700 viewers per hour.[69] And in their grandest,

[64] United States Information Agency, *USIA Correspondent* 4, no. 5 (May 1962): 7, RG 306, p. 191, Records of the United States Information Agency, National Archives and Records Administration, College Park.

[65] A. Snyder, *Warriors of Disinformation: How Lies, Videotape, and the USIA Won the Cold War* (New York: Arcade, 1995), 106.

[66] S. B. Hafsteinsson and T. Gréétarsdóóttir, "Screening Propaganda: The Reception of Soviet and American Film Screenings in Rural Iceland, 1950–1975," *Film History* 23, no. 4 (2011): 366.

[67] United States Information Agency, *USIA Correspondent* 2, no. 4 (March 1960): 1, RG 306, p. 191, Records of the United States Information Agency, National Archives and Records Administration, College Park.

[68] See United States Information Agency, *Building Circarama*, film, ca. 1960, 306.6935, Records of the United States Information Agency, National Archives and Records Administration, College Park.

[69] United States Information Agency, *USIA Correspondent* 3, no. 10 (November 1961): 1, RG 306, p. 191, Records of the United States Information Agency, National Archives and Records Administration, College Park.

though eventually unrealized, plan, the USIA laid the groundwork to exhibit a Cinerama-shot feature about the potential of nuclear energy, called *The Eighth Day*, on an aircraft carrier moving from port to port.[70]

Netflix's distribution and exhibition strategies echo many elements of the USIA's, especially as it seeks to create a tangible and socialized encounter with the content central to its brand. For example, the company plans to create a "fever dream" of an experience with their new "Netflix House" venues, which will open in 2025 in Dallas, Texas, and King of Prussia, Pennsylvania.[71] Not only will visitors be able to eat, shop, and play all things Netflix, the latter location will also feature a theater in which they will do screenings. It is easy to imagine that, if these two pilot locations prove successful, the company will internationalize the plan. Additionally, though most people view their content through the app, Netflix still releases its films in traditional commercial theaters. For the company, it is not about selling tickets. As with USIA's rare theatrically released narrative productions, Netflix's "capillary collaboration" with theaters is designed to "keep a presence in specific territories" and maintain a channel of political and cultural access.

However, the heart of Netflix's distribution and exhibition strategy, like USIA's, is predicated on reach and multimodality. Its fluid and legible platform, available on nearly every device with a screen, has been the defining factor of the brand since 2007. In what Lobato calls the company's "infrastructural ontology" or the "complex and dynamic metasystem made up of hundreds of different software processes that relies on *hard* and *soft* technological infrastructure, *open* and *closed* knowledge systems, and *public* and *private* investment,"[72] Netflix has entrenched itself as the most popular transnational streamer. The cohesiveness and ease of the platform is what gives Netflix the footing to negotiate things like access to foreign productions and the skirting of national media quotas. But this cohesiveness has especially given the company power in dealing with government investment in internet technologies, which only serve to reinforce the platform's reach. Echoing the USIA's relationship to satellite infrastructure and coverage,

[70] J. Hinkelman, "The Eighth Day: Presenting a Nuclear Age in Cinerama and Stereophonic Sound," paper presented at the Society for Cinema and Media Studies Conference, March 14–18, 2018, Toronto, Canada.

[71] H. Goldblatt, "Netflix House Will Let You Experience Your Favorite Shows, Movies in Real Life," Tudum, June 17, 2024, https://netflix.com/tudum/articles/netflix-house.

[72] Lobato, *Netflix Nations*, 101.

Netflix supplements its political lobbying for high-speed broadband in more remote regions, as Elkins claims, with its *own* development of infrastructure.[73] And it is in this sense—in Netflix's "infrastructural ontology"—that the "platform imperialism" to which Jin speaks becomes particularly clear. Netflix constructs the means by which to access content and then saturates that space with films and shows that speak to and across multiple cultural idioms.

Netflix's technological infrastructure, multimodality, and seamless platform all must be reinforced, of course, by the content itself—some of which is distributed globally and other content being targeted for particular regions or countries. According to Stuart Davis, this blended strategy, much like USIA's, establishes a supranational dominance.[74] Citing Ernest Mandel, Davis characterizes the monopolizing process as an "international amalgamation"[75] that creates cohesion, stability, and protection from market competitors. Differently from the USIA, Netflix's imperialism is not tethered to its home nation. Many of its flagship shows that are about or produced in the United States, indeed, are among their most popular content globally, such as *Stranger Things*, *The Night Agent*, and *Dahmer*. However, some programs that are shot abroad, such as *Wednesday* (shot in Romania), and others that are produced and take place abroad, like *Squid Game* (South Korea), also top this list—underscoring Netflix's growing transnational ethos.[76] These globally watched programs, then, are complemented by what Davis calls a "localization process." For the company, this involves both the "licensing of content distribution for within certain national markets" and the production of several "Netflix Originals" intended for regional or national distribution outside of the United States.[77] By seeking to invoke dually cosmopolitan and idiomatic relationship to the content, Netflix can sell its transnational branding and infrastructural aspirations to markets throughout the world.

73 Elkins, "Streaming Diplomacy," 155.
74 S. Davis, "What Is Netflix Imperialism? Interrogating the Monopoly Aspirations of the 'World's Largest Television Network,'" *Information, Communication, and Society* 26, no. 6 (2023): 1143–58.
75 E. Mandel, "International Capitalism and 'Supranationality,'" in *International Firms and Modern Imperialism*, ed. H. Radice (Harmondsworth, UK: Penguin, 1975), 146, quoted in Davis, "What Is Netflix Imperialism?," 1151.
76 Netflix, "Top 10," Tudum, https://www.netflix.com/tudum/top10/most-popular.
77 Davis, "What Is Netflix Imperialism?," 1151.

Conclusion

The USIA and Netflix are temporally and technologically distinct actors with different incentive structures operating in tremendously different contexts. However, their comparable reach, breadth of content, multinational network of labor, and investments in media infrastructure present a valuable opportunity for historicizing the contemporary streaming era. Each organization—speaking through a cosmopolitan discourse of a shared humanity that can exist locally and globally at the same time—seeks a form of control and dominance. For the USIA, operating in the context of the Cold War, that meant facilitating American interests and policies in perpetuity. For Netflix, like most streaming corporations today, it means predictable and consistent growth for its investors. Branding a familiar, accessible, but engaging media network across cultures is essential to fostering consent for Netflix's hegemonic aspirations.

This chapter aims to provide an introduction to a potentially fruitful historiographic framework through which to unpack the many obfuscated details of Netflix's operations today. But larger questions about power, politics, and identity remain, as fewer media companies oversee larger pieces of our cultural consumption. Despite some setbacks, such as the so-called "Great Netflix Correction" in 2022, in the past few years Netflix feels more embedded than ever in media cultures throughout the world. And, as scholars like John Cheney-Lippold have wondered, can the underlying algorithmic logics to Netflix lead to more uniform "algorithmic identities," in which our "corporeal and societal forms" are "de-essentialize[d]" and then "re-essentializ[ed] as a statistically-related, largely market research-driven category"?[78] In other words, will culture congeal into the borderless "taste cluster" to which Netflix assigns us?

This concern over the potential of sweeping, borderless hegemonies underlay the Cold War era as well. When we talk about the Cold War as simply the battle between two "superpowers," we can discount the histories of transcultural action, like the Non-Aligned Movement, which helped the world remain richly plural—despite the encroaching hegemonies of global capitalism. The USIA archive, while undeniably the product of a top-down bureaucracy serving a larger neoimperial project, helps activate this more complicated and heterogenous picture of the Cold War. Moreover, its content includes novel filmic styles,

[78] J. Cheney-Lippold, "A New Algorithmic Identity: Soft Biopolitics and the Modulation of Control," *Theory, Culture & Society* 28, no. 6 (2011): 170.

surprising dialectics, and diverse media artifacts that capture place and cultural perspective at given moments in time. With time, perhaps Netflix's transnational media and outreach will also inadvertently foster similar disruptive, diversifying externalities. Perhaps it will not.

Bibliography

Albornoz, L., and M. T. García Leiva. "Netflix Originals in Spain: Challenging Diversity." *Critical Studies in Television* 37, no. 1 (2022): 63–81.

Allen, G. V. "Information Is a Two-Way Street." *Challenge* 7, no. 5 (1959): 48–53.

Asmar, Axelle, Tim Raats, and Leo Van Audenhove. "Streaming Difference(s): Netflix and the Branding of Diversity." *Critical Studies in Television* 18, no. 1 (2022): 24–40.

Cheney-Lippold, J. "A New Algorithmic Identity: Soft Biopolitics and the Modulation of Control." *Theory, Culture & Society* 28, no. 6 (2011): 164–81.

Cull, N. J. *The Cold War and the United States Information Agency: American Propaganda and Public Diplomacy, 1944–1989*. Oxford: Oxford University Press, 2009.

Davis, S. "What Is Netflix Imperialism? Interrogating the Monopoly Aspirations of the 'World's Largest Television Network.'" *Information, Communication, and Society* 26, no. 6 (2023): 1143–58.

Dizard W. *The Strategy of Truth: The Story of the U.S. Information Service*. Washington, DC: Public Affairs Press, 1961.

Elder, R. *The Information Machine: The United States Information Agency and American Foreign Policy*. Syracuse, NY: Syracuse University Press, 1967.

Elkins, Evan. "Streaming Diplomacy: Netflix's Domestic Politics and Foreign Policy." In *The Routledge Handbook of Digital Media and Globalization*, edited by Dal Yong Jin, 150–7. New York: Routledge, 2021.

Gharabaghi, H. "American Mice Grow Big! The Syracuse Audiovisual Mission in Iran and the Rise of Documentary Diplomacy." PhD diss., New York University, 2018.

Gharabaghi, H."The Syracuse Mission to Iran during the 1950s and the Rise of Documentary Diplomacy." *Journal of Cinema and Media Studies* 60, no. 4 (2021): 9–36.

Gharabaghi, H., and Bret Vukoder. "The Motion Pictures of the United States Information Agency: Studying a Global Film and Television Operation." *Journal of e-Media Studies* 6, no. 1 (2022).

Goldblatt, H. "Netflix House Will Let You Experience Your Favorite Shows, Movies in Real Life." Tudum, June 17, 2025. https://www.netflix.com/tudum/articles/netflix-house.

Govinfo.gov. "H.R. 3342—80th Congress (1947–1949), United States Information and Educational Exchange Act of 1948." January 27, 1948. https://www.govinfo.gov/content/pkg/COMPS-1091/pdf/COMPS-1091.pdf.

Hafsteinsson, S. B., and T. Gréétarsdóóttir. "Screening Propaganda: The Reception of Soviet and American Film Screenings in Rural Iceland, 1950–1975." *Film History* 23, no. 4 (2011): 361–75.

Hallinan, B., and T. Striphas. "Recommended for You: The Netflix Prize and the Production of Algorithmic Culture." *New Media & Society* 18, no. 1 (2016): 117–37.

Harrington, Curtis, dir. *Prelude: Dog Star Man*. Film, 1961. United States Information Agency, 306.8489. Records of the United States Information Agency, National Archives and Records Administration, College Park.

Hinkelman, J. "The Eighth Day: Presenting a Nuclear Age in Cinerama and Stereophonic Sound." Paper presented at the Society for Cinema and Media Studies Conference, March 14–18, 2018, Toronto, Canada.

Jin, D. Y. *Digital Platforms, Imperialism, and Political Culture*. New York: Routledge, 2015.

Kim, H. S. "Uneven Screens, Contested Identities: USIS, Cultural Films, and the National Imaginary in South Korea, 1945–1972." PhD diss., Seoul National University, 2013.

Lobato R. *Netflix Nations: The Geography of Digital Distribution*. New York: NYU Press, 2019.

Lobato, R., A. Scarlata, and S. Cunningham. "Conceptualizing the National and the Global in SVOD Original Production." In *Streaming Video: Storytelling across Borders*, edited by A. Lotz and R. Lobato, 37–53. New York: NYU Press, 2023.

Mandel, E. "International Capitalism and 'Supranationality.'" In *International Firms and Modern Imperialism*, edited by H. Radice, 27–41. Harmondsworth, UK: Penguin, 1975.

Myung, Chun Sun, dir. *The Four Elements*. Film, 1965. United States Information Agency, 306.4585. Records of the United States Information Agency, National Archives and Records Administration, College Park.

Myung, Chun Sun, dir. *Island Doctor*. Film, 1965. United States Information Agency, 306.3318. Records of the United States Information Agency, National Archives and Records Administration, College Park.

Netflix. "About Netflix." 2024. https://about.netflix.com/en.

Netflix. "Netflix's View: Streaming Entertainment Is Replacing Linear TV." https://ir.netflix.net/ir-overview/long-term-view/default.aspx.

Netflix. *Netflix Tech Blog*. https://netflixtechblog.com.

Netflix. "Top 10." Tudum. https://www.netflix.com/tudum/top10/most-popular.

Netflix. "Top Investor Questions." 2024. https://ir.netflix.net/ir-overview/top-investor-questions/default.aspx.

Osgood, K. *Total Cold War: Eisenhower's Secret Propaganda Battle at Home and Abroad*. Lawrence: University Press of Kansas, 2006.

Parker, Morten, dir. *Fayrouz*. Film, 1972. United States Information Agency, 306.5084. Records of the United States Information Agency, National Archives and Records Administration, College Park.

Park-Primiano, S. "The Tentacular Reach of USIA/S in Postwar South Korea." *Journal of e-Media Studies* 6, no. 1 (2022). https://doi.org/10.1349/PS1.1938-6060.A.477.

Peters, J. "All the News about Netflix's Gaming Efforts." *Verge*, December 2024. https://www.theverge.com/22772589/netflix-video-games-app-news-updates.

Snyder A. *Warriors of Disinformation: How Lies, Videotape, and the USIA Won the Cold War.* New York: Arcade, 1995.

Steinhart, D. *Runaway Hollywood: Internationalizing Postwar Production and Location Shooting*. Oakland: University of California Press, 2019.

Tsuchiya, Y. *Occupying Eyes, Occupying Voices: CIE/USIS Films and VOA Radio*. Tokyo: University of Tokyo Press, 2012.

United States Information Agency. "4th Report to Congress, January 1–June 30, 1955," 8. https://www.google.com/books/edition/Report_to_Congress/yoTPDTsbXhgC?hl=en.

United States Information Agency. "7th Report to Congress, July 1–December 31, 1956," 30. https://www.google.com/books/edition/Report_to_Congress/yoTPDTsbXhgC?hl=en.

United States Information Agency. *Building Circarama*. Film, ca. 1960. 306.6935. Records of the United States Information Agency, National Archives and Records Administration, College Park.

United States Information Agency. "Facts about USIA," 1965: 5. https://www.google.com/books/edition/Facts_about_USIA_Audience_the_World/pO5GAQAAIAAJ?hl=en&gbpv=0.

United States Information Agency. *Nuestro Barrio*. Television series, 1962–8. 306-NB. Records of the United States Information Agency, National Archives and Records Administration, College Park.

United States Information Agency. *School of Assassins* (*Escuela de Assesinos*). Film, 1967. 306.2997. Records of the United States Information Agency, National Archives and Records Administration, College Park.

United States Information Agency. *USIA Correspondent* 1, no. 6 (May 1959): 3. RG 306, p. 191. Records of the United States Information Agency, National Archives and Records Administration, College Park.

United States Information Agency. *USIA Correspondent* 1, no. 6 (May 1959): 8. RG 306, p. 191. Records of the United States Information Agency, National Archives and Records Administration, College Park.

United States Information Agency. *USIA Correspondent* 2, no. 4 (March 1960): 1.// RG 306, p. 191. Records of the United States Information Agency, National Archives and Records Administration, College Park.

United States Information Agency. *USIA Correspondent* 3, no. 10 (November 1961): 1. RG 306, p. 191. Records of the United States Information Agency, National Archives and Records Administration, College Park.

United States Information Agency. *USIA Correspondent* 4, no. 5 (May 1962): 7.// RG 306, p. 191. Records of the United States Information Agency, National Archives and Records Administration, College Park.

United States Information Agency. *USIA Correspondent* 6, no. 1 (January 1964): 12. 306-USC, Box 1. Records of the United States Information Agency, National Archives and Records Administration, College Park.

United States Information Agency. *USIA Correspondent* 6, no. 12 (December 1964): 11. 306-USC, Box 3. Records of the United States Information Agency, National Archives and Records Administration, College Park.

United States Information Agency. *USIA World* 1, no. 6 (November 1967): 5. 306-USW, Box 10. Records of the United States Information Agency, National Archives and Records Administration, College Park.

United States Information Agency. *USIA World* 9, no. 9 (May 1976): 8. 306-USW, Box 25. Records of the United States Information Agency, National Archives and Records Administration, College Park.

United States Senate. "Overseas Information Programs of the United States: Hearings before a Subcommittee of the Committee on Foreign Relations," 82nd Congress, Second Session, November 20–21, 1952. https://www.google.com/books/edition/Overseas_Information_Programs_of_the_U_S/pdVEAQAAMAAJ?hl=en&gbpv=0.

US Department of Health Education and Welfare. *The Stuff We Throw Away*. Film, 1970. United States Information Agency, 306.10876. Records of the United States Information Agency, National Archives and Records Administration, College Park.

US Department of State. "Transcript of Proceedings, Film Advisory Committee, Monday, September 24, 1951." Washington, DC, 1951, C-32, P 218, RG 306. Records of the United States Information Agency, National Archives and Records Administration, College Park.

Washburn, A. "U.S. Information Agency." *World Affairs* 119, no. 4 (1956): 101–3.

Wayne, M. L. "Place in Netflix Original Police Drama: Local Signifiers and Global Audiences." In *Streaming Video: Storytelling across Borders*, edited by A. Lotz and R. Lobato, 54–70. New York: NYU Press, 2023.

3

Netflix's Double-Edged Sword for Local Storytelling: A Brazilian Perspective

Daniel Rios, Melina Meimaridis, and Daniela Mazur

Since 2016, Netflix has been producing original Brazilian content as part of its expansion in Latin America, introducing titles like *3%* (2016–20) and *Invisible City* (2021–3), which circulate both within Brazil and internationally. With a library now surpassing sixty original Brazilian productions, spanning series, reality shows, films, and documentaries, Netflix reaffirmed its commitment to fostering the local market in 2023.[1] This commitment materialized through the release of new original content developed in Brazil and a substantial financial injection of one billion *reais* (approximately two hundred million US dollars) into the country's audiovisual sector between 2023 and 2024. In general, transnational subscription video-on-demand (SVOD) services have been increasing their investments in locally developed original content to expand and maintain their international subscriber bases.[2] Estimates suggest that about 8 percent of Netflix's 238 million subscribers hail from Brazil.[3] Netflix's move to increase Brazilian production demonstrates its recognition of the importance of Brazil's market and its efforts to target and engage local viewers.

In examining Brazil's current audiovisual landscape, an analysis of Netflix's localization strategies becomes crucial, particularly as these strategies result in challenging the established dominance of Grupo Globo, the largest media conglomerate in Latin America and one of the largest in the world. Through

1 Luiz Pacete, "Netflix aposta no longo prazo para ampliar produção local no Brasil," *Forbes*, January 11, 2023.

2 Catalina Iordache, "Netflix in Europe: Four Markets, Four Platforms? A Comparative Analysis of Audio-Visual Offerings and Investment Strategies in Four EU States," *Television & New Media* 23, no. 7 (2022): 721–42.

3 Gustavo Ravache, "Exclusivo: Netflix faz investimento recorde no país, apesar da crise dos streamings," *Valor Econômico*, September 28, 2023.

its television network, TV Globo, this conglomerate has traditionally dominated the representation of Brazil and Brazilians on TV, both nationally and internationally.[4] Its depictions are often limited, predominantly mirroring the experiences of a more affluent, southeastern, White, and urban segment of Brazilian society.[5] For decades, TV Globo has offered such a restricted view of Brazilian national identity. In contrast, Netflix has strategically channeled investments into narratives that embrace a broader range of geographical and cultural perspectives, thereby potentially expanding and diversifying the ways Brazilian identity is portrayed on screen. The streaming company fosters varied storytelling opportunities, giving voice to media professionals previously sidelined in Brazil's audiovisual market.[6]

Yet, akin to developments observed in South Korea[7] and Mexico,[8] Netflix's Brazilian originals are fraught with contradictions. Khoo contrasts traditional TV's conservative approach, driven by advertisers, with Netflix's strategy relying on a vast, transnational subscriber base.[9] While Netflix champions diversity to appeal to a wide range of viewers, this inclusivity is often curated to maximize global appeal and profit. Lobato's view that Netflix's libraries "might be better understood through the prism of transnational class formations rather than the frame of the national audience" is valid.[10] Nevertheless, applying this view to understand Netflix's original productions in global South countries like Brazil can also risk oversimplification and may overlook crucial nuances. If in the past Brazilian narratives were mostly limited by Grupo Globo and aimed primarily at Brazilian audiences, now local producers catering to global streaming platforms like Netflix find themselves compelled to create stories that "feel" local

4 Maria I. V. Lopes, "Telenovela brasileira: uma narrativa sobre a nação," *Comunicação & Educação* 26 (2003): 17–34; Mauro Porto, "Telenovelas and Representations of National Identity in Brazil," *Media, Culture, and Society* 33, no. 1 (2011): 53–69.

5 Antonio La Pastina, Joseph Straubhaar, and Lirian Sifuentes, "Why Do I Feel I Don't Belong to the Brazil on TV?," *Popular Communication* 12, no. 2 (2014): 104–16.

6 Daniel Rios, "Diversidade e Dependência: as estratégias de comissionamento de séries da Netflix no Brasil," PhD diss., Universidade Federal Fluminense, 2024.

7 Dal Yong Jin, "Netflix's Effects on the Korean Wave: Power Relations between Local Cultural Industries and Global OTT Platforms," *Asian Journal of Communication* 33, no. 5 (2023): 452–69.

8 Juan Llamas-Rodríguez, "The Return of Indigenismo in Netflix Mexico," in *Streaming Video: Storytelling across Borders*, ed. Amanda Lotz and Ramon Lobato (New York: NYU Press, 2023), 225–47.

9 Olivia Khoo, "Picturing Diversity: Netflix's Inclusion Strategy and the Netflix Recommender Algorithm (NRA)," *Television & New Media* 24, no. 3 (2023): 281–97.

10 Ramon Lobato, "Rethinking International TV Flows Research in the Age of Netflix," *Television & New Media* 19, no. 3 (2018): 252.

but also have broad international appeal. The company's Western universalist approach to diversity within storytelling can ironically dilute locally produced representations of Brazilian culture and experiences designed to fit its platform.[11]

Through an examination of current literature, scrutiny of local media coverage, and analysis of Netflix's original series and miniseries in Brazil from 2016 to 2023, this study seeks to understand how Brazil and Brazilianness are represented in popular culture. Concentrating on the company's creations, we address the following questions: (1) What Brazilian narratives and storytelling strategies has Netflix privileged? (2) How do these narratives confirm or complicate dominant stereotypes about Brazil in cultural representation? And (3) what challenges arise as Netflix adapts its content to meet demand among Brazilian viewers? Our aim is to enrich discussions on how Netflix shapes and mediates popular perceptions of Brazil and Brazilians, both domestically and internationally.

We contend that while Netflix promotes greater cultural diversity in its Brazilian productions, it simultaneously reinforces US dominance by imposing US production practices and themes onto Brazil's cultural industry, which, for better or for worse, displaces local norms, conventions and aesthetics of Brazilian audiovisual works. Brazil's audiovisual industry has unique genres and formats, like telenovelas, developed over decades. These culturally specific genres and formats are at risk when a major company like Netflix imposes genre structures and storytelling approaches, such as the US series format, on local content production. While this concern applies also with regard to other foreign SVODs, we focus on Netflix due to its dominant presence and synonymity with streaming in Brazil.

Disrupting the Monopoly of TV Globo in the Streaming Era

The Brazilian television industry began in 1950. Initially, TV was restricted to wealthy urban elites due to high costs and limited signal coverage, but it spread geographically and socially in later decades. Notably during the military regime (1964–85), the state invested in the television industry to foster a national identity, particularly through TV Globo.[12] Owned by a private entity but bolstered by

[11] Daniela Mazur, "South Korea and Hallyu: An Alternative Globalization in the Multipolar World," PhD diss., Universidade Federal Fluminense, 2023.

[12] Esther Hamburger, *O Brasil antenado: A sociedade da novela* (Rio de Janeiro: Zahar, 2005).

governmental backing, TV Globo underwent significant expansion, wielding both political and cultural influence.[13] By the late 1960s, it introduced the so-called *Padrão Globo de Qualidade* (Globo Quality Standard), a framework aimed at crafting content emphasizing superior technical and visual finesse,[14] in relation to other broadcasters. This standardization not only established a benchmark for high-quality cultural output but also shaped a model of refined popular culture tailored for a wide national viewership,[15] across geographical, class, and racial divides.[16]

In the 1970s, some media analysts expressed concern about US influence on Brazilian television, reflecting the cultural imperialism perspective, which emphasized the asymmetries between the North and South, particularly evident in the Brazilian TV industry's reliance on imported US content.[17] In contrast, media analysts like Straubhaar observed a decline in US imports (dubbed films and TV shows) following the implementation of the national television broadcast system in the mid-1970s and the rise in popularity of domestic content.[18] In this context, TV Globo started to emphasize local productions and genres, especially telenovelas, establishing them as the primary storytelling format. By the late 1970s, the network started exporting telenovelas, thereby spreading Brazilian content and representations globally.

Despite its often ethically troubling, class-privileged representation of Brazilian nationhood, the enduring prevalence of broadcast television in Brazil, via TV Globo, stands as a distinctive national trait in contrast to other television industries around the world.[19] The broadcaster continues to hold significant importance and centrality in the country, its technological strength allowing Globo's signal to reach nearly the entire Brazilian territory. Additionally, as it strives to meet audience demands and to address gaps in

[13] During the military dictatorship, the regime imposed strict media censorship while simultaneously leveraging television to mold public opinion. It cultivated an illusion of democracy by permitting commercial television networks to operate but coerced them into aligning with government narratives.

[14] Ana P. G. Ribeiro and Igor Sacramento, "A renovação estética da TV," in *História da televisão no Brasil: Do início aos dias de hoje*, ed. Ana P. G. Ribeiro, Igor Sacramento, and Marco Roxo (São Paulo: Contexto, 2010), 109–36.

[15] Melina Meimaridis, "Disrupting Brazilian Television: Streaming and the Decline of Globo's Hegemony in Video Cultures," *International Journal of Cultural Studies* 27, no. 1 (2024): 9–27.

[16] Hamburger, *O Brasil antenado*.

[17] Sérgio Caparelli, *Televisão e capitalismo no Brasil* (Porto Alegre: L&PM, 1982).

[18] Joseph Straubhaar, "Brazilian Television: The Decline of American Influence," *Communication Research* 11, no. 2 (1984): 221–40.

[19] Meimaridis, "Disrupting Brazilian Television."

representation, TV Globo maintains a large national viewership, unlike in other countries where pay TV has diminished the dominance of free-to-air networks. In Brazil, pay TV penetration has been historically low, peaking at only 19.6 million subscriptions—about 10 percent of the population—in 2014.[20]

Within Brazil, TV Globo's telenovelas have wielded significant influence in shaping Brazilians' conception of nationhood and national identity.[21] Celebrated examples like *Roque Santeiro*[22] (1985) establish a "unified national public space that provides diverse audiences with a common experience that cuts across regional, class and other social boundaries."[23] However, these productions typically feature Caucasian protagonists, hailing from privileged socioeconomic backgrounds, situated in urban landscapes of major cities like Rio de Janeiro and São Paulo, and engaging in relationships that adhere to expectations of heteronormativity. These representations therefore fall short of reflecting the lived experiences of the broader Brazilian populace. They provide a narrow audiovisual narrative that fails to capture Brazil's cultural richness and diversity. Recently, the demand for video streaming services has emerged in response to growing viewer dissatisfaction and desire for more varied content. This shift has fragmented the audience, challenging Globo's dominance for the first time in decades.[24]

As a forerunner of this emerging technology, Netflix arrived in Brazil in 2011. Despite initial challenges, it successfully established itself as a leading player in the country. Alongside Netflix, other platforms such as Amazon Prime Video (established in 2016) and Disney+ (launched in 2020) have garnered some traction. In response to the competition created by foreign companies, Brazilian media conglomerates have launched their own streaming services. Among these, Grupo Globo's Globoplay, introduced in 2015, emerged as the foremost national platform in the streaming market. What sets Globoplay apart is its extensive telenovela library and its ability

[20] Since 2014, this number has steadily declined. By the end of 2024, subscriptions had dropped to 9.3 million. See https://www.abta.org.br/, accessed on December 12, 2024.

[21] Lopes, "Telenovela brasileira."

[22] *Roque Santeiro* was a popular telenovela about the residents of a small town who centered their lives around the supposed miracles of a martyr. It captivated over 80 percent of the audience nationwide.

[23] Porto, "Telenovelas and Representations," 55.

[24] Meimaridis, "Disrupting Brazilian Television."

to broadcast live sports events, thanks to Grupo Globo's long-standing investments in sports broadcasting.

Since the late 2000s, YouTube has gained popularity in Brazil, particularly among younger groups. Unlike SVOD services, YouTube doesn't require subscriptions or high-speed internet, making it accessible to a broader audience. YouTube offers content not available on traditional TV and hosts original Brazilian productions, including cartoons, kid YouTubers, and short sketch comedies. The continued success of the YouTube channel Porta dos Fundos, for example, leverages the platform's technological affordances and, according to Carter, disrupts TV Globo's dominance by offering audiences innovative aesthetics and tackling taboo subjects such as racism and misogyny.[25] As per Lunardi, YouTube represents an innovative sphere where Brazilian content creators can explore themes of cultural diversity and include greater representation by highlighting elements of national culture frequently overlooked by hegemonic media entities and top-down corporate streaming media.[26]

On the whole, the rise and adoption of video streaming platforms have led to renewed investment within the Brazilian audiovisual sector. This recent surge has prompted some diversification across formats, genres, and, notably, within modes of representing national identity, or exporting Brazilianness. This is not entirely unprecedented: foreign entities like HBO Latin America have previously produced narratives about Brazil for a Brazilian audience.

However, now the situation is more complex, as Netflix not only produces content for Brazilian audiences but also, in the process, positions itself as a "global storyteller."[27] In other words, Netflix, like other US media companies, positions itself as a global narrative authority by both distributing US content internationally and producing local content under its brand. While creating content that must balance local and international appeal, the company imposes its operational model on local markets, potentially undermining local cultural industries.

25 Eli Carter, "Representing Blackness in Brazil's Changing Television Landscape: The Cases of *Mister Brau* and *O Grande Gonzalez*," *Latin American Research Review* 53, no. 2 (2018): 344–57.

26 Gabriela Lunardi, "The 'Brazilian-ness' of Brazilian YouTube: New Voices, New Platforms, New Television?" PhD diss., Queensland University of Technology, 2022.

27 Melina Meimaridis, Daniela Mazur, and Daniel Rios, "Original by Deception: Netflix's Mixed Blessings to Hallyu in Brazil" (paper presented at "Is Netflix Riding the Korean Wave or Vice Versa?" Conference, Seoul, 2022), 19–32.

Netflix's Brazilian Odyssey

Multiterritory SVOD services like Netflix actively reconfigure established media flows and contraflows.[28] Freed from a strict twenty-four-hour programming schedule, they operate within regulatory limits and the capital available for acquiring titles and creating original content.[29] This opens new opportunities for distributing domestic productions but also raises concerns about unpredictable outcomes, particularly in Brazil's audiovisual sector.

However, gaining precise insights into Netflix's presence in Brazil is difficult. The company does not share detailed, up-to-date data on its library size, individual viewership statistics,[30] or subscriber breakdowns by country. Available information comes from sporadic academic research, private entities, and government agencies, offering only partial glimpses. For example, Lotz, Eklund, and Soroka found that Netflix's Brazilian library had about five thousand titles in 2021, with around 40 percent of the content emanating from the United States.[31] Another study noted that less than 5 percent of the titles were Brazilian, while over half came from other international markets.[32] Additionally, in 2021, the Conselho Administrativo de Defesa Econômica[33] (Administrative Council for Economic Defense) inadvertently disclosed that Netflix had 19 million subscribers in Brazil.[34] Although library size and subscription numbers vary over time, these data provide a snapshot of the company's scale and reach in the country.

[28] Thussu defines media flows as the circulation of media content across different regions and cultures. They are typically dominated by content from powerful Western countries (especially the United States and the United Kingdom). Media contraflows refer to the reverse movement, where media from less dominant regions (often from the global South) spreads into and influences the global North. Contraflows challenge the hegemony of Western media by introducing alternative perspectives, narratives, and cultural products into the global landscape. Daya Thussu, *Media on the Move: Global Flow and Contra-Flow* (London: Routledge, 2007).

[29] Amanda Lotz, *Netflix and Streaming Video: The Business of Subscriber-Funded Video on Demand* (Cambridge: Polity, 2022).

[30] In 2021, Netflix launched the "Top 10 Global" website, highlighting the most-watched titles each week. However, these reports lack detailed descriptions. Therefore, drawing significant conclusions from this data is challenging.

[31] Amanda Lotz, Oliver Eklund, and Stuart Soroka, "Netflix, Library Analysis, and Globalization: Rethinking Mass Media Flows," *Journal of Communication* 72, no. 4 (2022): 515.

[32] Amanda Lotz and Oliver Eklund, "Beyond Netflix: Ownership and Content Strategies among non-US-Based Video Streaming Services," *International Journal of Cultural Studies* 27, no. 1 (2023): 119–40

[33] Federal entity responsible for investigating economic abuses and ensuring free market competition.

[34] This disclosure occurred following an unintended faux pas wherein a confidential document was mistakenly appended to a legal proceeding.

In Brazil, Netflix has employed various strategies to expand its content library. Netflix shows a commitment to acquiring and producing national content, including scripted series, films, reality shows, documentaries, and stand-up comedy specials. However, the company's investments in the domestic market face challenges due to the ambiguity of the "Netflix Original" label. Corresponding to practices observed in other regions,[35] Netflix employs this classification to encompass both productions it directly commissions and titles licensed from other studios wherein Netflix simply holds exclusive distribution rights, leading to charges of platform imperialism.[36] Notably, the visual and textual elements used to denote the "Netflix Original" label, such as Netflix's brand identity in the production's thumbnail and the use of the official Netflix vignette (characterized by an animation with the company's logo and its trademark soundtrack), remain consistent across both the commissioned and licensed categories. Penner and Straubhaar note that the lack of differentiation in the "Netflix Original" label can "inflate the number... of content understood as 'original production.'"[37] Mazur, Meimaridis, and Rios take this further, arguing that Netflix's loose use of the "Original" label "contributes to the erasure of the cultural industries behind these works,"[38] as the Western company often claims credit as the creator and exclusive owner of titles that actually originate from non-Western countries.

Exclusive licensing and content financing can similarly seem to initially reinforce Netflix's service exclusivity. Yet it is in the commissioning of new works that it becomes possible to more specifically identify the formats, genres, and themes that are valuable to the service in a particular national context. Essentially, it is through this strategic approach that Netflix attains increased control over the content creation process, reducing its reliance on titles obtained from external entities within the industry, especially those that may be considered, to some extent, as competitors.

As of 2023, Netflix had commissioned thirty-six Brazilian scripted series and miniseries. It is important to note that in Brazil, the service has invested primarily

[35] Karen Petruska and Faye Woods, "Traveling without a Passport: 'Original' Streaming Content in the Transatlantic Distribution Ecosystem," in *Transatlantic Television Drama: Industries, Programs and Fans*, ed. Michele Hilmes, Matt Hills, and Roberta Pearson (Oxford: Oxford University Press, 2019), 49–68.

[36] Dal Yong Jin, *Digital Platforms, Imperialism and Political Culture* (London: Routledge, 2015); Mazur, Mazur, and Rios, "Original by Deception."

[37] Tomaz Penner and Joseph Straubhaar, "Netflix Originals and Exclusively Licensed Titles in Brazilian Catalog: A Mapping Producing Countries," *MATRIZes* 14, no. 1 (2020): 136.

[38] Mazur, Mazur, and Rios, "Original by Deception," 20.

in fully funded productions rather than co-commissions.[39] Netflix typically hires a local audiovisual production company, which in turn is responsible for assembling the team involved in developing the production. Furthermore, collaborations are pursued with esteemed Brazilian actors,[40] directors, and other industry professionals. These dynamics shape the commissioning of scripted series in Brazil: the narratives are crafted by Brazilian professionals and facilitated by domestic production companies, yet the final decision-making authority for original projects rests with the executives of the US-based company.[41] This allows Netflix to maintain control over productions across various media industries, imposing its own conditions and perspectives on foreign cultures and their representations.

Accounts from professionals involved in the production of the Brazilian original *3%* underscore Netflix's close oversight from its US headquarters during the series' development. The series was produced in the wake of Netflix's strategy in releasing non-English-language commissions, the first being the Mexican sitcom *Club de Cuervos*, introduced in 2015, about a local soccer team. This year also marked the launch of *Narcos* (2015–17), an exploration of the cocaine trade in Colombia under the leadership of Pablo Escobar. Despite being produced by a US company, *Narcos* aimed to resonate with the Latin American audience, as it was filmed in Colombia, featured a Brazilian actor, and incorporated dialogue in both English and Spanish. The following year the company released *3%*, which echoes elements from other global media franchises such as *The Hunger Games* saga by operating within a postapocalyptic dystopian backdrop and navigating issues of societal inequality and the pursuit of social ascent via perilous processes orchestrated by the state. During this period, Erick Barmack, serving as the director of international content, oversaw the few non-US commissions from Netflix, operating primarily from the company's Los Angeles office. De Marchi and Ladeira highlight the genesis of discussions between Barmack and the creator of *3%*, Pedro Aguilera, dating back to 2014.[42] In this context, the researchers claim that the decision to begin commissioning works was made at

[39] Melina Meimaridis, Daniela Mazur, and Daniel Rios, "From São Paulo to Seoul: Netflix's Strategies in Peripheral Markets," *Comunicación y Sociedad* 18 (2021): 1–26.

[40] The company has sought for its productions former stars from TV Globo, such as Wagner Moura.

[41] Rios, "Diversidade e Dependência."

[42] Leonardo De Marchi and João Ladeira, "Originais Netflix: um panorama da produção audiovisual da Netflix no Brasil 2016–2019," *FAMECOS* 30, no. 1 (2023): 6.

the same time for both the United States and other countries, thereby making them part of the same content strategy.

By 2018, Netflix Brasil had expanded its offerings to include the political drama *The Mechanism* (2018–19), based on a true government corruption scandal, and the comedy *Samantha!* (2018–19), which follows a former 1980s child star planning her comeback. Concurrently, Netflix announced the commissioning of two dramas, *Girls from Ipanema* (2019–20), about four women struggling with gender biases in the 1950s and 1960s, and *Sintonia* (2019–), focused on three teens from a São Paulo favela. This period marked a key moment for Brazil's operations, with the recruitment of executives to manage title commissioning for the Brazilian audience. Between 2016 and 2023, the rate of new series and seasons has been consistent: nine in 2019, seven in 2020, two in 2021 (smaller likely due to the Covid-19 pandemic), eleven in 2022, and seven in 2023.

The Netflix library showcases wide-ranging productions across diverse genres. Notably, police dramas and law enforcement series emerge prominently, featuring dramas like *Good Morning, Verônica* (2020–), *Brotherhood* (2019–22), and *Criminal Code* (2023–). While initial ventures included *Samantha!* and *Nobody's Looking* (2019), the development of comedies gained momentum starting in 2022 with the release of *Smother-in-Law* (2022–), *Unsuspicious* (2022), *Time Hustler* (2022–), *Reporting for Duty* (2023–), and *#NoFilter* (2023). Equally significant are productions targeting the teenage audience, exemplified by *Back to 15* (2022). Catering to teenagers aligns with research indicating that this demographic has historically not been privileged as TV consumers in linear ecosystems. Streaming video, as emphasized by Potter, has the capability to address this gap.[43]

Following the success of *3%*, Netflix introduced another science fiction series, *Omniscient* (2020). Noteworthy additions also include *Invisible City*, a fantasy series interwoven with a police investigation, and the horror series *Reality Z* (2020), a Brazilian adaptation derived from the British production *Dead Set*. In this sense, we observe how Netflix strategically invests in narrative genres, notably science fiction and fantasy, that occupy a comparatively limited space within the realm of Brazilian television, and which are too niche in comparison with Globo's telenovelas.

[43] Anna Potter, "SVODs' Innovation in Children's Content," in *Streaming Video: Storytelling across Borders*, ed. Amanda Lotz and Ramon Lobato (New York: NYU Press, 2023), 264–79.

Netflix has also introduced a new model of cross-national collaboration, as evidenced by the police drama *Santo* (2022–). Although developed by the Spanish production company Nostromo Pictures, the series serves as a collaborative effort between Spain and Brazil. The narrative unfolds as two detectives, one Spanish and one Brazilian, unite their efforts in a criminal investigation. Filmed in both Madrid and Salvador, the series incorporates scenes in Spanish and Portuguese. Mirroring Netflix's approach in *Narcos*, this is a bilingual production linked to esteemed personalities in both nations. The rising prominence on the global stage of Spanish productions—once a lesser-known player in television flows—and the longstanding influence of Brazilian media across Latin America have both contributed positively to Netflix's transnational expansion. This reinforces the company's goal of producing content with a more genre-based as well as more transnational appeal.

Localizing Content: Netflix's Strategies and Stereotypes in the Brazilian Landscape

Netflix's decision to invest in Brazilian productions coincided with a turbulent political period in the domestic cultural sphere that negatively affected the development of content outside the control of major national media conglomerates. Thus, the prevailing discourse surrounding Netflix described the service as the ideal space for cultivating content distinct from the TV Globo's conventions, in terms of narrative construction, character development, and representation frameworks. Professionals associated with Brazilian Netflix Originals often emphasize the uphill battle these narratives faced to secure any space on Brazilian television.[44] While they have found this space via streaming platforms, these narratives now face the pressure of incorporating stereotypes and standardized notions of cosmopolitanism into their narrative and character constructions. For example, a screenwriter for *3%* stated that Netflix's executives "had a requirement that was closely connected with the [international] market reality, guided by the narrative structure of [US] series."[45]

[44] This phenomenon extends beyond Brazil and can also be seen with the South Korean hit *Squid Game* (2021–), which faced a decade-long rejection from the domestic television industry. Evan Williams, "Squid Game's script was rejected for 10 years before coming to Netflix," *Early Game*, October 6, 2021.

[45] Rios, "Diversidade e Dependência," 123.

In contrast to TV Globo's narratives rooted in Rio de Janeiro and São Paulo, the service's originals, like *The Chosen One* (2019) and *Time Hustler*, offer an array of settings across Brazil, unveiling the cultural tapestry woven across different states. The former immerses viewers in the distinct landscapes of the Amazon rainforest, in Brazil's North Region, while the latter takes a comedic approach to show historical characters of the Northeast Region. Both regions are economically disadvantaged in the country and have had limited representation on Brazilian television. Yet these regions offer exotic landscapes—the rainforest in the former and the *sertão* (hinterlands) in the latter. While Netflix challenges certain representational norms, it also subtly reinforces stereotypes about Brazil's regions by leveraging "cultural peculiarities and local tastes"[46] from an external perspective.[47]

Regardless of these new avenues for representing Brazil's diverse regional landscapes, an underlying concern arises regarding the persistent concentration of production resources within the Southeast Region.[48] So, despite the narrative shift away from urban-centric portrayals, established production companies like Los Braga and O2—predominantly situated in the Southeast—continue to remain dominant. This reliance on Rio de Janeiro- and São Paulo-based experienced production houses and professionals, driven by Netflix's risk-averse strategy, inadvertently reinforces the Southeast's already dominant role in shaping narratives. This situation prompts important questions regarding the continuing need for decentralization of the industry and for fair distribution of opportunities among diverse regional talents and markets across Brazil.

Furthermore, it is noteworthy that a number of series, including *Samantha!* and *Good Morning, Veronica*, continue to employ Rio de Janeiro and São Paulo as their main settings. That said, the portrayal of these cities in these works does diverge significantly from their depiction on traditional broadcast television. A notable example is *Sintonia*, which delves into the funk music industry and the relationships the characters maintain with organized crime and religious groups. *Sintonia* originated from the creative vision of Kondzilla, a production company that has emerged as a prominent platform for funk artists, achieving widespread success in Brazil through its audiovisual content on YouTube.

46 Jin, "Netflix's Effects on the Korean Wave," 461.

47 Daniel Rios, Melina Meimaridis, and Daniela Mazur, "À sombra da dominância: produção audiovisual nacional à moda Netflix," *Intexto* 56 (2024): 1–23.

48 De Marchi and Ladeira, "Originais Netflix."

Lunardi underscores that Netflix's collaboration with Kondzilla represents a deliberate tactic aimed at engaging novel niche audiences within the country. A product of this tactic, *Sintonia* "aimed to fill a gap in Brazilian television culture—the representation of the *favela* youth."[49] The series is set in São Paulo, but it showcases the most marginalized population and forgotten part of the city. Netflix sought with *Sintonia* to replicate the international success of the Brazilian film *City of God* (2002), using this film as a model for reaching transnational audiences, and reviving in the process well-known stereotypes about Brazil, such as urban poverty and gang violence. In this way, Netflix at once actively engages niche audiences, addresses gaps in Brazil's television culture, and signals a shift toward more inclusive (even if often stereotypical) storytelling.

The shift to more inclusive storytelling has presented challenges for some production companies. These challenges were made visible with *Invisible City*, situated in Rio de Janeiro, which depicts paranormal interactions between individuals and entities intricately linked to Brazil's indigenous peoples' legends. After facing criticisms regarding the appropriation of indigenous narratives without proper acknowledgment, and also the absence of indigenous representation in the cast, the show's second season was compelled to relocate the action from Rio de Janeiro to Pará, a state encompassing the Amazon Rainforest, and also to incorporate actors of indigenous ancestry.

Also exemplary of the shift to inclusive storytelling are Netflix original productions like *Omniscient* and *Spectros* (2020) that have strategically ventured into reimagining the urban landscapes of São Paulo and Rio de Janeiro, often via dystopian and fantasy narrative frameworks, as well as narratives intertwined with indigenous and urban Brazilian folklore, thereby presenting Brazil from diverse perspectives. The former presents São Paulo as an integral component of a worldwide system employing drones for continuous citizen surveillance, while the latter assigns a supernatural dimension to Liberdade, a district in São Paulo recognized for its Japanese-Brazilian community, interweaving it with vengeful spirits. Yet the most popular example of this genre continues to be *3%*, which unfolds as a dystopian narrative wherein São Paulo represents the destitute sector, denoted as "our side" of the continent—Brazil's imagination of a future marked by profound social inequality amid a global capitalist crisis.

[49] Lunardi, "The 'Brazilian-ness' of Brazilian YouTube," 109.

The Double-Edged Sword in the Production of Brazilian Originals

Although Netflix and other SVOD services have made substantial investments in Brazil's audiovisual industry and prompted shifts to more inclusive forms of storytelling, their localization strategies have also exerted a coercive force on narrative production. In November 2021, Netflix organized the "More Brazil on Screen" event in São Paulo, dedicated to promoting original local productions and commemorating the company's decade-long presence in the country. During the event, Elisabetta Zenatti, vice president of content at Netflix Brasil, emphasized the desire of Brazilians for narratives that represent their lives, roots, and ancestry.[50] This statement underscores the importance of embracing Brazil's diversity and the necessity to decentralize production in this vast, continent-sized nation. However, the production of local stories by US streaming services can inadvertently lead to culturally flattening narratives, undermining the very diversity they claim to celebrate. They also impose their own methods of content creation and storytelling, which often clash with Brazil's audiovisual production norms.

The sitcom *Smother-in-Law* stands out as a notable case within this context. The series centers around Dona Isadir (Rodrigo Sant'Anna), a humble, outspoken Black woman who relocates to live with her son and his family at the onset of the pandemic, much to the chagrin of her daughter-in-law. Despite being set in Brazil, specifically in Rio de Janeiro, and featuring Brazilian actors, the series emulates the aesthetics of American situational comedies such as *The Cosby Show* (NBC, 1984–92) and *The Fresh Prince of Bel-Air* (NBC, 1990–6). Evident through its use of a multicamera setup and a living room-centric set design, the series presents an idealized depiction of an upper-middle-class Black experience, along American lines.[51] But these aesthetic choices represent a striking departure from Brazil's own well-established television comedy formats.

Many critics have bemoaned the awkwardness of *Smother-in-Law*'s viewing experience due to the humor's poor quality.[52] This sense of strangeness is more

50 Jamie Lang and John Hopewell, "Netflix Celebrates Five Years of Brazilian Production, Unveils Upcoming Slate in More Brazil on Screen Presentation," *Variety*, November 23, 2021, https://variety.com/2021/streaming/news/netflix-brazil-2022-slate-presentation-1235118087/.

51 Bambi Haggins, "The Black Situation Comedy," in *African Americans and Popular Culture*, ed. Todd Boyd (Westport, CT: Praeger, 2008), 217–43.

52 André Zuliani, "A Sogra que Te Pariu: Série da Netflix mira em Sai de Baixo, mas erra feio," *Notícias da TV*, April 13, 2022, https://noticiasdatv.uol.com.br/noticia/series/sogra-que-te-pariu-serie-da-netflix-mira-em-sai-de-baixo-mas-erra-feio-78987?cpid=txt.

pronounced in the realm of comedy compared to drama, as humor relies on specific local references and character types, which contrast with the more universally relatable themes and characters of melodrama. Cultural specificity can significantly impact comedy's effectiveness as well as translatability across borders. For instance, appreciating humor about Brazilian political dynamics necessitates a nuanced understanding of the country's civic culture; without this contextual knowledge, the comedic nuance may be lost on international viewers. Each culturally specific reference has the potential to alter a scene's comedic resonance profoundly. Melodrama, on the other hand, often explores universal themes—such as the archetypal battle between good and evil or the profound grief of bereavement—which tend to resonate more broadly across diverse audiences.

Shows like *Smother-in-Law* suggest a coercive imposition of the US hegemonic production chain on other cultural industries. Similar trends can be observed in other markets, as exemplified by South Korea, where Netflix is influencing aspects such as episode length, the number of episodes in a series, and even the adoption of multiple seasons, aligning more closely with the conventions of US scripted series, despite its continued production of K-dramas.[53] In the Turkish drama and Nollywood industries as well, similar sentiments arise concerning Netflix's role as a foreign agent that is coercively imposing production norms. While the streaming service provides a global stage and introduces new narrative and format possibilities, it simultaneously instigates changes in domestic industrial dynamics.[54]

In other words, this phenomenon is not exclusive to Brazil, and it extends to other countries in the global South. Netflix's approach in these regions involves meeting a demand for localized content, albeit within frameworks considered "universal." In this way it can cater to both a local audience's desire for more diverse representation and an international audience's cosmopolitan taste, seeking differentiated content from various cultural perspectives. Sometimes, the attempt to appeal very broadly leads to instances of Netflix Originals

[53] Hyejung Ju, "Deterritorialisation of Korean TV Dramas in 'Netflix Originals,'" *Critical Studies in Television* 19, no. 4 (2023): 429–49; Mazur, "South Korea and Hallyu."

[54] Ana Esteves, "O lugar da Netflix no campo do poder: reflexões sobre o impacto da empresa no audiovisual africano a partir da Nigeria," *Mídia e Cotidiano* 17, no. 2 (2023): 1–21; Asli Ildir and Ipek Celik Rappas, "Netflix in Turkey: Localization and Audience Expectations from Video on Demand," *Convergence* 28, no. 1 (2022): 255–71; Mazur, "South Korea and Hallyu."

featuring orientalist content, as seen in the Turkish production *The Protector*.[55] At other times, the streamer's original content might cater to particular local political imperatives. This was the case with Netflix's Brazilian drama *The Mechanism*, which portrays a romanticized view of the judiciary while casting leftist politicians in an unfavorable light. Compounding the issue, the drama was released just a few months before the 2018 presidential elections in Brazil. As Albuquerque, Meimaridis, and Quinan argue, the series served as an agent of biased mediatization of the ongoing Brazilian political and institutional crisis.[56]

Notwithstanding their perceived coercive strategies, certain industry insiders express gratitude for the emergence of foreign SVODs in the Brazilian market. In a 2021 interview on *Roda Viva*, renowned actor Wagner Moura remarked, "Brazil's audiovisual [sector] still exists thanks to streaming."[57] Conversely, Lazaro Ramos, a prominent TV Globo actor, characterized the Brazilian audiovisual market as "abandoned" and in need of rescue by international streaming platforms.[58] While it is unfortunate that both actors seem to be susceptible to imperialistic rhetoric that frames the national industry as desperately requiring foreign intervention, it is also true that Brazilian actors have found new opportunities through streaming services.

Screenwriters and creators, however, have been criticizing these businesses for their abusive practices. Services like Netflix, Amazon Prime Video, HBO Max, and Disney+ in Brazil have consistently restricted creators from retaining their intellectual property rights, a concern previously identified in other markets.[59] The Associação Brasileira de Autores Roteiristas (Brazilian Screenwriters Association) highlights that contracts presented by national production companies to Brazilian screenwriters are often dictated by foreign platforms, essentially being translations of foreign contractual arrangements into Brazilian Portuguese. In this process, streaming services deviate from established local production practices and laws.

[55] Yasemin Özkent, "Reflection of Orientalist Discourse on Netflix Turkey: *The Protector*," in *Handbook of Research on Contemporary Approaches to Orientalism in Media and Beyond*, ed. Isil Tombul (Hershey, PA: IGI Global, 2021), 939–52.

[56] Afonso de Albuquerque, Melina Meimaridis, and Rodrigo Quinan, "O mecanismo e a midiatização ficcional da Operação Lava-Jato," *Questões Transversais* 8, no. 16 (2021): 58–64.

[57] Wagner Maura on *Roda Viva*, retrieved from https://www.youtube.com/watch?v=FvOO3Gysd7Q. Accessed September 19, 2024.

[58] Carolina Moraes, "Lázaro Ramos, na Amazon, diz que em 2022 Brasil tem de conversar consigo mesmo," *Folha*, January 8, 2022, https://www1.folha.uol.com.br/ilustrada/2022/01/lazaro-ramos-na-amazon-diz-que-em-2022-brasil-tem-de-conversar-consigo-mesmo.shtml.

[59] Ji Hoon Park, Kristin April Kim, and Yongsuk Lee, "Netflix and Platform Imperialism: How Netflix Alters the Ecology of the Korean TV Drama Industry," *International Journal of Communication* 17 (2023): 72–91.

Some of the abusive practices streaming companies engage in include (1) denying screenwriters ownership of their work, (2) pressuring creatives to relinquish future rights and potential revenue sources, and (3) designating the contracting production company as the sole author of the project. These practices not only represent setbacks to the rights acquired by Brazilian screenwriters in recent decades but also contain provisions incompatible with Brazilian law, particularly regarding copyright. The professional landscape of Brazilian audiovisual production already has its own precarious conditions, and the introduction of foreign streaming companies' work models in Brazil has further worsened these conditions.

Ultimately, these issues surrounding the production of Netflix Originals can be understood within the framework of what Jin defines as platform imperialism—another type of media imperialism that reinforces Western supremacy while consolidating capital.[60] The United States, through digital intermediaries like Google, Netflix, and Facebook, continues to enhance its dominance. These platforms serve as the essential routes that industries from the global South must navigate to reach audiences and markets worldwide. Consequently, southern industries become reliant on these northern intermediaries for the production and dissemination of their cultural artifacts. Netflix does not merely provide a platform. By imposing its work model and creative processes on other cultural industries, it shapes how local producers craft narratives and representations of their own countries and cultures.

When examining Netflix, Elkins suggests that the company portrays its brand as cosmopolitan, attempting to present its extensive content library as a champion of diversity.[61] In doing so, it tries to legitimize "globally expanding business and technological practices as humanistic and cosmopolitan rather than faceless, mathematical, and all-consuming."[62] This analysis reveals Netflix's ongoing endeavor to satisfy the cosmopolitan preferences of what Lobato terms the "transnational class."[63] The streaming giant's content strategy often aligns with—and perpetuates—generalized expectations that shape international perceptions of countries deemed "exotic." This approach frequently relies on lenses of otherness and, at times, even orientalist perspectives. However, this trend is not exclusive to Netflix, as the imperialistic strategies of US-based SVODs persistently reinforce

[60] Jin, *Digital Platforms, Imperialism and Political Culture.*

[61] Evan Elkins, "Algorithmic Cosmopolitanism: On the Global Claims of Digital Entertainment Platforms," *Critical Studies in Media Communication* 36, no. 4 (2019): 376–89.

[62] Ibid., 2.

[63] Lobato, "Rethinking International TV Flows Research in the Age of Netflix," 252.

existing asymmetries that favor Anglophone audiovisual flows, particularly those originating from the United States. What is most concerning is how these services position themselves as arbiters of global narratives.[64]

Conclusion

In recent years, Brazilian audiovisual production has demonstrated a noteworthy capacity for diversification, thereby expanding its understanding and portrayal of Brazil and Brazilianness. This trend has facilitated the creation and dissemination of more diverse content, catering to the demand for nonhegemonic narratives and delving into themes that were previously underexplored in the national audiovisual industry, particularly on television. The expansion of production diversity conforms to international trends, aiming not to replace established norms but rather to broaden the spectrum of narratives accessible to the public.[65]

Within this context, video streaming services have emerged as crucial spaces for experimentation and creation, representing yet another influential agent influencing the content landscape. What is depicted as an essential aspect of Brazilian culture and identity in Netflix originals has a global resonance, reaching audiences worldwide through the platform's extensive libraries. The monopoly held by Globo over the portrayal of Brazilian identity in national fiction has introduced limitations, paving the way for other entities to address local demands. Netflix, operating as an additional agent, introduces novel possibilities.

Nevertheless, our concern centers on the gradual loss of media sovereignty over our own narratives and, specifically, the ways we articulate and produce them. In various ways, we can observe that media companies from the global North continue to dominate transnational content, perpetuating a system of cultural and economic dependence in countries of the global South. When series like *Smother-in-Law* adopt imported language and aesthetics to portray something as particular and culturally specific as comedy, they risk sacrificing artistic integrity and also homogenizing cultural industries from the South, including Brazil's.

[64] Meimaridis, Mazur, and Rios, "From São Paulo to Seoul."

[65] Amanda Lotz and Ramon Lobato, eds., *Streaming Video: Storytelling across Borders* (New York: NYU Press, 2023).

Bibliography

Albuquerque, A., M. Meimaridis, and R. Quinan. "O mecanismo e a midiatização ficcional da Operação Lava-Jato." *Questões Transversais* 8, no. 16 (2021): 58–64.

Caparelli, S. *Televisão e capitalismo no Brasil*. Porto Alegre: L&PM, 1982.

Carter, E. "Representing Blackness in Brazil's Changing Television Landscape: The Cases of *Mister Brau* and *O Grande Gonzalez*." *Latin American Research Review* 53, no. 2 (2018): 344–57.

De Marchi, L., and J. M. Ladeira. "Originais Netflix: um panorama da produção audiovisual da Netflix no Brasil 2016–2019." *FAMECOS* 30, no. 1 (2023): e42989. https://doi.org/10.15448/1980-3729.2023.1.42989.

Elkins, E. "Algorithmic Cosmopolitanism: On the Global Claims of Digital Entertainment Platforms." *Critical Studies in Media Communication* 36, no. 4 (2019): 376–89.

Esteves, A. "O lugar da Netflix no campo do poder: reflexões sobre o impacto da empresa no audiovisual africano a partir da Nigeria." *Mídia e Cotidiano* 17, no. 2 (2023): 1–21.

Haggins, B. "The Black Situation Comedy." In *African Americans and Popular Culture*, edited by T. Boyd, 217–43. Westport, CT: Praeger, 2008.

Hamburger, E. *O Brasil antenado: A sociedade da novela*. Rio de Janeiro: Zahar, 2005.

Ildir, A., and İ. Celik-Rappas. "Netflix in Turkey: Localization and Audience Expectations from Video on Demand." *Convergence* 28, no. 1 (2022): 255–71.

Iordache, C. "Netflix in Europe: Four Markets, Four Platforms? A Comparative Analysis of Audio-Visual Offerings and Investment Strategies in Four EU States." *Television & New Media* 23, no. 7 (2022): 721–42.

Jin, D. Y. *Digital Platforms, Imperialism and Political Culture*. London: Routledge, 2015.

Jin, D. Y. "Netflix's Effects on the Korean Wave: Power Relations between Local Cultural Industries and Global OTT Platforms." *Asian Journal of Communication* 33, no. 5 (2023): 452–69.

Ju, H. "Deterritorialisation of Korean TV Dramas in 'Netflix Originals.'" *Critical Studies in Television* 14, no. 9 (2023): 429–49.

Khoo, O. "Picturing Diversity: Netflix's Inclusion Strategy and the Netflix Recommender Algorithm (NRA)." *Television & New Media* 24, no. 3 (2023): 281–97.

Lang, J., and J. Hopewell. "Netflix Celebrates Five Years of Brazilian Production, Unveils Upcoming Slate in More Brazil on Screen Presentation." *Variety*, November 23, 2021. https://variety.com/2021/streaming/news/netflix-brazil-2022-slate-presentation-1235118087/.

La Pastina, A. C., J. Straubhaar, and L. Sifuentes. "Why Do I Feel I Don't Belong to the Brazil on TV?" *Popular Communication* 12, no. 2 (2014): 104–16.

Llamas-Rodríguez, J. "The Return of Indigenismo in Netflix Mexico." In *Streaming Video: Storytelling across Borders*, edited by A. Lotz and R. Lobato, 225–47, New York: NYU Press, 2023.

Lobato, R. "Rethinking International TV Flows Research in the Age of Netflix." *Television & New Media* 19, no. 3 (2018): 241–56.

Lopes, M. I. V. "Telenovela brasileira: uma narrativa sobre a nação." *Comunicação & Educação* 26 (2003): 17–34.

Lotz, A. D. *Netflix and Streaming Video: The Business of Subscriber-Funded Video on Demand*. Cambridge: Polity, 2022.

Lotz, A. D., and O. Eklund. "Beyond Netflix: Ownership and Content Strategies among Non-US-Based Video Streaming Services." *International Journal of Cultural Studies* 27, no. 1 (2023): 119–40. https://doi.org/10.1177/13678779231196314.

Lotz, A. D., Eklund, O., and Soroka, S. "Netflix, Library Analysis, and Globalization: Rethinking Mass Media Flows," *Journal of Communication* 72, no. 4 (2022): 511–21. https://doi.org/10.1093/joc/jqac020.

Lotz, A. D., and R. Lobato. *Streaming Video: Storytelling across Borders*. New York: NYU Press, 2023.

Lunardi, G. "The 'Brazilian-ness' of Brazilian YouTube: New Voices, New Platforms, New Television?" PhD diss., Queensland University of Technology, 2022.

Mazur, Daniela. "South Korea and Hallyu: An Alternative Globalization in the Multipolar World." PhD diss., Universidade Federal Fluminense, 2023.

Mazur, Daniela, Melina Meimaridis, and Daniel Rios. "Original by Deception: Netflix's Mixed Blessings to Hallyu in Brazil." Paper presented at "Is Netflix Riding the Korean Wave or Vice Versa?" Conference, Seoul, 2022.

Meimaridis, Melina. "Disrupting Brazilian Television: Streaming and the Decline of Globo's Hegemony in Video Cultures." *International Journal of Cultural Studies* 27, no. 1 (2024): 9–27. https://doi.org/10.1177/13678779231197.

Meimaridis, Melina, Daniela Mazur, and Daniel Rios. "From São Paulo to Seoul: Netflix's Strategies in Peripheral Markets." *Comunicación y Sociedad* 18 (2021): 1–26. https://doi.org/10.32870/cys.v2021.8038.

Moraes, C. "Lázaro Ramos, na Amazon, diz que em 2022 Brasil tem de conversar consigo mesmo." *Folha*, January 8, 2022. https://www1.folha.uol.com.br/ilustrada/2022/01/lazaro-ramos-na-amazon-diz-que-em-2022-brasil-tem-de-conversar-consigo-mesmo.shtml.

Özkent, Y. "Reflection of Orientalist Discourse on Netflix Turkey: *The Protector*." In *Handbook of Research on Contemporary Approaches to Orientalism in Media and Beyond*, edited by I. Tombul, 939–52. Hershey, PA: IGI Global, 2021.

Pacete, L. "Netflix aposta no longo prazo para ampliar produção local no Brasil." *Forbes*, January 11, 2023. https://forbes.com.br/forbes-tech/2023/01/netflix-aposta-no-longo-prazo-para-ampliar-conteudo-local-nobrasil/.

Park, J. H., K. A. Kim, and Y. S. Lee. "Netflix and Platform Imperialism: How Netflix Alters the Ecology of the Korean TV Drama Industry." *International Journal of Communication* 17 (2023): 72–91.

Penner, T. A., and J. Straubhaar. "Netflix Originals and Exclusively Licensed Titles in Brazilian Catalog: A Mapping Producing Countries." *MATRIZes* 14, no. 1 (2020): 125–49. https://doi.org/10.11606/issn.1982-8160.v14i1p125-149.

Petruska, K., and F. Woods. "Traveling without a Passport: 'Original' Streaming Content in the Transatlantic Distribution Ecosystem." In *Transatlantic Television Drama: Industries, Programs and Fans*, edited by M. Hilmes, M. Hills, and R. Pearson, 49–68. Oxford: Oxford University Press, 2019.

Potter, A. "SVODs' Innovation in Children's Content." In *Streaming Video: Storytelling across Borders*, edited by A. Lotz and R. Lobato, 264–79. New York: NYU Press, 2023.

Porto, M. "Telenovelas and Representations of National Identity in Brazil." *Media, Culture, and Society* 33, no. 1 (2011): 53–69.

Ravache, G. "Exclusivo: Netflix faz investimento recorde no país, apesar da crise dos streamings." *Valor Econômico*, September 28, 2023. https://valor.globo.com/opiniao/guilherme-ravache/coluna/exclusivo-netflix-faz-investimento-recorde-no-pais-para-enfrentar-a-crise-do-streaming.ghtml.

Ribeiro, A. P. G., and I. Sacramento. "A renovação estética da TV." In *História da televisão no Brasil: Do início aos dias de hoje*, edited by A. P. G. Ribeiro, I. Sacramento, and M. Roxo, 109–36. São Paulo: Contexto, 2010.

Rios, D. "Diversidade e Dependência: as estratégias de comissionamento de séries da Netflix no Brasil." PhD diss., Universidade Federal Fluminense, 2024.

Rios, D., M. Meimaridis, and D. Mazur. "À sombra da dominância: produção audiovisual nacional à moda Netflix." *Intexto* 56 (2024): 1–23.

Straubhaar, J. D. "Brazilian Television: The Decline of American Influence." *Communication Research* 11, no. 2 (1984): 221–40.

Thussu, D. *Media on the Move: Global Flow and Contra-Flow*. London: Routledge, 2007.

Williams, E. "Squid Game's Script Was Rejected for 10 Years before Coming to Netflix." *Early Game*, October 6, 2021. https://earlygame.com/entertainment/squid-games-script-rejected-10-years-netflix.

Zuliani, A. "A Sogra que Te Pariu: Série da Netflix mira em Sai de Baixo, mas erra feio." *Notícias da TV*, April 13, 2022. https://noticiasdatv.uol.com.br/noticia/series/sogra-que-te-pariu-serie-da-netflix-mira-em-sai-de-baixo-mas-erra-feio-78987?cpid=txt.

4

Streaming Platforms in Lebanon: A Tale of Business and Politics

Wissam Mouawad and Pamela Nassour

The media system in Lebanon, recognized as a pioneer in the region and renowned for its established regional and international reach, has always been influenced by the volatile socioeconomic and political context of the country. From the first newspapers to be published in Beirut in the second half of the nineteenth century,[1] to the emergence of radio stations in the late 1930s,[2] to the establishment of the first TV stations in the late 1950s to early 1960s by private investors,[3] to today's social media and digital platforms, media in Lebanon has consistently been shaped by the interaction of international, regional, and local politics and finance.

In fact, Lebanon's strategic geopolitical location, which made it a focal point for the ambitions of major imperial, colonial, and regional powers, along with its diverse religious and ethnic composition have ensured that its history and current situation remain deeply intertwined with broader geopolitical dynamics and rivalries. A "buffer state," as Georges Corm describes it,[4] and "a key arena"[5]

1 Launched on June 29, 1858, by Khalil Al-Khuri, *Hadiqat Al Akhbar* is the first newspaper to be published in Lebanon, then under Ottoman rule. For a brief history of Arab press, see Nabil Dajani, "Arab Press," in *Arab Media: Globalization and Emerging Media Industries*, ed. Noha Mellor, Mohammed Ayish, Nabil Dajani, and Khalil Rinnawi (Cambridge: Polity Press, 2011), 45–66.

2 Originally specialized in cable and radio communications in French-mandated eastern states, Radio-Orient began broadcasting in 1938 to counter the Arabic-language radio propaganda of the Italians and the Germans. It is identified as the first radio in Lebanon. Douglas A. Boyd, *Broadcasting in the Arab World: A Survey of the Electronic Media in the Middle-East* (AmesAmew: Iowa State University Press, 1993), 69.

3 Ibid., 76.

4 Georges Corm, "What Lebanon Needs," *Le Monde Diplomatique*, October 2020, https://mondediplo.com/2020/10/03lebanon.

5 Rachel Bessette, "In the Heart of the Fire: Lebanon's Place in a Regional Rivalry," *Georgetown Security Studies Review*, September 16, 2016, https://georgetownsecuritystudiesreview.org/2016/09/16/in-the-heart-of-the-fire-lebanons-place-in-a-regional-rivalry/.

for regional and international competition, this small Arab country is also the last to maintain an active front with Israel (except for the Gaza strip[6] and, very rarely, Syria). Lebanon also hosts more refugees per capita than any other country in the world,[7] the refugees fleeing mainly from Syria and Palestine. The country's strategic location, its complex sociopolitical composition marked by a culturally and religiously diverse population, and a sectarian political system "tightly interwoven with geopolitical allegiances"[8] offer a fertile ground for foreign influences of all kinds, which make their way into the media. In fact, this dynamic has significantly influenced the content and political orientation of media outlets, which often reflect the interests and agendas of political powers and financial backers.

The neoliberal turn that the country took at the end of the civil war in 1990 under Prime Minister Rafic Hariri[9] exacerbated the problematic effects of the country's already established "free" economic system promoted in the constitution.[10] This turn, coupled with the failure of state institutions particularly in the last fifteen years,[11] has further opened the door to foreign economic and political influence and pressure, while encouraging investment in media outlets and the production of media content aimed at local, regional, and international audiences.

In this context, the study of streaming platforms in Lebanon acquires multidimensional significance. As Ramon Lobato has shown, transnational streaming platforms often try to adapt to the tastes of previously existing audiences and to the contexts of specific territories.[12] This is true in Lebanon as

[6] Fabrice Balanche, "The Geopolitics of Lebanon in the Renewed Regional Tensions," in *Atlas of Lebanon: New Challenges*, ed. Eric Verdeil, Ghaleb Faour, and Mouin Hamze (Beirut: Presses de l'IFPO, CNRS Liban, 2019), 30–1.

[7] UNICEF, "Knowledge, Attitude and Practices of the Lebanese; Syrian and Palestinian Refugees in Lebanon, 2017," June 2016, https://www.unicef.org/lebanon/reports/knowledge-attitude-and-practices-lebanese-syrian-and-palestine-refugees-lebanon-2017.

[8] Tamirace Fakhoury and Allison McCulloch, "How Do Consociations Craft Asylum Policy? Lebanon's Response to Conflict-Induced Displacement as an Exploratory Case," *International Studies Quarterly* 67, no. 3 (September 2023): sqad057, https://doi.org/10.1093/isq/sqad057.

[9] Hannes Baumann, *Citizen Hariri: Lebanon's Neo-liberal Reconstruction* (Oxford: Oxford University Press, 2016).

[10] "The economic system is free and ensures private initiative and the right of private property" (Lebanese Constitution, promulgated 1926, accessed January 16, 2024, https://www.presidency.gov.lb/English/LebaneseSystem/Documents/Lebanese%20Constitution.pdf).

[11] For example, the Parliament elected in 2009 for four years extended its own mandate until 2018 due to the deadlock over the electoral law, and in the past fifteen years, there were four years of presidential vacancy.

[12] See Ramon Lobato, *Netflix Nations: The Geography of Digital Distribution* (New York: NYU Press, 2019), 61–2.

well. But due to its rich history and ongoing contributions to the regional media industry, this small country plays a pivotal role in also shaping the content of many of these platforms. In turn, these platforms, which operate within a deregulated environment vulnerable to piracy, political funding, and various external influences, exert a profound impact on society.

In this chapter, we examine the various international, regional, and local streaming services utilized today by viewers in Lebanon. Through an exploration of the operating models, funding sources, and infrastructure of these platforms, coupled with textual analysis of select original works in which Lebanese professionals and viewers play an important role, we explore the financial as well as political and diplomatic underpinnings of these platforms in this volatile part of the world. We argue that the story of the rise of streaming in Lebanon is significant because it challenges conventional narratives about media globalization by highlighting instead the unique challenges faced by smaller nations in asserting their presence within a globalized media ecosystem.

The Troubled History of Lebanon's TV Broadcasting

When streaming platforms reached Lebanon in 2007, they entered a market that had its own history and dynamics. As Benjamin Burroughs shows in his article "A Cultural Lineage of Streaming,"[13] streaming does not represent a pure technological break but rather a cultural and technological evolution, constantly shaped by historical and social contexts. Therefore, it is important to look back at the different dynamics that have shaped the broadcasting industry in Lebanon over the years in order to understand how these platforms have been influenced by and have adapted to Lebanon's unique media landscape and audience expectations. In "The Domestication of Netflix in the Gulf," Zaid, Benmoussa, Ayyad, Ibahrine, and El Kadoussi write: "The history of Arab television can be divided into three main time periods. They refer to, respectively, state-controlled broadcasting, Pan-Arab satellite TV, and multiplatform content distribution with the introduction of online streaming, IPTV, and SVoD."[14] While these three time periods somewhat apply to Lebanon as well, the Lebanese context is distinct

[13] Benjamin Burroughs, "A Cultural Lineage of Streaming," *Internet Histories* 3, no. 2 (2019): 147–61.

[14] Bouziane Zaid, Mohammed Benmoussa, Khayrat Ayyad, Mohamed Ibahrine, and Abdelmalek El Kadoussi, "The Domestication of Netflix in the Gulf," *International Journal of Communication* 17 (2023): 6615.

in that, unlike other Arab countries where state control was initially absolute and still holds significant influence today, the Lebanese state has shared control of television with private entities from the outset and now appears completely withdrawn, perhaps even overtaken by the private sector.

The first Lebanese television was established on the initiative of Lebanese investors, Alex Aridi and Wissam Izzedine, who proposed a project for a TV station that would be financed through advertising revenue. Their project envisioned two channels: one broadcasting in Arabic and another predominantly focused on international programming. Launched in 1959, La Companie Libanaise de Télévision (CLT), with its channels 7 (broadcasting in Arabic and targeting the majority of the Lebanese population) and 9 (broadcasting in French and catering to the francophone elite), became the first nongovernmental commercial television in the Arab world. The same year, an organization partly owned by the American Broadcasting Company (ABC) reached an agreement with the Lebanese government to start operating another TV station, the Compagnie de Télévision du Liban et du Proche-Orient (Télé-Orient), which started broadcasting effectively in 1962.[15] So, from the very beginning, television in Lebanon was a private initiative, even though the state maintained tight censorship of programming and had a monopoly on news bulletins. This model remained stable for a period of time, as television became more and more popular. A survey conducted in 1974 showed that 75 percent of Lebanese households owned television sets,[16] a remarkably high penetration rate for the time, especially when compared to other Arab countries.

When the civil war erupted in 1975, television channels initially remained politically neutral in the face of the intercommunal conflict. However, the situation changed dramatically when Brigadier Aziz Al Ahdab, a Muslim general in the Lebanese army, seized control of CLT during a news broadcast to announce a (failed) coup d'état. This act was perceived by part of the Christian population as an aggression, prompting Christian factions to occupy Télé-Orient and use it for their own propaganda. The Lebanese audience found itself amid a televised information war.[17] This situation continued until 1977 when, following a fragile return to calm caused by the entry of the Syrian Army into the country, and consequent to the reluctance of the private owners of both stations to finance

15 Boyd, *Broadcasting in the Arab World*, 72.

16 Nabil Dajani, *The Media in Lebanon: Fragmentation and Conflict in the Middle-East* (London: I. B. Taurus, 2019), 86.

17 Boyd, *Broadcasting in the Arab World*, 76.

the restoration of damaged infrastructure that could have resulted in the closing of the television stations, the state decided to use its capital and merge the two stations into a single entity. This new station, named Télé-Liban, was jointly owned, with half controlled by the Lebanese government (51 percent) and the other half by Lebanese investors, along with the French government company Sofirad.[18] Thus, while the Lebanese state eventually came to own and control the national television company, it seems to have been pushed into doing so. Its weak management of television, stemming from a lack of strategic vision, and its overall fragility due to the expansion of the civil war, ultimately led to the public sector's fragmentation and disintegration. It is within this fragmented landscape that streaming platforms would later emerge.

When Télé-Liban was created, it was initially granted exclusive broadcasting rights for twenty-five years.[19] However, this did not help enhance quality standards nor broaden the appeal of programming across the different sectors of the population. The television directors took the path of least resistance by offering programs with no genuine local grounding, attempting to promote this approach as an openness to the international market. A study conducted on the programs of early Télé-Liban showed that only 41 percent of the afternoon programs were in Arabic (mainly classical, not even the spoken Lebanese Arabic); 53 percent of entertainment programs were produced in the United States and 23 percent in France; and 50 percent of the educational programs were American and 25 percent French. Dr. Charles Rizk, the first director-general of Télé-Liban, boasted that "the Lebanese viewer can 'now' watch the same programs shown 'now' in New York and Paris."[20] The colonial undertones of this assertion hardly conceal the fact that the real intention was to avoid at all costs programming that dealt with the complex historic events the country was going through. It is often remarked, for instance, that Lebanese series of that era avoided characters with names that could suggest sectarian affiliation.

With the intensification of the war, the gradual disintegration of the country into various territories under partisan and sectarian governance, the Israeli invasion of 1982, and the weakening of infrastructure, particularly electricity and telecommunications, Télé-Liban's monopoly was ultimately undermined. The interruption of broadcasts in certain territories and the use of antennas capable

[18] Ibid., 80.
[19] Dajani, *Media in Lebanon*, 88.
[20] Ibid., 90.

of a more powerful reception pushed the Lebanese audience to turn to channels from Syria, Israel, Cyprus, Jordan, and even Egypt, depending on the regions they lived in and the reach of their antennas. This happened in a context when demand for TV content was becoming higher, especially with the escalation of war. As *New York Times* journalist Thomas Friedman wrote in 1983, "People have retreated into their neighborhoods or living rooms for relaxation [and] Lebanese Television… suddenly found itself with a massive new audience."[21] This rise in demand for audiovisual entertainment spurred the emergence of unofficial television stations, whose number and diverse affiliations reflect the fragmentation of the country.

In 1979, the Christian right-wing party Phalange decided to launch its own TV. This project came to fruition in August 1985 with the launch of the Lebanese Broadcasting Corporation (LBC), which was managed by the Lebanese Forces, a militia that emerged from the Phalange. At the time, 80 percent of LBC's programs were American and French.[22] Simultaneously, a Christian channel named Middle East Television (METV), owned by the American Christian Broadcasting Network (CBN), was set up in southern Lebanon as early as 1982 with the support of Israel, which had occupied part of the country that same year. METV's programs were American and could be received in Lebanon and Israel. Other initiatives also took place around the same time, including Fihe Television, which broadcast sporadically from the village of Fihe, a village disputed by the Syrian Social Nationalist Party and former president Sleiman Frangie's Marada movement. Additionally, a television project initiated by the Hezbollah began in 1987 with promised French support, but it only materialized in 1991 under the name Al Manar, without French support.[23] It is estimated that when the war ended in 1990, fifty-four TV channels were operating illegally in Lebanon, and many served small geographic areas and were run from small apartments.[24] This is mirrored today by the decentralized nature of the use of streaming platforms in Lebanon. Much like those early illegal TV stations catered to hyperlocal

[21] Thomas Friedman, "Living with the Violence of Beirut," *New York Times Magazine*, National Edition, July 17, 1983, 12, https://www.nytimes.com/1983/07/17/magazine/living-with-the-violence-of-beirut.html.

[22] Douglas A. Boyd, "Lebanese Broadcasting: Unofficial Electronic Media During a Prolonged Civil War," *Journal of Broadcasting and Electronic Media* 35, no. 3 (1991): 282, https://doi.org/10.1080/08838159109364125.

[23] Ibid., 283.

[24] Zahera Harb, *Channels of Resistance in Lebanon: Liberation Propaganda, Hezbollah and the Media* (London: I. B. Tauris, 2011), 92–100.

audiences with minimal resources, today's streaming platforms benefit from a similarly flexible and fragmented media ecosystem.

The rise in illegally operating TV channels posed a significant problem for the resurgent Lebanese state, pushing it to adopt Media Law 382 in 1994, which aimed at reorganizing the sector.[25] Media Law 382 granted licenses to six TV stations out of sixty-three applications. The logic behind the licensing was both political (all these channels belong to powerful politicians or political groups) and sectarian, and programming on these channels was also determined by their respective political and sectarian agendas, and by the audiences to which they catered. For example, in the 1990s while LBC regularly scheduled William Wyler's *Ben Hur* (1959) every Easter and Franco Zefirelli's *Jesus of Nazareth* (1977) every Christmas for many years, Future TV (now operating from the UAE) marked Ramadan and the important Islamic Eids with special programming, and Al Manar marked prayer time five times a day. However, the mid-1990s was also a time when cable and satellite TV were becoming increasingly popular in the region. Eventually, most of these channels switched to satellite broadcasting and reached regional and sometimes international audiences as well. This expansion compelled the channels to consider the needs, culture, and viewing habits of broader audiences, particularly in Arab countries, which significantly influenced their programming. In particular, Ramadan shows gained increasing importance, even for a channel like LBC, all of which helped establish Lebanon as a major media powerhouse in the region.

From Satellite TV to the Internet

In 1967, Arab ministers of information outlined a plan for a satellite network that would be designed to bring together the cultural and social efforts of the Arab League. Consequently, the Arab States Broadcasting Union (ASBU) was established in 1969. This union was formed without the participation of Saudi Arabia, however, due to the tense Saudi–Egyptian relations during that period. After Saudi Arabia joined ASBU in 1974, efforts led to the establishment of the Arab Satellite Communications Organization (Arabsat) under the jurisdiction of

[25] Marwan M. Kraidy, "Broadcasting Regulation and Civil Society in Postwar Lebanon," *Journal of Broadcasting and Electronic Media* 42, no. 3 (1998): 391–2, https://doi.org/10.1080/08838159809364457.

the Arab League.[26] Arabsat launched in 1985 the first Arab satellite, Arabsat-1A, which remained largely underused until the first Gulf War. Following Kuwait's invasion by Iraq, Egypt aligned with Syria, Saudi Arabia, and other Gulf states in the US-led coalition to remove Iraqi forces from Kuwait. Egyptian troops, stationed in the Gulf, encountered Iraqi propaganda that criticized US actions. To counter this, in December 1990, the Egyptian Radio and Television Union (ERTU) leased an Arabsat transponder to launch the Egyptian Space Channel (ESC). This channel, the first satellite television station in the Arab world, was intended to provide news and entertainment to Egyptian soldiers and Gulf viewers.[27] After the war, more satellites capable of covering Southwest Asia were launched, including Eutelsat (Europe) and Asiasat (Hong Kong), and more Arab satellite television stations started emerging: Middle-East Broadcasting Center (MBC), launched in London by Saudi investors in 1991; Emirates Dubai Television (EDTV), a state-owned Emirati channel launched in 1992; Jordan Satellite Channel (JSC), established by Jordan's national broadcaster in 1993, and so on.

By 1996, when Qatar launched Al-Jazeera, all Arab countries had their own satellite television stations, except for two countries: Iraq, then under international embargo, and Lebanon, which was struggling to rebuild its institutions after the war. Despite an early attempt at satellite broadcasting by Future TV in October and November 1994, it was only in late 1996 that Lebanon entered the realm of satellite TV (law 531) with the launch of LBCSAT, followed by Future TV International, then in 1999 Télé-Liban, and Hezbollah's Manar TV, which started broadcasting the day before the liberation of South Lebanon in May 2000. MTV and Al-Jadeed followed suit the same year.[28] The entry of Lebanese channels into the satellite television market transformed the media landscape in the Arab world. Their politically and culturally free-spirited tone, coupled with a bold approach to entertainment, set them apart. These channels were unafraid to broadcast content uncommon in other Arab countries, such as depictions of relative nudity, challenging traditional norms in the region. The success of Lebanese channels also rested on the positive reputation that Lebanese entertainment already enjoyed. This included areas such as music,

[26] Marwan M. Kraidy, "Arab Satellite Television between Regionalization and Globalization," *Global Media Journal* 1, no. 1 (2002), https://www.globalmediajournal.com/open-access/arab-satellite-television-between-regionalization-and-globalization-.php?aid=35056.

[27] Naomi Sakr, *Satellite Realms: Transnational Television, Globalization and the Middle-East* (London: I. B. Taurus, 2011), 10–11.

[28] Dajani, *Media in Lebanon*, 102.

theater, and cinema. In addition to all this, the media industry boom in Lebanon in the late 1990s to early 2000s established the country as a production hub for regional channels, many of which set up offices in Beirut and employed Lebanese professionals. This contributed to the lasting influence of Lebanese talent on content production, which continues to resonate even today in the era of streaming platforms.

Lebanese households were able to access—legally or illegally—hundreds of satellite television channels offering programs ranging from sport shows to religion, to films, to news and more. A large-scale study on the media industries in the MENA (Middle East and North Africa) region conducted by Northwestern University in Qatar in 2015–16 showed that there were 724 free-to-air channels available in the region, with twenty-seven headquartered in Lebanon, twenty-six of which were privately owned. The top ten channels by share of revenue in Lebanon were MBC 4 (headquartered in the UAE), MBC 2 (UAE), Zee Alwan (UAE), Al Mayadeen (Lebanon), Al Arabia (UAE), Fox Movies (KSA), MBC Action (UAE), Rotana Cinema (Bahrain), Zee Aflam (UAE), and Al Jazeera (Qatar). The analysis of the top twenty TV channels in Lebanon by share of audiences showed that Lebanese audiences turn mainly to channels specializing in entertainment (10 percent), then to channels specializing in films (9 percent), and thirdly to channels specializing in news (8 percent), with other types of channels (religion, music, sports) at 1 percent or less.[29] This reflects an environment conducive to the subsequent success of streaming platforms that are usually mainly oriented around entertainment, movies, and TV series.

The development of satellite TV in Lebanon occurred concurrently with the introduction of the internet, which was to become essential for the rise of streaming platforms. The internet was introduced as part of the Horizon 2000 reconstruction project, which was launched by the government in 1991. The telecommunications sector—successor to the postal and telegraph services established during the Ottoman rule at the end of the nineteenth century and further developed under the French mandate (1923–43)—was allocated 5.6 percent of the total reconstruction budget.[30] During the pre-civil-war period, the state held a monopoly over this sector, and telephony developed relatively well, achieving a penetration rate of 7 percent by 1974.

[29] Klaus Schoenbach, Robb Wood, and Marium Saeed, "Media Industries in the Middle East 2016," Northwestern University in Qatar, 2016, https://www.qatar.northwestern.edu/docs/publications/research/2016-middle-east-media-industries-report.pdf.

[30] Nabil Bukhalid, "Lebanese Internet Governance Landscape," MEAC-SIG 2016 (August 2016), https://icann-community.atlassian.net/wiki/spaces/MES/pages/112985710/MEAC-SIG+2016.

Although such growth continued during the civil war, reaching a penetration rate of 12.5 percent by the end of the war, half of the lines were out of service due to the considerable damage caused by the conflict.[31] When the war ended, the state endeavored to rehabilitate the sector in hopes of attracting foreign capital. Lebanon became the first Arab country to introduce the internet, thanks to an initiative taken by the American University of Beirut.[32]

Internet in Lebanon is relatively free. Law No. 81/2018 relating to Electronic Transactions and Personal Data stipulates that "the transmission to the public by digital means is free within the limits required for the respect of the constitution, laws, and public order" (article 65) and that service providers are not required to monitor the information that they transmit or stock, but are nonetheless required to remove it on the authorities' request (articles 69 and 70).[33] Thus, while on the one hand the internet is accessible for the vast majority, on the other hand human rights activists have recently observed a significant increase in governmental efforts to suppress or regulate free speech on the internet. At the same time, Lebanon is also witnessing an unprecedented number of outspoken human rights activists and NGOs, who mainly use the internet to express their criticism.[34] It is in this complex terrain that streaming platforms have made their entrance in recent years.

The Rise of Streaming Platforms in the MENA Region

By establishing a culturally diverse, commercially driven, relatively free, and somewhat deregulated media landscape shaped by political, sectarian, and foreign influences, earlier initiatives have shaped audience expectations while simultaneously fostering capabilities such as broad accessibility, transnational

[31] Dima Jamali, "Post-War Telecommunications Reform in Lebanon: Trends and Challenges," *Info* 5, no. 2 (2003): 37.

[32] See Mohamad M. Taha, "Le développement d'internet au Liban," in *Mondialisation et nouveaux médias dans l'espace arabe*, ed. Frank Mermier (Lyon: Maison de l'Orient et de la Méditerranée Jean Pouilloux, 2003), 261–72.

[33] "Law No. 81/2018," Lebanese University, http://77.42.251.205/Law.aspx?lawId=278573&language=ar.

[34] On the history of civil society and NGOs in Lebanon, see Khaldoun AbouAssi, "The Third Wheel in Public Policy: An Overview of NGOs in Lebanon," in *Public Administration and Policy in the Middle East*, ed. Alexander R. Dawoody (New York: Springer, 2015), 215–30, https://doi.org/10.1007/978-1-4939-1553-8_12, and Khaldoun AbouAssi, *Lebanese Civil Society: A Long History of Achievements, Facing Decisive Challenges Ahead of an Uncertain Future* (Johannesburg: CIVICUS, 2006), https://www.civicus.org/media/CSI_Lebanon_Country_Report.pdf.

reach, and a skilled technical workforce. Streaming platforms in Lebanon inherited this legacy of innovation as well as political-economic entanglement.

From the perspective of Lebanese law, streaming platforms are governed by rather liberal telecom laws. Consequently, as soon as a service becomes technically available in the MENA region, it becomes accessible to the Lebanese public. Khalil and Zayani explain that the foundation of the audience base for streaming platforms in the MENA region was, on one hand, those who consumed videos on sites like YouTube—some of whose channels were already producing original content popular among young people seeking more progressive and unconventional media content—and on the other hand, those who used pay-TV channels that catered to an affluent viewer class with more conservative and traditional content.[35] In this context, the satellite television channel MBC launched in 2007 the first video streaming platform in the region: Shahid. Shahid (which means "Watch" in Arabic) was initially designed to contain archives of MBC programs but later diversified its offering to also include international programs. Today, Shahid is a full-fledged OTT platform with international, regional, and original content.

The success of Shahid spurred the launch of numerous streaming platforms in the region, many of which were either based in Lebanon or relied on Lebanese talent and professionals, taking advantage of the country's diverse and open media tradition. These platforms, whether tied to preexisting channels or not, and whether subscription based or free, became available as websites or mobile applications. Among these platforms, Cinemoz, a Lebanese platform originally dedicated to Arabic and Lebanese content, launched in 2011 and has since undergone several reconfigurations. Its programs include shows like *Stand Up Baladi* (2020), a Lebanese stand-up comedy from Awk.word comedy club, and *Arabs in Space* (2018), a comedic science-fiction miniseries depicting "a crew of poorly picked Arab explorers who are sent into deep space to save the Middle East." GO, created in 2014 by the satellite channel OSN and later rebranded as Wavo, became OSN+ in 2022. OSN+ prides itself for offering the largest selection of content in Arabic in the region, including international programs dubbed into Arabic like the romantic Turkish series *Sharab Al Toot* (2022) or HBO's satirical drama series *Succession* (2018).

[35] Joe F. Khalil and Mohamed Zayani, "Streaming the Middle-East: From Legacy Media to Digital Platforms," in *Digital Platforms and the Global South: Reconfiguring Power Relations in the Cultural Industries*, ed. Philippe Bouquillon, Christine Ithurbide, and Tristan Mattelart (London: Routledge, 2024), 165–81.

Starzplay Arabia, an entertainment platform founded in the United Arab Emirates in 2015, offers a wide range of content, including famous TV shows like *Man Sa Yarbah Al Malioun* (the Arabic version of *Who Wants to Be a Millionaire*) and the Arabic-dubbed vintage Japanese anime *Captain Majed* (1982). M.Media, a platform dedicated to Lebanese content and notably featuring Ziad Al Rahbani's filmed theater plays *Bel Nesbeh La Bokra Chou* and *Film Ameriki Tawil*, appeared in 2017 and was suspended in 2020. And Aflamuna Online, a Lebanese nonprofit streaming platform supported by several nongovernmental funds and dedicated to sharing independent films, was created in 2020. Aflamuna Online currently curates weekly programs, featuring independent films like Jocelyne Saab's war documentary *Beirut Madinati* (1982) and Feiruz Serhal's short drama *Tashweesh* (2017). Many other platforms continue to be launched.

These regional platforms, which sometimes offer content dubbed into Arabic, coexist and compete with global platforms such as Netflix, which launched its MENA service in 2016; Amazon Prime Video launched in MENA in 2021 and Disney+ in 2022. This list is not exhaustive, as the landscape of streaming platforms is in constant flux: some platforms cease to operate (M.Media), others undergo name changes (GO, OSN), and some enter into collaborations (Netflix and OSN+, Jawwy TV and Starzplay, to name a few). A study by Ipsos in 2019 mentioned the existence of more than forty OTT platforms in the region.[36] However, the platforms commanding the largest market share in the MENA region are Netflix, Shahid, and Starzplay,[37] with some reports showing Shahid ahead of Netflix.[38]

Several studies have noted that media audiences usually prefer local and regional content to international content. Joseph Straubhaar, for instance, suggests that "cultural proximity" plays a significant role in influencing audience preferences.[39] Similarly, Jang, Kim, and Baek identify linguistic proximity and geographical distance as crucial factors in determining audience choices.[40] This

[36] Ipsos, "OTT and Content Streaming in MENA," November 25, 2019, https://www.ipsos.com/sites/default/files/ct/news/documents/2019-12/ott_and_content_streaming_2019.pdf.

[37] Khalil and Zayani, "Streaming the Middle-East," 173.

[38] Capital Market Authority, "MBC Group Prospectus," December 2023, 75, https://cma.org.sa/en/Market/Prospectuses/Documents/MBC_en.pdf.

[39] Joseph D. Straubhaar, "Beyond Media Imperialism: Asymmetrical Interdependence and Cultural Proximity," *Critical Studies in Mass Communication* 8, no. 1 (1991): 39–59, http://dx.doi.org/10.1080/15295039109366779.

[40] Moonkyoung Jang, Dokyung Kim, and Hyunmi Baek, "How Do Global Audiences of TV Shows Take Shape? Evidence from Netflix," *Applied Economics Letters* 30, no. 3 (2023): 285–91, https://doi.org/10.1080/13504851.2021.1983916.

is also true in the MENA region, where "most Arab audiences prefer to watch content in Arabic."[41] Therefore, platforms that cater to the taste of local audiences succeed in garnering a bigger number of subscribers. This is exemplified by Shahid's peak audience in Ramadan 2023, with a record number of 3.76 million subscribers due to content curated for the occasion, compared to Netflix's 2.2 million subscribers in Q1 of 2023.[42]

Lebanese viewers enjoy access to much of the same online video content as other audiences in the MENA region, thanks to the largely uncensored nature of the internet in Lebanon. However, two important nuances should be considered: first, streaming platforms can geo-restrict content, making some programs unavailable in certain countries; second, the relatively unrestricted internet access in Lebanon allows viewers to access content that may be censored elsewhere in the region. Nonetheless, the widespread use of VPNs, which often go undetected, complicates efforts to quantify the extent of these differences.

It is also important to note that very few studies have been conducted on the use of these platforms in Lebanon, for several reasons: First, the country's unstable economic and political situation, combined with limited resources provided to researchers by local universities and the state, makes it difficult to conduct serious large-scale studies of the phenomenon. Second, information on the use of platforms in Lebanon is often compromised by the fact that piracy is widespread, allowing the population to access platform content outside of any documented circuit. Finally, Lebanon represents a limited market relative to the region as a whole, making it less prominent in regional studies than countries like Egypt and the Gulf nations, particularly the United Arab Emirates, Saudi Arabia, and Qatar, which are much more studied and documented. The few available studies show that, in 2019, 48 percent of Lebanese viewers used streaming services (compared to 69 percent in the UAE and 25 percent in Egypt)[43] and that 15 percent used Shahid (33 percent in the UAE and 25 percent in Egypt) while 12 percent used Netflix (35 percent in the UAE and 4 percent in Egypt).[44] Studies also showed that, in 2016, 53 percent of Lebanese respondents to a survey answered yes to the prompt "More entertainment media should be based on my

[41] Zaid, Benmoussa, Ayyad, Ibahrine, and El Kadoussi, "Domestication of Netflix," 6625.

[42] Capital Market Authority, "MBC Group Prospectus," 97.

[43] Everette E. Dennis, Justin D. Martin, Elizabeth A. Lance, and Fouad Hassan, "Media Use in the Middle-East 2019: A Seven-Nation Survey," Northwestern University in Qatar, 2019, 37, https://www.mideastmedia.org/survey/2019/uploads/file/NUQ_Media_Use_2019_WebVersion_FNL_051219%5B2%5D(1).pdf.

[44] Ibid., 38.

culture and history" (64 percent in Egypt, 65 percent in the UAE, and 80 percent in Qatar; only Tunisia scored less with 52 percent); 51 percent answered yes to the prompt "I prefer to watch films that portray my own culture" (62 percent in Egypt, 73 percent in the UAE, 72 percent in Qatar, and 46 percent in Tunisia); and 58 percent answered yes to the prompt "People benefit from watching content from different parts of the world" (59 percent in Egypt, 69 percent in the UAE, 67 percent in Qatar, and 51 percent in Tunisia).[45] It is interesting to note that, overall in the Arab countries covered by this study, less respondents answered yes to this last prompt in 2016 than in 2014 (down 4 percent in the Arab world, with Lebanon in particular moving down by 16 percent from 74 to 58 percent), at a time when more international platforms were becoming available. This may have been linked to the intensification of political problems in the region, which may have exacerbated populist policies and nationalist sentiment. In Lebanon, the deadlock over the issue of the presidential vacancy heightened sectarian tensions, with the presidency, traditionally held by a Christian, becoming a pretext for an uninhibited expression of sectarian or racist sentiments.

Studies also show that the Lebanese audience's use of media content follows a pattern generally similar to that of other MENA audiences, and that, for example, the choice of genres by viewers is more determined by reception habits (time of day, context) than by the content of specific genres: when moving from TV to online content, MENA audiences tend to exhibit an explorative tendency at first, then go back to their previous viewing preferences,[46] challenging the dominant narrative that platforms are particularly popular solely because they offer a new kind of content.

All these studies and figures are to be considered with caution, given that the chosen macroscopic samples used might not perfectly represent the highly fragmented Lebanese population and its use of video content. For this reason, the importance of Lebanon for streaming platforms might not lie so much in reception and consumption of video content as in its creation. Indeed, while Lebanon represents a negligible and unstable market for these platforms, it nevertheless plays a very important role in content creation, which relies on its notable legacy in the realm of media entertainment and the existence of a skilled

[45] Northwestern University in Qatar, "Media Use in the Middle-East: A Five-Year Retrospection," 2018, https://www.qatar.northwestern.edu/docs/publications/research/2018-middle-east-media-retrospection.pdf.

[46] Klaus Schoenbach, Marium Saeed, and Rob Wood, "Audience Responses to Online Video in MENA: New Favorite Genres or Just More of the Same as Television," *International Communication Gazette* 80, no. 8 (2018): 697–713.

workforce. Even when this workforce is not employed within Lebanon, it plays a pivotal role in shaping the media landscape of key regional hubs, particularly the United Arab Emirates and, to a growing extent, Saudi Arabia. This is perfectly expressed in the original content that these platforms produce for the region. Hence, in the following pages the initial original content from three platforms—Netflix, Shahid, and M.Media—will be the focus of our attention. By representing respectively international, regional, and Lebanon-based players, they allow insight into the different types of negotiations platforms are compelled to make with audiences and stakeholders.

Netflix: A study of Netflix libraries in seventeen countries showed that while American productions were not necessarily dominant, local productions were not widely prevalent either, with an average of only 7.7 percent of domestic titles.[47] When Netflix began streaming in the MENA region in 2016, it encountered difficulties in acquiring popular local content due to the presence of well-established platforms and the fact that the local production and distribution network was already firmly in place. Netflix's initial content acquisitions included recorded theater performances,[48] mainly Egyptian, and festival films that were rarely broadcast by established broadcasters. Majid Al Ansari's *Zinzana* (2015) for instance, an Emirati festival film, was the first Arab film to be purchased by the platform.

An analysis of Netflix content available in Lebanon at the beginning of 2024 through the "browse by language" function available on the laptop version of the app revealed that there are 384 Arabic-language titles (including series, films, shows, etc.) out of a total of 6,568 titles (5.8 percent). There is also a "Lebanese" language category, which contains twenty-one titles. Interestingly, Lebanese is the only Arabic dialect to have a dedicated category, and this is not just the case when accessing the app from Lebanon but also from other Arab countries[49] such as the United Arab Emirates or Egypt, even though Egyptian productions are more numerous on the platform.

Netflix's first original production targeting the MENA audience was a stand-up comedy titled *Adel Karam: Live from Beirut* (2018). While stand-up comedy might not typically be the foremost choice for a platform, the bold and

[47] Amanda D. Lotz, Oliver Eklund, and Stuart Soroka, "Netflix, Library Analysis, and Globalization: Rethinking Mass Media Flows," *Journal of Communication* 72, no. 4 (August 2022): 511–21, https://doi.org/10.1093/joc/jqac020.

[48] Khalil and Zayani, "Streaming the Middle-East," 172.

[49] Not all. In Morocco, for instance, there is only an "Arabic" language category.

unfiltered humor famously embraced by Lebanese comedians likely encouraged the platform to invest in this genre. Adel Karam is a Lebanese actor, locally famous since the 1990s, who gained regional and international fame with his roles in award-winning films, notably Ziad Doueiri's Oscar-nominated feature film *The Insult* (2017), and with his live night TV show. However, *Adel Karam: Live from Beirut* did not achieve the anticipated success (rated 5.3/10 on IMDB), and after a second unsuccessful attempt to present stand-up comedy content featuring Saudi Arabian comedians Ibrahim Alkhairallah and Moayad Alnefaie (*Comedians of the World*, 2019), Netflix shifted its focus toward serial dramas by producing *Jinn* (2019), a paranormal series set in Jordan and directed by the Lebanese Mir-Jean Bou Chaya, who had made a name for himself a few years earlier with his independent feature film *Very Big Shot* (2015), which won Best Picture at the Marrakech International Film Festival. However, *Jinn*, released during Ramadan, when families across many Arab countries traditionally gather after Iftar to watch television series, also fell short of meeting audience expectations. Moreover, it proved controversial for its portrayal of teenagers smoking cannabis, forming romantic relationships, and displaying tattoos and piercings, which some deemed "immoral."[50]

The same kind of controversy occurred in response to the first feature-length fiction film produced by Netflix. Feature films are, numerically, the dominant content on Netflix. Out of the 384 Arabic-language items, 291 are feature films, with the Lebanese language category boasting thirteen feature films. The remaining items are series, limited series, or comedy shows. For its first original feature film, Netflix attempted to play it safe by choosing to produce a regional adaptation of a comedy film whose international success was already confirmed and to cast bankable actors with established reputations. *Ashab Wala Aaz* (*Perfect Strangers*, 2022) was filmed in Lebanon, directed by Lebanese filmmaker Wissam Smayra, and produced by Dubai-based and Lebanese-owned Front Row Filmed Entertainment, with Egypt's Film Clinic and Lebanon's Empire Entertainment. An adaptation of Paolo Genovese's Italian success *Perfetti Sconosciuti* (2016), which had already been adapted in more than fifteen countries, *Ashab Wala Aaz* featured an all-star cast including Adel Karam, Nadine Labaki—Lebanon's most prominent filmmaker and actress whose film *Capernaum* (2018) won the Jury Award at the Cannes Film

[50] Bruce Haring, "Netflix series 'Jinn' Sparks Uproar in Jordan over Alleged 'Immoral Scenes,'" *Deadline*, June 15, 2019, https://deadline.com/2019/06/netflix-jinn-uproar-jordan-immoral-scenes-1202633178/.

Festival and was also nominated at the Oscars—and Mona Zaki, an Egyptian superstar and L'Oréal's face in the MENA region. Despite this star-studded cast, the film became a subject of controversy in the region on its release, especially in Egypt. One of the scenes that sparked debate depicts the character played by Mona Zaki removing her underwear before leaving the house for an evening out. It is also revealed in the film that Mona Zaki's character, while going through a challenging period with her husband, has been exchanging text messages with another man. Another element deemed offensive was the inclusion of a homosexual character. These elements prompted conservative critics to accuse the film of trying to put "Western ideas in a conservative society" and to "encourage homosexuality." This pushed MP Mustafa Bakri to submit a statement to Egypt's Speaker of the House, stating that the film incites homosexuality and marital infidelity. Later, a lawsuit against the film was filed by Egyptian lawyer Ayman Mahfouz.[51] In Lebanon, by contrast, the film did not stir any controversy and was generally well received. The limited criticism it faced was aimed at its relatively mediocre cinematic qualities rather than its themes or content.[52]

These controversies and challenges highlight the difficulties faced by an international provider like Netflix in positioning itself within a regional market that has its own logics and where the audience's expectations are strongly shaped by cultural and even political factors. This is not to say that Netflix's original productions are always doomed to failure in the region. In Lebanon, for instance, seven out of the top ten most-watched productions at the time of writing were Netflix originals. Leading the chart was *Al Rawabi School for Girls* (2021) season 2, a high-school drama set in Jordan, with its third season securing the third spot. While a Jordanian series, it features significant contributions from Lebanese professionals in key creative roles, including cinematography and sound design. The remaining five Netflix originals in the top ten are Western productions. Interestingly, the list also includes another Arab title, a non-Netflix production: *It's OK* (2024), an intimate documentary series on Lebanese singer Elissa, which held a place in the top ten for seven weeks.

[51] Alex Ritman, "Netflix's First Arabic Original Film 'Perfect Strangers' Accused of 'Moral Degradation' in the Middle-East Due to Gay Character," *Hollywood Reporter*, January 23, 2022, https://www.hollywoodreporter.com/movies/movie-news/perfect-strangers-netflix-middle-east-egypt-controversy-lgbt-1235079663/.

[52] Chafic Tabbara, "'Ashab Wala Aaz.' Film mansoukh wa zawbaa fi fenjan: satazhab 'al fadiha' wa la yabka ella thiyab al moumasilin" ["Perfect strangers"… a Copied Film and a Squall in a Cup: The "Scandal" Will Disappear and Only the Actors' Clothes Will Remain]. *Al Akhbar*, January 25, 2022, https://al-akhbar.com/Cinema/328704.

Shahid: The composition of Shahid's content and the way it is presented on the website reflects its origin as an extension of a satellite television channel deeply rooted in the Arab market.

Indeed, while the Netflix home page primarily showcases recommendations tailored to viewers based on their viewing habits, followed by a wide range of generic categories with predominantly Western content, Shahid's home page highlights Arabic or Turkish series and films, with a category for Western films placed much farther down the page. Additionally, it is worth noting that, at the time of writing, one month before the start of Ramadan, Shahid has dedicated a "Ramadan 2024" category on its home page, a feature not (yet) present on Netflix. Another interesting point to note is the presence of an "Egyptian Movies" category on Shahid—and the absence of a Lebanese film or Lebanese language category. This aligns with historical market trends in the region, where Egyptian cinematic production has consistently been the most dominant and the most popular.

Shahid's original productions—not considering those of its parent company MBC but focusing solely on productions branded as Shahid Originals—began in 2019 with a series titled *Al Diva* that focuses on the behind-the-scenes action of a reality TV singing competition similar to *Arabs Got Talent*. *Al Diva* is set in Lebanon and features, in the main role, Lebanese singer and actress Cyrine Abdelnour (The Diva), accompanied in more secondary roles by Egyptian actress Boussy and Saudi actor Yaqoub Al-Farhan. Choosing actors from these three nationalities reflects Shahid's intention to appeal to audiences in the three main markets it targets: the Levant, Egypt, and the Gulf countries. But giving the main role to a Lebanese star and setting the series in Lebanon both reflect the country's importance in the regional entertainment market. It is worth noting that the Maghreb countries are rarely represented in Shahid's original productions, although *Al Diva* does feature a contestant who speaks Arabic with a Moroccan accent, certainly with an eye toward luring another subsection of the regional audience.

A key feature of *Al Diva* is the explicit product placement carried out in the show. In fact, the appearance of the product, often in close-up and shown under a glowing light, creates a sort of pause in the narration. The introduction of the product is always narratively justified (for example, the Diva, wanting to take a moment to relax, savors a Galaxy chocolate bar), but the way it makes the product prominent while suspending the narrative is reminiscent of television advertisements. This relationship with the product illustrates how

Shahid's television origins can manifest in its streaming content. However, while audience figures are lacking, it appears that *Al Diva* did not achieve the expected success. Despite the fact that the final scene of season 1 ended with a cliffhanger that could have led to a second season, a second season never materialized. This highlights the complexity and unpredictability of the market, even for a well-established platform with extensive knowledge of audience expectations.

Following *Al Diva*, Shahid has produced a significant amount of original content (73 productions currently appear on the website branded as Shahid Originals). These productions span a broad range of genres, including dramatic series, movies, children's programs, reality shows, talk shows, cooking shows, and sports programs. While these productions, all in Arabic, may include a wide range of dialects, and appeal to audiences in a number of Arabic-speaking countries—from Saudi Arabia to Morocco—the majority of these productions are in Egyptian Arabic. Lebanese productions are not numerous among Shahid Originals. They tend to be characterized by a slightly more liberated or "bling-bling" aspect, as seen in the series *Arrabat Beirut* (The Godmother of Beirut, 2023), which tells the story of a woman managing an elite nightclub, or by explorations of the flashy world of celebrities as in *Al Diva* and *The Fixer* (2020), or by the depiction of an apocalyptic world as in the series *Hell's Gate* (2021) or *Beirut 6:07* (2020), a series of fifteen dramatic short films directed by fifteen young Lebanese filmmakers which narrates the events of the massive explosion that occurred in the port of Beirut on August 4, 2020. *Beirut 6:07* was nominated for the Best Short-Form Series Award at the 2021 International Emmy Awards, marking one of Shahid's rare appearances on the international stage, beyond its regional presence. The critical success of *Beirut 6:07* did not, however, translate into audience success.

It is worth noting that Shahid's original productions rarely rank among the most-watched programs. At the moment of writing, no Shahid Original production was part of the Lebanese top ten (not counting the two MBC productions), and even in the UAE, where Shahid is headquartered, only one Shahid original production (*Al Naseem Highschool*, a high-school series released in 2024 and set in Saudi Arabia) made it to the top ten. This does not mean that Shahid's original programs are not successful. It simply means that Shahid's originals are not as successful as the other programs acquired by the platform. This can be explained by the fact that, due to its establishment within the industry and the network it inherited from MBC, Shahid is capable of acquiring programs that are themselves very successful and whose success surpasses that of its originals. Conversely, if Netflix Originals manage to make it to the top ten on Netflix more often, it is probably because the platform struggles to acquire successful nonoriginal content from the region. As Joe Khalil writes, "Netflix encountered limitations in its capacity to acquire popular content, largely due to producers being bound by long-term contracts or preferring to remain affiliated with their long-established traditional venues."[53] Shahid, on the contrary, did not face such issues, especially thanks to MBC's preestablished network.

Counting on its originals, but also on an established network that allows it to buy high-quality regional and local content, Shahid appears to be much more in tune with the regional and local markets than Netflix. This is to be expected, as Shahid only has this market to consider, whereas for Netflix, this market is just a portion, albeit a significant one, of its global market.

Shahid and its parent company MBC are more exposed to a dynamic of "permutations of global media transformation" that Joe Khalil describes as "business push/global pull," which influences their decision-making processes.[54] The "business push" can be described as the collective force of diverse cultural industries at global, regional, and local levels, which are focused on the standardization and commercialization of their entertainment offerings. The "audience pull" centers around local symbolic elements, including tastes, interests, and communities. This dynamic pushes regional and local entertainment outlets to recognize more seriously the viewers' specificity. As Khalil explains, "All regional channels have either recognized location as a niche-audience criterion, or have broken away from localization principles that assume Arab audiences are homogeneous."[55] This is illustrated by Shahid, which aims to diversify its content to cater to the varied tastes of its regional audience, all while being held by global business trends and norms.

M.Media: Although M.Media was suspended in the aftermath of the Beirut port explosion and due to the challenging political and economic situation in Lebanon, its story is worth revisiting because it serves as a perfect illustration of the challenges faced by local Lebanese platforms. Launched in 2017 in Beirut with the aim of serving a localized niche audience, M.Media, whose slogan was "Entertainment with a twist from Beirut to the world," primarily specialized in

[53] Khalil and Zayani, "Streaming the Middle-East," 172.

[54] Joe F. Khalil, "The Business Push and Audience Pull in Arab Entertainment Television," *International Journal of Communication* 10 (2016): 3638.

[55] Ibid., 3642.

Lebanese content and targeted Lebanese audiences, not only in Lebanon but also in the diaspora. One should note here that while the Lebanese population in Lebanon is estimated at about four million citizens, the Lebanese diaspora is estimated at three to four million according to conservative estimates[56] and up to fourteen million when including the descendants of emigrants scattered across the globe.[57] M.Media, therefore, was potentially viable. However, as we will see, the country's instability would ultimately lead to the downfall of the platform, which suspended its operations in 2020.

M.Media was created as part of Lebanese investor Eli Khoury's Quantum Group, which includes advertising agencies and communication firms involved in, among other things, political propaganda (for instance, in Iraq in 2004 and Lebanon in 2005).[58] M.Media adopted a rather interesting launch strategy that illustrates its deep understanding of the Lebanese market. Several years before its launch, M.Media embarked on the restoration of a filmed rehearsal of a 1978 play by Ziad Rahbani, *Bel Nesbeh La Bokra Chou* (What About Tomorrow?). Ziad Rahbani is the son of composer Assi Rahbani and Feyrouz, the greatest star of Lebanese song. A musician, an actor, and a playwright, Rahbani is idolized by a large portion of the population, and his plays are cultural phenomena, with some of their lines being repeated by Lebanese people in their daily lives. These plays, *Sahrieh* (Evening, 1973), *Nazl El Sourour* (Happiness Hotel, 1974), *Bel Nesbeh La Bokra Chou* (What About Tomorrow?, 1978), *Film Ameriki Tawil* (An American Motion Picture, 1980), and *Shi Feshil* (Failure, 1985), are widely known and have been listened to through audio recordings from generation to generation. It would not be an exaggeration to say that almost every Lebanese household owns a cassette recording of at least one of these plays. However, while a visual of *Sahrieh* does exist in the form of a performance filmed in a television studio—which has been regularly broadcast by Télé-Liban since it no longer had the means to invest in new and original programming—the visuals of the other plays were completely nonexistent and mysterious. The extreme popularity of Rahbani's plays, coupled with the total absence of visuals, transformed these

[56] Eric Verdeil and Bruno Dewailly, "International Migration and the Lebanese Diaspora," in *Atlas of Lebanon: New Challenges*, ed. Eric Verdeil, Ghaleb Faour, and Mouin Hamze (Beirut: Presses de l'IFPO, CNRS Liban, 2019), 42–3.

[57] Lisa Barrington, "Lebanon Gets a New Government, Now It Needs a New Economy," Reuters, January 19, 2017, https://www.reuters.com/article/idUSKBN1530MT/.

[58] Michael Karam, "Man of Many Parts Aims to Ignite a Media Revolution," *National*, December 19, 2012, https://www.thenationalnews.com/business/man-of-many-parts-aims-to-ignite-a-media-revolution-1.435758.

works into mythical objects fantasized about by the Lebanese. So when M.Media managed to persuade Rahbani to restore and release the filmed recordings—first of *Bel Nesbeh La Bokra Chou*, then of *Film Ameriki Tawil*—it did not just obtain audiovisual content; it secured cultural artifacts of profound significance for Lebanese audiences.

The filmed plays, which are said to have cost hundreds of thousands of dollars in rights and restoration, were not directly launched on the platform. In the "Pirate's Paradise"[59] that is Lebanon, it would have been reckless to rely on revenues generated by a digital platform that could have been illegally accessed. Instead, M.Media—which, in early 2016, did not yet exist as a platform but as a content production company—chose to release these works in cinemas after a major teasing campaign. The results were phenomenal. On the very first day of its release in theaters in January 2016, *Bel Nesbeh La Bokra Chou* attracted 28,000 viewers, setting an absolute record in the country. This is more than what *Star Wars: The Force Awakens* did in its entire opening weekend.[60] The "film" ultimately became the biggest box office hit of 2016, earning more than $1,350,000, far ahead of its closest competitors *London Has Fallen* and *Batman v Superman*, which garnered around $859,000 and $843,000, respectively.[61] Released in October 2016, *Film Ameriki Tawil* was also very successful, although not as much as the former.

Encouraged by these commercial and symbolic successes, the M.Media platform was launched in the first quarter of 2017, targeting Lebanese audiences both in Lebanon and abroad. The platform's catalog consisted primarily of popular contemporary Lebanese films, such as Nadine Labaki's *Caramel* (2007) and *Where Do We Go Now* (2011), and Amin Dora's *Ghadi* (2013); "vintage" Lebanese films and series like Gary Garabedian's *Abu Salim in Africa* (1965), Mohammed Salman's *Guitar El Hobb* (The Guitar of Love, 1973) and *Badawia Fi Baris* (A Bedouin in Paris, 1964); filmed popular theater like Ziad Rahbani's *Bel Nesbeh La Bokra Chou*, Marwan Najjar's *Aarisein Midri Min Wein* (Suitors of Unknown Origins, 1986) and *Jawz al Jawz* (The Pair's Husband, 1984); and documentaries about Lebanese public figures (such as Bachir Gemayel, Kamal

[59] Peter Speetjens, "Pirate's Paradise: When Stealing Is the Norm," *Journal of Middle-East Broadcasters* 4 (November–December 2005).

[60] Ali Jaafar, "Lebanese Film 'What about Tomorrow?' Crushes Box Office Records in the Country," *Deadline*, January 22, 2016, https://deadline.com/2016/01/lebanese-film-ziad-rahbani-what-about-tomorrow-crushes-box-office-records-in-lebanon-fairuz-civil-war-1201688154/.

[61] Box Office Mojo by IMDbPro, "Lebanese Box Office of 2016," January 15, 2024, https://www.boxofficemojo.com/year/2016/?area=LB&grossesOption=totalGrosses.

Jumblatt, and Rafic Hariri), or about topics of Lebanese interest such as the civil war, old Beirut, or the defunct railway.

This all came to an abrupt halt the day after the Beirut port explosion in August 2020. The company was located in the Ashrafieh district, which was severely damaged by the massive explosion. This explosion, considered one of the largest nonnuclear explosions in history,[62] damaged the company's servers, effectively ending its operations.[63] Thus, all that is left of a platform that was designed for a niche audience and that aimed to position itself in the gap left by regional and international streamers, are its Instagram and Facebook pages in ruins and a website promising, "We'll be back soon." In a way, its collapse mirrors that of the country itself. Quantum Group, which had benefited both from the unregulated financial system and political money, eventually suffered the repercussions of the state's failure on which it had built part of its power.

Final Words

Lebanon's media ecosystem, shaped by a history of mostly private-sector initiatives, has paved the way for the emergence of streaming platforms and content that reflect the country's dynamic interaction of local, regional, and global forces. Streaming platforms in Lebanon operate at the crossroads of a deregulated and piracy-prone environment, leveraging the country's skilled workforce, rich media legacy, and relatively free internet access to create and distribute content in the region. While international platforms like Netflix may grapple with regional cultural sensitivities, and while regional players such as Shahid and local initiatives like M.Media can be more in tune with the expectations of local audiences, all of these platforms are involved in simultaneously reshaping media consumption patterns and engaging in complex negotiations with local and regional markets to produce content that resonates with their audiences. But it is Lebanon's role as not just a consumer but also a contributor to streaming content that underscores its pivotal position in the MENA region's streaming landscape, despite the country's small size and volatile political context.

62 Bassem Mroue and Karim Chehayeb, "Investigation into Beirut's Massive 2020 Port Blast Resumes," Associated Press, January 24, 2023, https://apnews.com/article/politics-beirut-lebanon-03eb5ff04f6ad86b3d0d6fb2ab25bdd6.

63 NowLebanon, "About Us," accessed January 15, 2024, https://nowlebanon.com/about-us/.

Bibliography

AbouAssi, Khaldoun. *Lebanese Civil Society: A Long History of Achievements, Facing Decisive Challenges Ahead of an Uncertain Future*. Johannesburg: CIVICUS, 2006. https://www.civicus.org/media/CSI_Lebanon_Country_Report.pdf.

AbouAssi, Khaldoun. "The Third Wheel in Public Policy: An Overview of NGOs in Lebanon." In *Public Administration and Policy in the Middle East*, edited by Alexander R. Dawoody, 215–30. New York: Springer, 2015. https://doi.org/10.1007/978-1-4939-1553-8_12.

Balanche, Fabrice. "The Geopolitics of Lebanon in the Renewed Regional Tensions." In *Atlas of Lebanon: New Challenges*, edited by Eric Verdeil, Ghaleb Faour, and Mouin Hamze, 30–1. Beirut: Presses de l'IFPO, CNRS Liban, 2019.

Barrington, Lisa. "Lebanon Gets a New Government, Now It Needs a New Economy." Reuters, January 19, 2017. https://www.reuters.com/article/idUSKBN1530MT/.

Baumann, Hannes. *Citizen Hariri: Lebanon's Neo-liberal Reconstruction*. Oxford: Oxford University Press, 2016.

Bessette, Rachel. "In the Heart of the Fire: Lebanon's Place in a Regional Rivalry." *Georgetown Security Studies Review*, September 16, 2016. https://georgetownsecuritystudiesreview.org/2016/09/16/in-the-heart-of-the-fire-lebanons-place-in-a-regional-rivalry/.

Boyd, Douglas A. *Broadcasting in the Arab World: A Survey of the Electronic Media in the Middle-East*. Ames: Iowa State University Press, 1993.

Boyd, Douglas A. "Lebanese Broadcasting: Unofficial Electronic Media during a Prolonged Civil War." *Journal of Broadcasting and Electronic Media* 35, no. 3 (1991): 269–87. https://doi.org/10.1080/08838159109364125.

Box Office Mojo by IMDbPro. "Lebanese Box Office of 2016." January 15, 2024. https://www.boxofficemojo.com/year/2016/?area=LB&grossesOption=totalGrosses.

Bukhalid, Nabil. "Lebanese Internet Governance Landscape." MEAC-SIG 2016 (August 2016). https://icann-community.atlassian.net/wiki/spaces/MES/pages/112985710/MEAC-SIG+2016.

Burroughs, Benjamin. "A Cultural Lineage of Streaming." *Internet Histories* 3, no. 2 (2019): 147–61.

Capital Market Authority. "MBC Group Prospectus." December 2023. https://cma.org.sa/en/Market/Prospectuses/Documents/MBC_en.pdf.

Corm, Georges. "What Lebanon Needs." *Le Monde Diplomatique*, October 2020. https://mondediplo.com/2020/10/03lebanon.

Dajani, Nabil. "Arab Press." In *Arab Media: Globalization and Emerging Media Industries*, edited by Noha Mellor, Mohammed Ayish, Nabil Dajani, and Khalil Rinnawi, 45–66. Cambridge: Polity Press, 2011.

Dajani, Nabil. *The Media in Lebanon: Fragmentation and Conflict in the Middle-East*. London: I. B. Taurus, 2019.

Dennis, Everette E., Justin D. Martin, Elizabeth A. Lance, and Fouad Hassan. "Media Use in the Middle-East 2019: A Seven-Nation Survey." Northern University in Qatar, 2019. https://www.mideastmedia.org/survey/2019/uploads/file/NUQ_Media_Use_2019_WebVersion_FNL_051219%5B2%5D(1).pdf.

Fakhoury, Tamirace, and Allison McCulloch. "How Do Consociations Craft Asylum Policy? Lebanon's Response to Conflict-Induced Displacement as an Exploratory Case." *International Studies Quarterly* 67, no. 3 (September 2023): sqad057. https://doi.org/10.1093/isq/sqad057.

Friedman, Thomas. "Living with the Violence of Beirut." *New York Times Magazine*, National Edition, July 17, 1983. https://www.nytimes.com/1983/07/17/magazine/living-with-the-violence-of-beirut.html.

Harb, Zahera. *Channels of Resistance in Lebanon: Liberation Propaganda, Hezbollah and the Media*. London: I. B. Tauris, 2011.

Haring, Bruce. "Netflix Series 'Jinn' Sparks Uproar in Jordan over Alleged 'Immoral Scenes.'" *Deadline*, June 15, 2019. https://deadline.com/2019/06/netflix-jinn-uproar-jordan-immoral-scenes-1202633178/.

Ipsos. "OTT and Content Streaming in MENA." Accessed November 25, 2019. https://www.ipsos.com/sites/default/files/ct/news/documents/2019-12/ott_and_content_streaming_2019.pdf.

Jaafar, Ali. "Lebanese Film 'What about Tomorrow?' Crushes Box Office Records in the Country." *Deadline*, January 22, 2016. https://deadline.com/2016/01/lebanese-film-ziad-rahbani-what-about-tomorrow-crushes-box-office-records-in-lebanon-fairuz-civil-war-1201688154/.

Jamali, Dima. "Post-War Telecommunications Reform in Lebanon: Trends and Challenges." *Info* 5, no. 2 (2003): 34–44.

Jang Moonkyoung, Dokyung Kim, and Hyunmi Baek. "How Do Global Audiences of TV Shows Take Shape? Evidence from Netflix." *Applied Economics Letters* 30, no. 3 (2023): 285–91. https://doi.org/10.1080/13504851.2021.1983916.

Karam, Michael. "Man of Many Parts Aims to Ignite a Media Revolution." *National*, December 19, 2012. https://www.thenationalnews.com/business/man-of-many-parts-aims-to-ignite-a-media-revolution-1.435758.

Khalil, Joe F. "The Business Push and Audience Pull in Arab Entertainment Television." *International Journal of Communication* 10 (2016): 3632–46.

Khalil, Joe F., and Mohamed Zayani. "Streaming the Middle-East: From Legacy Media to Digital Platforms." In *Digital Platforms and the Global South: Reconfiguring Power Relations in the Cultural Industries*, edited by Philippe Bouquillon, Christine Ithurbide, and Tristan Mattelart, 165–81. London: Routledge, 2024.

Kraidy, Marwan M. "Broadcasting Regulation and Civil Society in Postwar Lebanon." *Journal of Broadcasting and Electronic Media* 42, no. 3 (1998): 387–400. https://doi.org/10.1080/08838159809364457.

Kraidy, Marwan M. "Arab Satellite Television between Regionalization and Globalization." *Global Media Journal* 1, no. 1 (2002). https://www.globalmediajournal.com/open-access/arab-satellite-television-between-regionalization-and-globalization-php?aid=35056.

Lebanese University. "Law No. 81/2018." http://77.42.251.205/Law.aspx?lawId=278573&language=ar.

Lobato, Ramon. *Netflix Nations: The Geography of Digital Distribution*. New York: NYU NYU Press, 2019.

Lotz, Amanda D., Oliver Eklund, and Stuart Soroka. "Netflix, Library Analysis, and Globalization: Rethinking Mass Media Flows." *Journal of Communication* 72, no. 4 (August 2022): 511–21. https://doi.org/10.1093/joc/jqac020.2413000889000.

Mroue, Bassem, and Karim Chehayeb. "Investigation into Beirut's Massive 2020 Port Blast Resumes." Associated Press, January 24, 2023. https://apnews.com/article/politics-beirut-lebanon-03eb5ff04f6ad86b3d0d6fb2ab25bdd6.

Northern University in Qatar. "Media Use in the Middle-East: A Five-Year Retrospection." 2018. https://www.qatar.northwestern.edu/docs/publications/research/2018-middle-east-media-retrospection.pdf.

NowLebanon. "About Us." Accessed January 15, 2024. https://nowlebanon.com/about-us/.

Presidency of the Republic of Lebanon. "The Lebanese Constitution." https://www.presidency.gov.lb/English/LebaneseSystem/Documents/Lebanese%20Constitution.pdf.

Ritman, Alex. "Netflix's First Arabic Original Film Perfect Strangers Accused of 'Moral Degradation' in the Middle-East Due to Gay Character." *Hollywood Reporter*, January 23, 2022. https://www.hollywoodreporter.com/movies/movie-news/perfect-strangers-netflix-middle-east-egypt-controversy-lgbt-1235079663/.

Sakr, Naomi. *Satellite Realms: Transnational Television, Globalization and the Middle-East*. London: I. B. Taurus, 2011.

Schoenbach, Klaus, Mariam Saeed, and Rob Wood. "Audience Responses to Online Video in MENA: New Favorite Genres or Just More of the Same as Television." *International Communication Gazette* 80, no. 8 (2018): 1–17.

Schoenbach, Klaus, Robb Wood, and Marium Saeed. "Media Industries in the Middle East 2016." Northwestern University in Qatar, 2016. https://www.qatar.northwestern.edu/docs/publications/research/2016-middle-east-media-industries-report.pdf.

Sciacchitano, Francesco. "Assessment of Media Legislation in Lebanon." MedMedia, 2015. http://data.infopro.com.lb/file/AssessmentofMediaLegislationMedmedia2015.pdf.

Speetjens, Peter. "Pirate's Paradise: When Stealing Is the Norm." *Journal of Middle-East Broadcasters* 4 (November–December 2005).

Straubhaar, Joseph D. "Beyond Media Imperialism: Asymmetrical Interdependence and Cultural Proximity." *Critical Studies in Mass Communication* 8, no. 1 (1991): 39–59. http://dx.doi.org/10.1080/15295039109366779.

Tabbara, Chafic. "'Ashab Wala Aaz.' Film mansoukh wa zawbaa fi fenjan: satazhab 'al fadiha' wa la yabka ella thiyab al moumasilin" ["Perfect strangers"… a Copied Film and a Squall in a Cup: The "Scandal" Will Disappear and Only the Actors' Clothes Will Remain]. *Al Akhbar*, January 25, 2022. https://al-akhbar.com/Cinema/328704.

Taha, Mohamad M. "Le développement d'internet au Liban." In *Mondialisation et nouveaux médias dans l'espace arabe*, edited by Frank Mermier, 261–72. Lyon: Maison de l'Orient et de la Méditerranée Jean Pouilloux, 2003.

UNICEF. "Knowledge, Attitude and Practices of the Lebanese; Syrian and Palestinian Refugees in Lebanon, 2017." https://www.unicef.org/lebanon/reports/knowledge-attitude-and-practices-lebanese-syrian-and-palestine-refugees-lebanon-2017.

Verdeil, Eric, and Bruno Dewailly. "International Migration and the Lebanese Diaspora." In *Atlas of Lebanon: New Challenges*, edited by Eric Verdeil, Ghaleb Faour, and Mouin Hamze, 42–3. Beirut: Presses de l'IFPO, CNRS Liban, 2019.

World Bank. "Individuals Using the Internet (% of the Population)." Accessed January 24, 2024. https://data.worldbank.org/indicator/IT.NET.USER.ZS.

Zaid, Bouziane, Mohammed Benmoussa, Khayrat Ayyad, Mohamed Ibahrine, and Abdelmalek El Kadoussi. "The Domestication of Netflix in the Gulf." *International Journal of Communication* 17 (2023): 6613–32.

5

Streaming Feminism? South Asian TV Series by/about Women

Valentina Vitali

This essay takes off from research undertaken by a small network of film scholars and filmmakers on the spread of streaming in South Asia.[1] We were interested in assessing the reach of corporate streaming video platforms like Netflix and Amazon Prime Video across the region, their competition with large regional providers like Zee5, and the undergrowth of hyperlocal platforms in every South Asian nation, from Afghanistan to Myanmar. Unlike existing research, our network focused on the impact of VOD (video on demand) or OTT (over-the-top) platforms' operation (as exhibitors, distributors, and producers) on the work of independent filmmakers. Two years of exchange between network members—most based in a South Asian country—generated a picture of the new streaming economy that is very different from the dazzling sight publicized by corporate operators. As much as their shareholders would like us to believe otherwise, so-called global platforms do not yet have actual global reach. Nor is there consistent evidence that corporate streaming companies' operation as producers or distributors has opened up significant opportunities for emerging talent, as these companies tend to claim. But the landscape is changing, also in regionally powerful mediascapes like India's, where the workings of both corporate and hyperlocal platforms have begun to intervene into long-established film-industrial patterns.

What is changing? First, streaming enables content to circulate more or less legally within South Asia across national borders, also in countries between which movement of goods and people remains at best difficult. Second, the

[1] This research was funded by the Arts and Humanities Research Council (AHRC) (UK). I would like to thank the members of the South Asian Cinema and VOD Network—a group that I was able to establish with AHRC funding—for their time, dedication, and input into the project. Information about each network member can be found at https://southasiavod.com/who-we-are-2.

proliferation of small, independent platforms in every South Asian nation—and their provision of content in local languages—is beginning to make a dent in the regional dominance of Hindi films. Established distributors and their networks are no longer the only game in town. In India, where distributors have historically controlled finance and, with it, film production, streaming seems to have opened up opportunities for small producers, though crucially not with global operators but with local or hyperlocal streaming platforms. Netflix and Amazon, for instance, tend to commission content from creatives who are already well established locally. Smaller operators like OHO Gujarati or Samuh Bhutan, on the other hand, tend to show openness to new creatives. Whether these openings with smaller operators are sustainable remains to be seen. Early signs are that hyperlocal platforms offering independent content in local languages struggle to survive on subscriptions only. Financiers and advertisers remain, for these platforms, very hard to come by. Lastly, by bringing content directly into the home and mobile phones, streaming impacts the exhibition sector, which in turn influences aspects of production such as casting. In India and other parts of South Asia, where the film star is key to theatrical box office sales, streaming has facilitated the production of films with lesser-known actors—some by now veritable streaming superstarlets.

The landscape that is beginning to emerge may be a prelude to a better cinema. In principle this juncture is ideal to promote cross-border exchange in a region that has historically experienced tensions and wars between neighboring nations. Setting up a research network of South Asian filmmakers and producers enabled us to start mapping the international connections made possible by a select number of noncorporate South Asian streaming platforms. The resulting map can be seen on our network's website (at www.southasiavod.com, under a tab tentatively called "findings"). Although the ground is extremely fluid and, for many of the local platforms we identified, precarious, the possibilities of cultural exchange are there, because by far the majority of the local platforms listed on our project's website are accessible from anywhere. Many also provide subtitled content. What obstructs exchange is neither the technology nor the reach of government control (which is very real even when it can be bypassed with VPNs). The difficulties lie, rather, in the fact that nonlocal audiences are unaware of the existence of independent, local, or hyperlocal platforms. Only a small minority of (regional and global) viewers venture out to seek content from, say, Sri Lanka or Bhutan—unless they are based in these countries or part of the diaspora. When curious viewers do look for such content, they search for it on corporate providers like Netflix or Amazon Prime Video, which hardly offer anything of the sort. Audiences are largely oblivious of local platforms like CinemaCeylon and Samuh, both of which provide eminently affordable Sri Lankan and Bhutanese independent films via pay-per-view options. Local platforms do not as yet have the capacity to bypass the marketing reach and algorithmic clamp of the major corporate players.

In theory this juncture is also ideal to promote emerging talent and films or series made by historically marginalized identities, including women. The rest of this essay will focus on the impact of the streaming economy on women filmmakers and producers, and on the increasing visibility of women-centered content in South Asia. I argue that these developments have significant implications for a transformative feminist politics within the realms of cinema and cinematic television.

Women in the New Streaming Economy

After #MeToo, globally influential platforms like Netflix began to claim commitment to the promotion of women creatives. Yet more than fifteen years since Netflix's launch as a streaming service, in most countries the situation for women in media remains dire. In the UK, from where I write, less than 23 percent of directors were women in 2018—5 percent less than ten years earlier.[2] Some platforms have recently appointed women in top roles, including programming and commissioning: in South Asia, Shailja Kejriwal at Zee5 and Aparna Purohit at Amazon India. But beyond these high-profile figures, how do women filmmakers fare in India, the country in the region with the largest industry and one of the largest cinemas in the world?

The report *Shift Focus: Women Shaping the Narrative in Media and Entertainment* by Ammu Joseph, Bina Paul, Asiya Shervani, Putul Mahmood, and Mimi Sukumar is damning. It notes that

> currently the environment in film industries is far from conducive for women. Professional contracts, let alone gender-sensitive ones, that take into account existing laws that guarantee women workers' rights and the need for proper

[2] BFI Research and Statistics Unit, *BFI Statistical Yearbook*, 2019, https://www.bfi.org.uk/industry-data-insights/statistical-yearbook.

> implementation of legislation relating to sexual harassment at the workplace, maternity leave and childcare facilities and/or allowances are almost completely absent.[3]

The result is that the number of women producers in India is "dismally low." Further,

> [the] onus to hire more women professionals in film projects is mainly on the few female producers and directors, [when] there ought to be a platform or mechanism that allows for regular conversations with male producers, too, to explore ways to produce and promote movies directed by women.[4]

There is a lack of training and rights for women screenwriters; lack of training for, and access to, technical jobs for women, including assistant director[5]; "and most producers do not invest time and effort in structured interaction with film schools… to share knowledge and tools related to finance and other production-related skills."[6] Nor is there data on how film companies and producers comply with existing labor legislation on maternity leave, pay equality or nondiscriminatory recruitment.[7] In fact,

> the "locker room culture" that prevails on sets in terms of language, topics of conversation, attitudes and behavior towards women, sharing of pornographic videos and material, inappropriate humor and behavior, etc., contributes towards making shoots an uncomfortable experience for most women. Women who dare to object to such everyday indignities are labelled "troublemakers" and are unlikely to be hired again.[8]

This situation is not unique to India. The report points to "a dearth of coproduction opportunities," nationally and internationally, and advocates more programming of women's films at festivals to promote women's work in cinema.[9] Crucially for our purposes, it notes that "the few female producers in southern Indian film industries have so far not been able to tap potential revenue sources from the OTT media market, which was projected to grow 45 percent by the

[3] Ammu Joseph, Bina Paul, Asiya Shervani, Putul Mahmood, and Mini Sukumar, *Shift Focus: Women Shaping the Narrative in Media and Entertainment*, Women in Cinema Collective and Sakhi's Women Resource Centre, 2019, 24, https://wccollective.org/wp-content/uploads/2022/01/Shift-Focus-Women-Shaping-the-Narrative-in-Media-and-Entertainment.pdf.

[4] Ibid., 23.

[5] Ibid., 34–5.

[6] Ibid., 24.

[7] Ibid., 50.

[8] Ibid., 24.

[9] Ibid., 24, 27.

end of fiscal year 2023."[10] One of the report's recommendations to "encourage the creation of global pipelines that could help women in cinema rise above existing constraints" is thus to create content for alternative film distribution avenues such as the variety of new OTT platforms that are relatively free from the constraints imposed by traditional, commercial, movie-theater-based distribution and therefore allow greater diversity in terms of themes, stories, and characters. Already OTT platforms appear to have considerably more women-oriented stories than theatrical films. OTT content is also varied in form—features, shorts, series, docuseries, and more—and can therefore offer a wider range of options to both female filmmakers and female audiences. As relatively new platforms, they are also likely to be less entrenched in patriarchal attitudes and behavior as well as related structures and systems.[11]

Shift Focus surveys women in the Indian film and media industry. Elsewhere in the region pockets do exist where women creatives appear to fare better. Typically, this is in countries where cinema is a relatively young industry, and thus more open. In Bhutan, for instance, the premier streaming platform, Samuh, was founded and is run by a woman, Nyema Zam. In the last few years Samuh has started its own content production. This is resulting in new opportunities for women creatives, including for one of the country's most prominent film directors, Dechen Roder, herself a woman.[12] Admittedly, changes and opportunities take time to materialize. In the long term streaming may well effect radical change in the gender composition of the film industry worldwide. For now, though, not much appears to be changing. What change there seems to be is painfully slow and often contested. And yet, as the authors of *Shift Focus* note, "OTT platforms appear to have considerably more women-oriented stories than theatrical films," and the range of diversified streaming platforms does facilitate the circulation of content that is "varied in form."[13] This is, to date, the most visible and only real sign of a potential for improvement: the number, quality, and high visibility of women-centred content, mostly series, produced or distributed by large platforms across Asia. In South Asia these include *Bombay Begums* (Netflix, 2021, created by Alankrita Shrivastava and Bornila Chatterjee), *Qatil Haseenaon ke Naam* (Zee5, 2021, created by Meenu Gaur),

[10] Ibid., 23–4.

[11] Ibid., 76.

[12] Valentina Vitali, "Interview with Dechen Roder," *BioScope: South Asian Screen Studies* 11, no. 1 (2020): 96–100.

[13] Joseph, Paul, Shervani, Mahmood, and Sukumar, *Shift Focus*, 76.

Churails (Zee5, 202, created by Asim Abbasi), *Scoop* (Netflix, 2023, created by Hansal Mehta), *Dahaad* (Amazon, 2023, created by Reema Kagti and Zoya Akhtar), *Delhi Crime* (Netflix, 2019, created by Richie Mehta), *Leila* (Netflix, 2019, created by Urmi Juvekar and Deepa Mehta), *Made in Heaven* (Amazon, 2019, created by Reema Kagti and Zoya Akhtar), *The Broken News* (Zee5, 2022, directed by Vinay Waikul), and *Hush Hush* (Amazon, 2022, created by Tanuja Chandra), to name but a few. While the centrality of female protagonists in these Indian and Pakistani works is clear, what is less clear is whether these or similar series can, as the *Shift Focus* report anticipates, "offer a wider range of options to both female filmmakers and female audiences."[14]

Feminism and Women-Centred OTT Series

> Sing softly of thy love, or silent be,
> O Makhfi, lest the hunter secretly
> Shall come and hear thy voice, and capture thee.[15]

Can this OTT content be seen to mark a real change for women's voices in cinema and media? On social media and within academic circles some of the series mentioned above have had a fair degree of visibility. While a few have been discussed in terms of their potential to transform dominant discourses about women in society, there has been no systematic comparative analysis across OTT platforms in any region. Nor have sporadic debates on this or that series' ideological positioning been brought to bear on the all-important question raised by *Shift Focus*: these series' potential transformative impact on the position of women creatives in the film and media industry. Both perspectives are important. Exploring the transformative and especially feminist potential of these series is important because the rise of VOD in the last fifteen years has coincided with a new wave of global feminism within which South Asian feminists, tapping into a long history of regional and intersectional alliances, have played a central role. But the interface between economics and aesthetics often complicates the progressive potential of OTT content. What follows is therefore an attempt to map—without falling into essentialist, deterministic, or reflectionist postulations—the interconnections between (a) the composition of

[14] Ibid., 76.
[15] Zeb-un-Nissa, *Diwan-i-Makhfi; The Diwan of Zeb-un-Nissa: The First Fifty Ghazals*, trans. Magan Lal and Jessie Duncan Westbrook (London: John Murray, 1913), XXXVII, 85.

the film-industrial terrain that generated these series as commodities and (b) the series as cultural artifacts mediating specific discursive positions that may have implications for a feminist politics.

Not every film or series made by a woman speaks progressive discourses about women, nor do industrial conditions favorable to women filmmakers necessarily indicate better conditions for women at large. To what extent, then, are the women-centred OTT series mentioned above "transformative" or simply "affirmative"? I use these terms in reference to a seminal essay in which feminist philosopher Nancy Fraser sought to disentangle a perceived contradiction in identity politics.[16] Gender, like race, Fraser argued, is a "paradigmatic bivalent collectivity." To begin with, gender discrimination has political-economic dimensions. Structuring the division, in film and media, between higher-paid, accredited creative roles for men and lower-paid, nonaccredited peripheral roles for women, gender falls within a problematic of distribution. But in this realm gender also structures cultural-valuational differentiations: sexually exploitative images, objectification of female figures, demeaning stereotypical depictions, and so forth tend to privilege traits associated with masculinity and devaluate things coded feminine. To this extent, gender falls within the problematic of recognition.[17] The bivalent character of gender is, for Fraser, the source of what she calls the "redistribution-recognition dilemma":

> In so far as women suffer at least two analytically distinct kinds of injustice [i.e., maldistribution and misrecognition], they necessarily require at least two analytically distinct kinds of remedy—both redistribution and recognition. The two remedies pull in opposite directions, however.... Whereas the logic of redistribution is to put gender out of business as such, the logic of recognition is to valorize gender specificity.... How can feminists fight simultaneously to abolish gender differentiation [i.e., in the labor force] and to valorize gender specificity?[18]

I am not interested here in revisiting Fraser's dilemma, partly because, as Judith Butler argued,[19] the two functions are more interlocked than Fraser, for theoretical ease, would have us believe. More useful is Fraser's conceptualization of the remedies that maldistribution and misrecognition call for. These are the

16 Nancy Fraser, "From Redistribution to Recognition? Dilemmas of Justice in a 'Post-Socialist' Age," *New Left Review* 212 (1995): 68–93.

17 Ibid., 78–9.

18 Ibid., 79–80.

19 Judith Butler, "Merely Cultural," *Social Text* 52–53 (1997): 265–77.

bivalent issues that still confront us today in cinema. Fraser distinguishes between two broad approaches to remedy "injustice that cuts across the redistribution-recognition divide": "affirmation" and "transformation":

> By affirmative remedies for injustice I mean remedies aimed at correcting inequitable outcomes of social arrangements without disturbing the underlying framework that generates them. By transformative remedies, in contrast, I mean remedies aimed at correcting inequitable outcomes precisely by restructuring the underlying generative framework.[20]

Fraser's differentiation is valuable, and not simply because it is indeed fairer and better to eliminate injustice at the root. The two concepts of transformation and affirmation are both useful because they enable us to disentangle contradictions and tensions in existing film-industry practices. If, as the WCC report suggests, the new streaming economy seems to offer better options for women, what combinations of remedies are at work in the South Asian series I consider here that minimize, if not altogether eliminate, maldistribution and misrecognition simultaneously?

In a sector characterized by the predominance of male directors and assumptions that spectators are, by default, male, "remedies" could be the appointment of a female director, producer, or screenwriter or the selection of women-centered themes or narratives. A tentative mapping of the OTT series listed above into the table originally designed by Fraser gives table 5.1. By this scheme, the only series to be transformative at both redistribution and recognition levels simultaneously would be *Dahaad*, an Indian show directed and created by women in an industry where women are structurally marginalized, which taps into a genre (crime drama) historically not aimed at the "female audience" market, and features a woman policeman, a profession not traditionally exercised by women. *Delhi Crime*, *Scoop*, and *The Broken News* also feature women characters in traditionally "male" occupations, but these shows are created by male directors. While they satisfy recognition remedies, they fail to remedy existing industrial dynamics privileging the allocation of creative roles to men. Similarly *Leila* and *Made in Heaven* remedy maldistribution, to the extent that they are directed-created by women, but leave dominant discourses about women more or less undisturbed because they feature women in the stereotypical roles of mother, home carer, wife, and wedding planner.

20 Fraser, "From Redistribution to Recognition?," 82.

Table 5.1 Mapping of select South Asian VOD/OTT series, based on Nancy Fraser's differentiation between "affirmative" and "transformative" remedies for "redistribution" and "recognition."

	AFFIRMATIVE	TRANSFORMATIVE
REDISTRIBUTION	male directed all genres *Scoop* *The Broken News* *Delhi Crime*	female directed women genres *Leila* *Made in Heaven* *Hush Hush* *Qatil Haseenaon ke Naam* *Bombay Begums* *Dahaad*
RECOGNITION	women genres male directed *Churails*	all genres female directed *Dahaad*

In the making of *Leila*, *Made in Heaven*, *Hush Hush*, *Qatil Haseenaon ke Naam*, *Bombay Begums*, and *Dahaad*, women were involved in a range of roles, as actors, screenwriters, directors, creators, and producers.[21] To use Louis Hjelmslev's terminology, women held these series together at the level of expression.[22] How does their collective female voice at the level of expression (or how does their "labor") translate at the level of content? As Fraser might ask, how do these women-directed/produced series handle (mis)recognition? Images are very complex objects to which viewers relate in even more complex ways. Yet Fraser's table allows us to begin to ask what may make these series transformative at both industrial and cultural levels.

[21] It is important to note here that *Made in Heaven* and *Dahaad*, both created-directed by the female duo Reema Kagti and Zoya Akhtar, were coproduced by Tiger Baby Films—the directors' own production company, and Excel Entertainment—one of India's largest production companies, owned by Ritesh Sidhwani and Farhan Akhtar. Farhan and Zoya Akhtar are Honey Irani and Javed Akhtar's son and daughter. As it is sadly often the case in the film industry worldwide, family connections and lineage in the cinema are instrumental to women's access to creative roles. From this perspective, *Dahaad* is hardly "transformative." Excel Entertainment has been very active in OTT, with Amazon Prime Video in particular, through which the producer has released also the sports and crime dramas *Inside Edge* (2017, 2019) and *Mirzapur* (2018). By contrast, as a "noir" series directed by a woman who is not Bollywood royalty in an industry where, until a few years ago, women would not normally be given the chance to direct crime movies, *Qatil Haseenaon ke Naam* is transformative. I return to this point below.

[22] Louis Hjelmslev, *Prolegomena to a Theory of Language*, trans. Francis J. Whitfield (Madison: University of Wisconsin Press, 1963). The Danish linguist Louis Hjelmslev conceptualized any cultural item as consisting of levels of expression (the raw materials, including technologies, capital, types of labor, etc.) and of content (the semiotic field, including ideologies, discourses, tropes, generic categories, etc.). See also Metz's reformulation of Hjelmslev's concepts in Christian Metz, *Language and Cinema* (The Hague: Mouton, 1974), 208–12. Here I consider the series at both levels of expression and content and situate them within the broader streaming and discursive landscape.

This is an urgent question for two reasons. First, in a streaming economy, content exhibition, distribution, and production are largely dictated by algorithms. As Nishant Shah explains, based on past consumption patterns, algorithms simultaneously predict and shape future use.[23] This is because executive decisions at commissioning and production levels are taken on the basis of consumption data. In other words, the more viewers watch male-directed content, the more companies are (profit-)driven to produce male-directed content, leaving little option to viewers but to watch male-directed content, and so on, in a self-fulfilling algorithmic cycle.

It follows that there is therefore more to exposing the paucity of women-directed content than creative jobs for women. In this new exhibition-distribution-production complex, the high "probability coefficient" that a film is directed or produced by a man—a coefficient arrived at because *that* is the current state of affairs in the film industry—not only predicts that a man will direct or produce the next film; it also "shapes the [future] behaviour" of viewers (and, with them, of the industry as a whole), who will continue watching mostly films directed or produced by men. To redress the dismally low proportion of women in creative roles in cinema and investigate whether women creatives fare better with OTT platforms is thus also to challenge the algorithmic logic of the industry as it grows and consolidates into the new era of digital distribution- and exhibition-led production.

Second, narratives of women are rarely only about women, when they are about women at all. Representations of women often function as a terrain to vent preoccupations about a wide range of issues affecting a society. As Partha Chatterjee noted in his seminal discussion of nineteenth-century Bengali women's autobiographies,

> The[ir] most striking feature is the ways in which the very theme of disclosure of self remains suppressed under a narrative of changing times, changing manners and customs, and changing values.... What held these [memoirs or "stories from memory," from *smirtikatha*] together into a single narrative was not the life story of the [female] narrator or the development of her "self", but rather the social history of the "times".[24]

[23] Nishant Shah, "The Insidiousness of Information Overload," in *Overload, Creep, Excess: An Internet from India*, ed. N. Shah, A. Rajadhyaksha, and N. A. Hasan (New Delhi: Left Word Books, 2023), 98–9.

[24] Partha Chatterjee, *The Nation and Its Fragments: Colonial and Post-Colonial Histories* (Oxford: Oxford University Press, 1994), 138–9.

Many of the seemingly women-centered South Asian OTT series, whether directed/created/produced by men or women, also "suppress" female subjectivity within a narrative of changing social values. In *Made in Heaven*, *Bombay Begums*, *Churails*, and *Dahaad* the characters are in essence social types whose inner desires and psychological turmoil are barely known to viewers. In *Made in Heaven*, *Bombay Begums*, and *Churails* especially, the women's gender enables them to navigate across class, caste, and religious divides. Unlike the male characters, they can connect, even if their social background is different, because they are women. Womanhood provides a bond between the women who populate these series over and above other social markers, but this female-led narrative comments primarily on class, caste, and religion in Indian and Pakistani society. It is not the women's subjectivity but their actions—what they can and cannot do—that function as our points of entry (of identification) into the narrative. The women's actions signal that times and values are changing, or should be.

What kinds of social change does the woman character, here an *object* of representation, mediate? In most of the OTT series mentioned above, women as types are offered up for broad-based, pluralistic, and inclusive identification. In another context, Indian film theorist Madhava Prasad has argued that "pluralism" and "inclusiveness" based on language and non-Brahminical, low-caste discourses sustained the populist pull of political party Dravida Munnetra Kazhagam (DMK) and its leader, film star Maruthur Gopalan Ramachandran (MGR)[25]—what Partha Chatterjee called "the sovereignty effect of the star as political leader": "Followers create relatively closed communities whose identity is anchored to the star."[26] This virtual community—"a virtual socio-political order within which subjects feel securely located"[27]—is not unlike the imagined community constructed by our women-centered OTT series, except that the marker of pluralistic, inclusive identification here is not language or caste but gender, or rather "woman-ness." OTT women characters seem designed to enable identification across the social spectrum. Their womanhood—that same gender-based marker that allows these characters to move across class, caste, religion—also enables broad-based spectatorial identification. The range

[25] Madhava M. Prasad, *Cine-politics: Film Stars and Political Existence in South India* (New Delhi: Orient Blackswan, 2014).

[26] Partha Chatterjee, *I Am the People: Reflections on Popular Sovereignty Today* (Ranikhet, India: Permanent Black, 2020), 99–100.

[27] Prasad, *Cine-politics*, 182.

of women featured in the series—from the cross-class coalitions of *Churails* and *Made in Heaven* to *Dahaad* and *Delhi Crime*, where caste is explicitly addressed, or even *Ghoul* (dir. Patrick Graham, 2018, Netflix) and *Beetal* (dir. Patrick Graham, 2020, Netflix), which feature the "ordinary" Indian woman (of action)—functions as a catchall, anchoring audience identification across class, caste, religion, and so forth.

Churails and *Made in Heaven* are in this sense the most interesting cases. Both series construct a virtual community that is firmly located in the contemporary. They inhabit a South Asian or Northern Indian sociopolitical terrain that, in the real world, is overinscribed by the populism of Indian Prime Minister Narendra Modi, whose upright, puritan, lower-middle-class address alienates, at the level of discourse, a very thin but affluent, urban, upper-middle-class, cosmopolitan, university-educated, and somewhat bohemian layer of society, which is represented in these shows—freethinking, *sharab*-drinking, party-going women and men who can afford fashionable clothes, expensive drugs, foreign holidays, and Netflix subscriptions. The OTT female figure anchoring identification in these series is a kind of *citoyenne*:[28] bourgeois, modern, sexually active, ostensibly progressive, and openly so in the face of a form of populism that promotes itself as morally, religiously, and politically "clean."

Churails was censored, albeit temporarily and by the Pakistani government, for showing female characters doing things deemed socially reprehensible for women. Yet, temporary banning of *Churails* notwithstanding, censorship is not the real question here, partly because series circulate beneath and above it.[29] For global corporate providers, the largest share of revenues is not from streaming content. For the likes of Amazon, OTT content is little more than the means by which large sections of the global population are scooped up for retail, advertising, and data extraction.[30] Netflix claims not to sell its user data. Subscriptions have been historically the company's predominant source of revenue. However, since the company started a lower-cost membership that includes advertising, Netflix

[28] Literally, the citizen-woman. While in this general sense the term was widely used since the French Revolution, *La Citoyenne* was a French feminist fortnightly magazine published in Paris by Hubertine Auclert between 1881 and 1891. International in its address, the magazine advocated women's enfranchisement and right to run for public office and demanded changes to the Napoleonic Code, which relegated women to an inferior status. Digitized issues of *La Citoyenne* can be accessed here: https://bibliotheques-specialisees.paris.fr/ark:/73873/pf0000538600/1891-03-15.

[29] Waseem Mushtaq, "Churails: Why a Feminist Detective Show Was Banned in Pakistan," BBC Monitoring, November 2, 2020, https://www.bbc.co.uk/news/world-asia-54629439.

[30] Kenio Fontes, "Prime Video's Profit Surge: A Game-Changer for Amazon Stock," *Street*, November 9, 2023, https://www.thestreet.com/amazon/media/prime-videos-profit-surge-a-game-changer-for-amazon-stock.

too has been bringing in revenues through that channel,[31] and this is likely to increase.[32] For Indian OTT service Zee5, subscriptions do bring in the bulk of income,[33] but Zee5 subscribers use access to stream a wide variety of content, with sport and news forming the largest share.

Indian economist Kalyan Sanyal has argued that in postcolonial capitalism it is the state, not capital itself, that shoulders the burden of "maintaining the conditions for securing the legitimacy of capitalist domination."[34] India and other states did this originally through national radio and television. In the new streaming economy, that ideological burden is now largely shouldered by private media companies like Zee5. Banned or not banned, generating surplus or not, an OTT series plays an important ideological function for digital giants and their wider surplus accumulation strategies, perhaps especially when it addresses a "minority." Women-oriented series, with their address to a discursively alienated (if politically influential and economically affluent) minority, may be loss leaders[35] but also help secure the legitimacy of global digital corporations. This too is a dimension of the algorithmic logic that drives corporate platforms, for algorithms enable them to target niche markets with precisely designed content, thus containing the risk of losses that come with catering to small audiences.

Streaming and Women's Media

The mirror of my heart I burnish bright
Until, reflected fair for my delight,
The self's eternal beauty greets my sight.[36]

It is impossible to say whether the appointment of women in leading creative roles at Amazon India, Netflix, or Zee5 is playing a role in the legitimation of streaming companies. It is certain that it did Amazon and Zee5 no harm. Nor

31 Nathan Reiff, "How Netflix Makes Money," *Investopedia*, September 26, 2023, https://www.investopedia.com/insights/how-netflix-makes-money/.

32 Netflix, "Q2 2023 Letter to the Shareholders," July 19, 2023, 5, https://s22.q4cdn.com/959853165/files/doc_financials/2023/q2/FINAL-Q2-23-Shareholder-Letter.pdf.

33 Best Media Info Bureau, "ZEEL's Ad Revenue down 4%; Total Revenue up 6.46% in Q1FY24," August 10, 2023, https://bestmediainfo.com/2023/08/zeel-s-ad-revenue-down-4-total-revenue-up-6-46-in-q1fy24.

34 Kalyan Sanyal, *Rethinking Capitalist Development: Primitive Accumulation, Governmentality and Postcolonial Capitalism* (London: Routledge, 2007), quoted in Chatterjee, *I Am the People*, 79–80.

35 In this context a loss leader is a product that although it may not generate a profit as a commodity enables a company to gain access to a market.

36 Zeb-un-Nissa, *Diwan-i-Makhfi*, XXIII, 64.

is it possible to ascertain whether there are connections between the lower wages historically earned by women creatives in this industry (as in others) and women's visibility among OTT series creators. But then again, employing creatives at lower costs is hardly likely to impact negatively on streaming companies' profits.[37] For all this, our question stands. Given that, at the level of expression, women hold together many of the OTT series considered here, how is the appointment of women in leading creative roles within streaming companies impacting kind of content produced?

I have argued that in *Churails*, *Made in Heaven*, *Dahaad*, and *Delhi Crime*, among other series, women feature as types and "doers." Each of the Pakistani show *Churails*' ten episodes focuses on one of the five protagonists and their story. Yet each episode is symptomatically titled after a quality or condition these women's stories are meant to illuminate: *wafa* (loyalty), *ghulami* (servitude), *izzat* (honor), *hudood* (limits, as dictated by *shari'ah*), *taaqat* (strength), and so forth. The official English titles, in turn, describe actions or situations: "The Fight for a Cause," "A Gruesome Crime," "Zubaida Goes Undercover." We are asked to look at the *churails* (literally "witches") and identify with them through their actions, not their thoughts and feelings. As in the nineteenth- and twentieth-century women autobiographies discussed by Partha Chatterjee, there is, in this series, an absence of personal female voice, in spite of the centrality of female characters. Consistently throughout *Churails* none of the characters is entrusted with narration.

However extraordinary the *churails* may be, subjective points of view (POVs) are few and far between. We mostly look *at* the women, we do not look *as* them, through the women's eyes, making it hard to valorize the gender specificity that Fraser talks about.

A scene in episode 1 of *Churails* makes this clear. The episode shows Sara as she is about to tell her husband Jameel that she would like to return to work. The couple is back from a work-related evening out, dinner with one of Jameel's clients. They both stand in front of mirrors in their bedroom, about to undress and go to bed. The camera is behind them. We see Jameel's reflection in the wardrobe mirror in the left half of the frame, cut across diagonally, as if to signal his duplicitous nature. In the right half is Sara's whole (uncut) reflection in the round mirror above her dressing table. While Jameel is visible nearly only as a reflection, on the right-hand corner of the frame is visible, in the foreground,

[37] Global operators have so far proved very reluctant to release any information on the subject of labor.

blurred, Sara's shoulder and part of her face (Figure 5.1). Jameel praises her for "looking perfect." Up to this point we looked at Sara's own reflection in the mirror if not quite through Sara's eyes, at least from her side of things. But on his cue the camera pans slowly to the right, toward Sara's reflection, and tracks in. We now see only Sara's reflection in the round mirror of her dressing table: a picture of "the perfect wife" (Figure 5.2). Extradiegetic music and the absence

Figure 5.1 *Churails* © Glasshouse, Motion Content Group, Asim Abbasi, Mo Azmi, and Shailja Kejriwal 2020.

Figure 5.2 *Churails* © Glasshouse, Motion Content Group, Asim Abbasi, Mo Azmi, and Shailja Kejriwal 2020.

of ambient diegetic sound signal that we are entering Sara's private dimension. This mise-en-scène alerts us to a split between how Sara looks and what she may think. But by the time the camera pulls out again, the real Sara is no longer there, as if the look into the round mirror at "the perfect wife" can no longer be allocated to the real Sara. Sara's feelings are transmigrated onto a voice-over, displaced onto an extradiegetic voice conveying a short interior monologue: "Fuck you, perfect wife." Now Jameel's reflection joins Sara in the dressing table mirror: he promises to take her on an expensive holiday; Sara tells him of her desire to work. But there is no escape: Jameel dismisses Sara's plans and demeans her skills as a qualified lawyer. Sara gets up from the table and walks across the room to sit on the bed. The camera follows her from a distance (Figure 5.3). Jameel is off frame, and we now look at Sara, small on the bed, from Jameel's side of things, nearly his POV.

Here and there in *Churails*, women are allocated subjective POV shots, but these have an instrumental function: they move the plot along and tell us a character's story, but hardly her feelings. They are exceptions rather than moments of an underlying subjective mode. The female perspective is often carried largely by the soundtrack, either as voice-over or as in the retrospective interrogation interludes scattered across each episode, where each central

Figure 5.3 *Churails* © Glasshouse, Motion Content Group, Asim Abbasi, Mo Azmi, and Shailja Kejriwal 2020.

character, whether male or female, is shown in heavily tilted, angled close-ups as they are made to admit their crime to the police. These are forced moments, of a confessional rather than subjective nature, which may explain why the stories they are intended to recount, in flashback, are rarely staged formally from the protagonists' own perspective.

To be clear, *Churails* is a hugely enjoyable and forward-thinking series. I remain to be convinced, however, that it can have transformative potential as far as women's position and representation in cinema is concerned. Aesthetically, it offers women affirmative spaces of identification but little room for enunciation. The women's look, their flamboyance, and, more generally, the contemporary glamour of the series' art direction take precedence over the women's filmic voice. The same occurs in *Dahaad* and other series featuring women as police officers and detectives. Affirmative identification is enabled by pointing the camera at the women as they ride a motorbike or hold a gun, rather than by the camera taking up their gaze as women detectives.

A key scene in episode 1 of *Dahaad* is a case in point. Two men are interviewed about a missing female relative by police officer Anjali, her colleague, and the chief inspector, Anjali's boss. The two men are sitting at a table in the police station. They face the chief inspector, also sitting at the table. Anjali and her colleague stand behind him. For most of the scene the camera takes up the inspector's POV. Occasionally we do look at the two men from Anjali or her colleague's perspectives. But when Anjali is handed over one of the men's mobile phones showing a photograph of the missing woman, we are shown Anjali looking at the mobile phone, rather than the missing woman's photo from Anjali's POV. We only see the photograph from the chief inspector's POV as Anjali hands over the phone to him. This sets up a voyeuristic perspective toward a female survivor/victim of sexual abuse. The problem is that the filmmaker seems to reproduce that voyeurism, rather than undermine or challenge it. Had the directors enabled the spectator to view the missing woman's photo from another woman's (Anjali's) POV, the relation between viewer and viewed would have been identificatory rather than voyeuristic and objectifying.

At the risk of sounding whimsical, one could say that women in *Churails*, *Dahaad*, *Delhi Crime*, and similar series are absent in their presence: they populate the diegesis but do not carry narration. I would not want to argue that this is to do with the fact that *Churails* and *Delhi Crime*'s creators or *Dahaad*'s coproducers are men.[38] It is an aesthetic choice, and one that anyone could have made, irrespective of sex, gender, or sexuality. I simply note that it was made and that as such it does little to transform what Laura Mulvey called women's "to-be-looked-at-ness," which Mulvey identified as the dominant and patriarchal mode of cinema. Mulvey argued that

> cinematic codes create a gaze, a world, and an object, thereby producing an illusion cut to the measure of desire. It is these cinematic codes and their relationship to formative external structures that must be broken down before mainstream film and the pleasure it provides can be challenged.[39]

Of the series considered here, *Qatil Haseenaon ke Naam* (*QHKN*) is closer to the mark. Written by Pakistani filmmaker Farjad Nabi and British-Indian Meenu Gaur, directed by Gaur, and produced, like *Churails*, by Indian media executive Shailja Kejriwal's Zindagi channel for Zee5, *QHKN* features seven women and their stories of revenge. Each episode is titled with the woman's name: Mehek, Kanwal, Zuvi, Massey Ma, Zehra, Anarkali, and Mai Malki. From the start, narration is emphatically subjective. The women's subjective POVs hold from the pretitle sequence through to the rest of the series. As when, at the end of episode 2, Kanwal can finally confront the man with whom she fell in love and whom she helped escape from the police, but who, now free, scorns her ingenuity and demeans her looks. Here, as in many other moments in the series, we are stitched into the story from the woman's point of view.

In *QHKN* flashbacks are carefully introduced as subjective retrospective narration, memories charged with feelings. Episode 3 opens with a woman feeding pieces of Zehra's corpse to a rabid dog. We cut to Mai and two other women sitting around a fire at night. They are all in frame, Mai in the middle, the camera in front of them, at a distance (Figure 5.4). "If this was for revenge alone"—the woman on the right, Anarkali, asks Mai—"you could have just buried everything and be done with it. But so much rage?" As Anarkali speaks, the frontal camera tracks in slowly: first toward Anarkali and Mai then to Mai only, and finally tilting down into the fire. The slow tilt signals the transition to

[38] *Churails* and *Qatil Haseenaon ke Naam* actors Sanam Saeed and Sarwat Gilani claim something along these lines in an interview for *Goodtimes*. They observe that Asim Abbasi felt he had to pass by them aspects of his script and dialogues for *Churails*, whereas Meenu Gaur, being a woman, did not need to do so, "instinctively" being able to relate to the situations *Qatil Haseenaon ke Naam*'s characters faced.

[39] Laura Mulvey, "Visual Pleasure in Narrative Cinema," *Screen* 16, no. 3 (1975): 17.

Figure 5.4 *Qatil Haseenaon ke Naam* © Zee Entertainment Enterprises Ltd. 2021.

Mai's memories, her flashback. Flashback over, the movement is reversed. As the three women are back into frame, Mai replies:

> My rage has extinguished the warmth of my heart. Now my every heartbeat spews poison, not blood. It runs through my veins. If I don't do this, I'll succumb to my own venom. This isn't a recent occurrence I speak of. A woman's rage and wrath, fury and temper are ancient. They precede her by thousands of years. When the time comes, that wrath seeps into the desolation inside, transforms itself into her lover.

As Mai speaks out her state of mind, the camera pulls out sufficiently to reveal the Mughal citadel that gives *QHKN* its distinctive gothic look and also evokes women's ancient fury and temper. This is not the gothic as English and European cinema know it. The series situates the women's lives in a precise moment of South Asian history, a time of syncretic Islamic rule when imperial court culture promulgated discourses of individuation through poetry and music. This is neither the glossy Mughal India of Bollywood cinema nor the polished glamour of Incredible India, the Ministry of Indian Tourism's advertising logo. Mai Malka and her women inhabit a dystopian world that evokes both past and future—a walled city where neon lights and motorbikes coexist side by side with the spectral grandeur of historic buildings, isolated from the present, as if that temporality, "our times," cannot accommodate the women's individuated voice. This is a gothic look that emphasizes the eeriness not of things dead but of possibilities unrealized. A past not quite wiped out, still speaking and, through Mai's and past women's rage, carried through into the present, at an oblique remove from the reality we know.

Conclusion

This is perhaps why Urdu poetry also features prominently in *QHKN*, as if prose was too implicated in modernity not to betray the women, thousands of years of them. The one poet actually mentioned in the series is a woman: in episode 1 a teacher (who is also a poet, lover, and contract killer) teaches a poem by Parveen Shakir.[40] Mai's lines are often delivered with a faint rhythmic cadence, and her imagery is heavy in poetic connotations. The series' plots, mise-en-scène and soundtrack ooze female desire—women's voices that are simultaneously individuated and covert, carrying the marks of suppression and making them their own.

Combined with the choice of location, the art direction, and references to Mughal India, *QHKN*'s poetic undertones evoke the *tawaifs* (courtesans), female musicians, dancers, and poetesses who populated cultural life at the Mughal court. These include enigmatic figures like Zeb-un-Nissa (1638–1702), daughter of Emperor Aurangzeb and author of many poems posthumously collected as *Diwan-i-Makhfi*, quoted above. Imprisoned by her father for the last twenty years of her life at Salimgarth Fort, Delhi, Zeb-un-Nissa was a polymath, a woman who defied social conventions and never married. In her award-winning *Makfhi: The Hidden One* (2018), filmmaker Prachee Bajania conveys the layeredness of Zeb-un-Nissa, her complexity as a historical figure quite literally hidden away, silenced. Bajania stages Zeb-un-Nissa's subjectivity as a diegetic absence and a powerful individuated voice all the more for that. The poems that survived her are a call against erasure and amnesia. Significantly, Bajania's film is also about, and was shot in and around, the small palace where Zeb-un-Nissa lived or might have lived. As in *QHKN*, the building is eerie and in ruin, now used as a shelter for goats. But its former grandeur and beauty, like that of its original

[40] Parveen Shakir (1952–94) was a Pakistani poet. She is best known for bringing a distinctive feminine voice to Urdu literature and her consistent use of the rare grammatical feminine gender for the word "lover." Shakir wrote ghazals and in *āzād nazm* (free verse). She used the latter to explore social issues and taboos, including gender inequality, discrimination, and prostitution. Along with themes of femininity and female sexuality, Shakir used free verse also to write about economic disparities, social exploitation of the weak and poor, and the harsh reality of low-income laborers.

occupier, Zeb-un-Nissa, and her poetry, survive the ravages of time and neglect. South Asian culture is replete with such women: from the sixteenth century all the way to present day—creative women in their time mostly removed from history.[41] In its own way, *QHKN* also works against amnesia by foregrounding female subjectivity at the levels of expression and of content, in both direction-production and filmic address, inviting viewers to see through its (angry, scorned, and rejected) women's eyes.

It is, at both these levels, a transformative example of filmmaking, one that simultaneously disrupts patterns of maldistribution and misrecognition. Whether this would have been possible outside or without the new streaming economy is too early to say. As I have outlined in my introduction, streaming is beginning to bring significant changes to the film industry in parts of South Asia. When decades-old dynamics are disrupted, openings materialize that may, for a short while, flash up opportunities for a change in direction. Whether this will lead to improvement in the long term is indeed too early to say. But we can say that while a few years ago directing a "noir" was considered a male director's job, *QHKN* was directed by a woman. Streaming makes it possible for it to circulate globally to an extent that earlier women's cinema would not have been able to do fifteen years back. This is happening at a time when feminism is gaining new ground globally. It is up to viewers now not to let algorithms decide whose films or series they watch.

Bibliography

Best Media Info Bureau. "ZEEL's Ad Revenue down 4%; Total Revenue up 6.46% in Q1FY24." August 10, 2023. https://bestmediainfo.com/2023/08/zeel-s-ad-revenue-down-4-total-revenue-up-6-46-in-q1fy24.

BFI Research and Statistics Unit. *BFI Statistical Yearbook*. 2019. https://www.bfi.org.uk/industry-data-insights/statistical-yearbook.

Bollywood Film Fame Canada. "What about the Female Gaze: About *Qatil Haseenaon ke Naam* with Meenu Gaur and Farjad Nabi." YouTube, January 25, 2022. https://www.youtube.com/watch?v=VFtLAOffylg.

Butler, Judith. "Merely Cultural." *Social Text* 52–53 (1997): 265–77. Also published in *New Left Review* 227 (1998): 33–44.

41 For a contemporary view of the history of this rich, women-based cultural lineage and its present condition, see Anna Morcom, *Illicit Worlds of Indian Dance: Cultures of Exclusion* (London: Hurst and Company, 2013).

Chatterjee, Partha. *I Am the People: Reflections on Popular Sovereignty Today*. Ranikhet, India: Permanent Black, 2020.

Chatterjee, Partha. *The Nation and Its Fragments: Colonial and Post-Colonial Histories*. Oxford: Oxford University Press, 1994.

Fontes, Kenio. "Prime Video's Profit Surge: A Game-Changer for Amazon Stock." *Street*, November 9, 2023. https://www.thestreet.com/amazon/media/prime-videos-profit-surge-a-game-changer-for-amazon-stock.

Fraser, Nancy. "From Redistribution to Recognition? Dilemmas of Justice in a 'Post-Socialist' Age." *New Left Review* 212 (1995): 68–93.

Goodtimes. "Meet the Qatil Haseenas: Sanam Saeed, Sarwat Gilani, Meenu Gaur." YouTube, January 8, 2022. https://www.youtube.com/watch?v=Lh2jpqShP_g.

Hjelmslev, Louis. *Prolegomena to a Theory of Language*. Translated by Francis J. Whitfield. Madison: University of Wisconsin Press, 1963.

Joseph, Ammu, Bina Paul, Asiya Shervani, Putul Mahmood, and Mini Sukumar. *Shift Focus: Women Shaping the Narrative in Media and Entertainment*. Women in Cinema Collective and Sakhi's Women Resource Centre, 2019. https://wccollective.org/wp-content/uploads/2022/01/Shift-Focus-Women-Shaping-the-Narrative-in-Media-and-Entertainment.pdf.

Metz, Christian. *Language and Cinema*. The Hague: Mouton, 1974.

Morcom, Anna. *Illicit Worlds of Indian Dance: Cultures of Exclusion*. London: Hurst and Company, 2013.

Mulvey, Laura. "Visual Pleasure in Narrative Cinema." *Screen* 16, no. 3 (1975): 6–18.

Mushtaq, Waseem. "Churails: Why a Feminist Detective Show Was Banned in Pakistan." BBC Monitoring, November 2, 2020. https://www.bbc.co.uk/news/world-asia-54629439.

Netflix. "Q2 2023 Letter to the Shareholders." July 19, 2023. https://s22.q4cdn.com/959853165/files/doc_financials/2023/q2/FINAL-Q2-23-Shareholder-Letter.pdf.

Prasad, Madhava M. *Cine-politics: Film Stars and Political Existence in South India*. New Delhi: Orient Blackswan, 2014.

Reiff, Nathan. "How Netflix Makes Money." Investopedia, September 26, 2023. https://www.investopedia.com/insights/how-netflix-makes-money/.

Sanyal, Kalyan. *Rethinking Capitalist Development: Primitive Accumulation, Governmentality and Postcolonial Capitalism*. London: Routledge, 2007.

Shah, Nishant. "The Insidiousness of Information Overload." In *Overload, Creep, Excess: An Internet from India*, edited by N. Shah, A. Rajadhyaksha, and N. A. Hasan, 23–66. New Delhi: LeftWord Books, 2023.

Vitali, Valentina. "Interview with Dechen Roder." *BioScope: South Asian Screen Studies* 11, no. 11 (2020): 96–100.

Zeb-un-Nissa. *Diwan-i-Makhfi; The Diwan of Zeb-un-Nissa: The First Fifty Ghazals*. Translated by Magan Lal and Jessie Duncan Westbrook. London: John Murray, 1913.

6

Television Drama in Mexico: From Broadcast to Streaming

Paul Julian Smith

Three Trends: Telenovela Subgenres, Crime, Multimedia Bio Series

This article treats very recent TV drama, both broadcast and streaming, in Mexico, the largest and most dynamic Spanish-speaking territory. The main body of the article gives an account of my chosen Mexican corpus, title by title, and its conclusion discusses the latest research on production and reception, which offers the opportunity to compare Mexico with other Spanish-speaking territories. In the absence of academic studies, beyond the invaluable annual handbooks produced by OBITEL (Observatorio Iberoamericano de Ficción Televisiva), I have drawn on little-known sources from the specialist and general press, especially bilingual trade journal *Produ*, to chart the production and reception of these new dramas.

The past decade was a time when Mexican media goliath Televisa was merging with top-rated US Spanish-language network Univision, a process completed in 2022. The joint company created a new and vast audience on its streaming app ViX, which is available in both countries in a basic free and a premium for payment version, known as ViX+. This article, then, aims to touch on general issues (television as vehicle for social and historical education; politics, philosophy, and documentary in the legal system; the private and public lives of famous sportsmen or actors) by offering an account of significant case studies. But it also calls attention to the paradox that at a time of much heralded crisis due to the transition between traditional or legacy media and new streaming or SVOD (subscription video on demand) platforms there should be a creative flourishing of television drama.

Telenovela Subgenres

My corpus of Mexican TV series of the past decade is divided into three parts, each exemplified by no more than four titles. The first part is on telenovela subgenres and the persistence of new iterations of the heritage serial. Didactic telenovela, a subgenre with a lengthy history, returned with the *Vencer* (Overcoming) franchise, which began in 2020. This franchise consists to date of five telenovelas all on the theme of gender violence and was broadcast in prime time on Televisa in Mexico and Univision in the United States before streaming on the merged networks' new platform. Created by distinguished showrunner Rosy Ocampo, also a longtime executive at Televisa, the franchise is unique in focusing in each edition on four female friends, rather than the familiar heterosexual couple. Moreover, unlike in traditional telenovela, characters and even locations recur in the various editions, creating a kind of *Vencer* "universe" in which different versions of gender violence coexist. While much of the *Vencer* phenomenon seems unprecedented, and most unexpected at a time when traditional telenovela and free-to-air television are thought to be in decline, there is in fact a decades-long tradition in Mexico of didactic telenovelas that aim to teach the audiences lessons about social issues. This tradition can be compared with a concept used by current TV scholars in Mexico: "social marketing" is the "selling" of controversial issues to an audience in a way analogous to that of the advertising of branded products.

The fifth edition of the franchise, *Vencer el pasado* (Overcoming the Past), which focuses on online violence against women in the workplace, is a clear example of that phenomenon. Offering production context, OBITEL tells us that there were fourteen new fiction titles on air in Mexico in 2021 (as many as twelve were from Televisa).[1] This is a fall from a high of twenty-seven in 2017, and over the same period the telenovela genre has slowly declined.[2] Even though its social-issue-led *unitarios* or anthology dramas, such as the much maligned *La rosa de Guadalupe* (The Rose of Guadalupe), have also lost some degree of popularity, Televisa still takes eight of the top ten titles most seen in the year.[3] Based on Nielsen ratings, *Vencer el pasado* comes in at number two. OBITEL breaks down its audience: more female than male (10.1 percent to 6.1 percent),

[1] OBITEL, *Observatorio Iberoamericano de Ficción Televisiva 2022* (Porto Alegre: Globo, 2022), 210.
[2] Ibid., 211.
[3] Ibid., 213.

more lower than upper class (DE 13.4 percent; ABC+ 6.1 percent), and older rather than younger (45+ at 16.2 percent; 19–29 at 11.8 percent).[4]

Turning to the digital platforms, OBITEL tells us that SVOD penetration has risen year on year, although it feels that Netflix is now threatened by competition from cheaper Mexican services, bundled in with ubiquitous cell phones.[5] OBITEL's most outstanding trend is the real-life feminist "disruption" that has shaken the country with massive antigovernment demonstrations in the capital. This trend produced, for the third year running, a *Vencer* telenovela "focused on speaking about the diverse forms of violence that run through the lives of Mexican women belonging to different generations." In spite of its overtly feminist agenda, the franchise's treatment of gender violence is, we are told, "widely accepted" by audiences: the three editions to date all figure in the top ten hit dramas of the last two years.[6]

Most telling is a lengthy interview with executive producer Rosy Ocampo by *Produ* on July 14, 2021.[7] This is helpfully divided into sections based on key words. The first is "always women." Whether the precise topic is gender violence, teen pregnancy, or cyberbullying, the general focus of the franchise is always on four strong female protagonists. While most Mexican telenovelas are now based on secondhand foreign formats, these original stories are created by a team of researchers whose task is to investigate emerging social trends. The second is "leadership in innovation." With her thirty years of experience, Ocampo is a pioneer in genres as diverse as children's programming and multimedia, quite apart from her role in "renovating" the telenovela. Next comes "attracting millions": according to *Produ*, the serial regularly draws over three million viewers in Mexico and in the United States and is number one in its time slot in both countries. These viewers include 1.37 million self-defined Mexican "housewives," a demographic Ocampo does not despise or neglect.

In the section called "more relevant than ever" Ocampo claims that the newly transformed genre of telenovela is influencing streaming series, rather than the other way around, and that this generic transformation is based on "research." As a "structured" producer (her own word), she talks constantly to her team, who inform her about the audience's reactions and preferences. Ocampo's franchise is

4 Ibid., 214.
5 Ibid., 215.
6 Ibid., 222.
7 Richard Izarra, "Rosy Ocampo de Televisa en #PRODUprimetime: Franquicia Vencer en preparación de cuarta y quinta temporadas y todas disponibles al mercado mundial," *Produ*, July 14, 2021.

thus "entertainment" first and foremost, but with what she calls "a social profile" (a key phrase). Vital here is a "new [narrative] structure": the use of flashbacks at the start of each episode and "epilogues" at the end, where the cast offers advice speaking as themselves and not as their characters. Further stylistic innovations cited by Ocampo in the interview are the "rapid rhythm" of the plot, much faster than traditional telenovela; the forthcoming year "2022," whose special preoccupations Ocampo's team are still investigating (they will eventually opt for postpandemic "loss" as the key concept); and finally, "tropicalization," the belief that those preoccupations of Latin American and US Latinx audiences are global problems that merit worldwide distribution. What is striking in this major interview is the ease with which Ocampo shifts between the social and the fictional spheres.

Similarly, the real-life issues identified by her meticulous researchers are then fed back to mass audiences (most especially to those with the "social profile" of older, poorer women) via carefully created TV characters and plots that the production team anticipates will both educate and entertain them.

My second revised subgenre is historical telenovela, a prestige format in legacy television. A major recent example is the remake as a short form series of *El extraño retorno de Diana Salazar* (The Strange Return of Diana Salazar, ViX, 2024). The original serial version of the same title, a rare historical and supernatural title broadcast by Televisa from 1988 to 1989, is one of the most celebrated telenovelas in the history of Mexican television. The production processes of the remake of *El extraño retorno* for binational streaming platform ViX in 2024 were complex. This is confirmed by the profiles of the new series' transnational executive producers, who tasked themselves with remaking a valuable and beloved piece of Mexican intellectual property.

On May16, 2023, *El extraño retorno* was the main title highlighted at the upfronts where TelevisaUnivision presented its new slate of "fiction contents" to the press with appearances by the series' celebrated stars, French-born Angelique Boyer and Argentine Sebastián Rulli. The pair are a real-life couple long familiar to audiences from free-to-air telenovelas, including the previously mentioned *Vencer el pasado*.[8] This celebrity-filled event took place even before Colombian-born Patricio Wills and Venezuelan Carlos Bardasano, both of independent company W Studios, announced the start of production on

[8] Vanessa Maldonado, "*El extraño retorno de Diana Salazar* y la serie animada de comedia Circo Gómez entre los contenidos de ficción de TelevisaUnivision," *Produ*, May 16, 2023.

the series on August 22.[9] During the public announcement, ViX's senior VP for development said this was a "new version of an emblematic *novela*... in a modern format." Its attractions include "high production values" and "talented stars." By January 24, 2024, Wills would announce that his W Studios was working on *El extraño retorno* simultaneously with a Mexican remake of the controversial Venezuelan format *Juana la virgen* (Jane the Virgin) and hoped to build on the success of their earlier *El gallo de oro* (The Golden Rooster, 2023), another period piece. ViX, he says, offers "great exposure" to W's products, and the aim will be to "keep up the level of quality, reception, and penetration in both Mexico and the US."[10]

Along similar lines, Bardasano claimed on May 15, 2024, that "our previous projects prepared us for the new version of *Diana Salazar*."[11] Referring back to the visually spectacular original serial, Bardasano said that this time there will be "fewer special effects and more of the human factor." Yet he stressed the "quality" that will appear on the screen, as shown by the lengthy shooting schedule of seventeen weeks, inconceivable in traditional telenovelas. Finally, Bardasano, who had made his career at NBC and Telemundo in the United States, cited the youth of the writing team and the "strong" and "adult" content they had created, euphemisms for the inclusions of scenes of sex and violence unacceptable on legacy TV in Mexico.

Amy E. Wright has recently offered a meticulous close reading of the original *El extraño retorno* in the wider context of Mexican seriality, across centuries of literary and visual culture. She also gives us the chance to reconsider in a new light the remake, which came out too late to be studied in her book. In a chapter called "History's Eternal Return," Wright distinguishes between the key terms "remediality" and "remediation." The first is defined as "the intense recycling of past elements," the second "the refashioning of prior elements to present needs."[12] This recycling of the past in the historical *novela* is shown in the original's first episode, a baroque backstory for the modern drama to come, in which a young woman with supernatural powers is burnt at the stake by the

9 Cynthia Plohn, "Patricio Wills y Carlos Bardasano de W Studios comenzaron producción de *El extraño retorno de Diana Salazar* para ViX," *Produ*, August 22, 2023.

10 Vanessa Maldonado, "Patricio Wills de W Studios: Trabajando en Juana la virgen y *El extraño retorno de Diana Salazar*," *Produ*, January 24, 2024.

11 Vanessa Maldonado, "Carlos Bardasano de W Studios: Nuestros proyectos anteriores nos prepararon para hacer la nueva versión de Diana Salazar," *Produ*, May 15, 2024.

12 Amy E. Wright, *Serial Mexico: Storytelling across Media from Nationhood to Now* (Nashville: Vanderbilt University Press, 2023), 159.

Inquisition alongside her lover. Significantly, the premiere of the remake is less challenging to its new audience, focusing almost exclusively on the couple's reencounter in the familiar present day, with only some brief flashbacks to the period lovers' naked tryst and their passage through the Inquisition torture chambers. The remake of *El extraño retorno* does, then, feature some graphic scenes of sex and violence that would have been impossible on broadcast television, especially in the 1980s. But it remains the case that an industrially original and artistically disruptive legacy telenovela, which was called "bizarre" at the time of its first broadcast, was transformed into a more aesthetically and thematically conventional series for the present-day viewer.

Crime and Punishment

After these revivals or rare telenovela subgenres, the second broad area I treat here as a current trend on Mexican television is crime and punishment. Two streaming thrillers made three years apart, *Ingobernable* (Uncontrollable, Netflix, 2017) and *El candidato* (The Candidate, Amazon Prime Video, 2020), both raise knotty questions of politics and philosophy. Although streamed on rival platforms, *Ingobernable* and *El candidato* have much in common.

Aesthetically and industrially, both profit from their creators' collaboration with transnational streamers Netflix and Amazon Prime Video, respectively, enjoying higher production values than most Mexican titles on free-to-air TV. Narratively, both labyrinthine plots suggest corruption and conspiracy theories at the highest level of government (the presidency and the mayoralty of Mexico City, respectively). And politically both point the finger at the United States' supposed covert and nefarious influence. They even exploit the same symbolically charged location (the main square or Zócalo of the capital, which houses both the National Palace and the Town Hall), and feature the same actress in a central role. Finally, both were well received by the local press, which highlighted their potentially explosive political effects.

Yet the two series also diverge from each other. The Spanish-language *Ingobernable* is Mexican and female centered, focusing on a first lady falsely accused of murdering her president husband (she is played by longtime telenovela star Kate del Castillo, who in real life had paid a controversial visit to notorious drug baron El Chapo). The bilingual *El candidato* is American and male centered, as its protagonist is a CIA operative (played by British actor James

Purefoy) investigating a Mexican master criminal allegedly linked to the young mayor. These textual differences may derive to some extent from the series' opposed production histories. *Ingobernable* is made by Argos, an indie Mexican producer that has long specialized in controversial content (its feminist-themed telenovela *Las Aparicio* ended with the first lesbian wedding on Mexican TV). Argos's CEO, Epigmenio Ibarra, was one of the highest profile opponents of the centrist president Enrique Peña Nieto, who, along with his then wife, a real-life telenovela actress, is fictionalized in his series.

Conversely, *El candidato* is produced by Televisa, Mexico's hegemonic broadcaster long allied with the federal government, but is billed as an "Amazon Prime Original." Its showrunner is the American Peter Blake, the creator of network and premium cable dramas such as *House* and *Billions*, and although the series was shot on location in Mexico City, most of its producers, writers, and directors are based in the United States.

Produ fully documents the lengthy creative process of *Ingobernable*, attesting from the start to the collective nature of TV authorship. Their first article mentioning the title, as early as July 24, 2015, cites Ted Sarandos, general content director at Netflix, as saying that the streamer is "committed to making quality series for Mexico, Latin America, the United States and the world."[13] The second, on the same date, interviews Epigmenio Ibarra, CEO of Argos, the series' indie production company. The combative Ibarra is said to have come from an event at the Guanajuato International Film Festival that addressed Ayotzinapa, a notorious case of the kidnapping and disappearance of students that came to define the presidency of Peña Nieto (the president re-created in the series). Ever the booster of the national industry, Ibarra claims that "Mexico must have more confidence in its TV series." He notes that his company Argos's series have always had "social subject matter," but now the audience has changed, and young people watch "other things." It is necessary, then, to "push the [boundaries of the] formats."[14] *Produ*'s third mention of *Ingobernable* comes in an interview with star Kate del Castillo on November 30, 2015. Where previously she had praised the "strong character" she plays who "shakes the foundations of the National Palace," here she focuses on her role as producer not actor. Castillo has bought the rights

[13] Produ, "Netflix: Kate del Castillo will be the star of our new original Mexican series *Ingobernable* produced by Argos," July 24, 2015.

[14] Vanessa Maldonado, "Epigmenio Ibarra de Argos: México debe tener más confianza en las series de TV," *Produ*, July 24, 2015.

to Argos's previous envelope-pushing hit *Las Aparicio* and hopes to launch an English-language version on the US market.[15]

Produ's coverage of Amazon Prime Video's *El candidato* is skimpier, although it documents once more a creative and industrial collaboration between Latin America and the United States. On July 31, 2020, Natalie Osma and Juan Rendón, the Colombian executive producers of the series who are based in Los Angeles, admit that collaborating with established US showrunner Peter Blake and serving as a "hinge" between the American and Mexican ways of working was a "challenge" and an "apprenticeship." From their earlier positions in the news division of Univision, the Spanish-language US network, the future producers came across what they call real-life IP (intellectual property) on corruption in Mexico and its relation to US agencies.

After two years of development and now with Amazon in the mix, the lavish shoot in Mexico involved 150 people, took 82 days, and used 20 locations. Osma and Redón praise the high production values of the series and the "professionalism" of their Mexican colleagues, which "can be seen on screen."[16]

Continuing and confirming this trend of crime and punishment narratives is the recent emergence of true crime in both documentary and drama series in Mexico. Paco Stanley, perhaps the best-known and most loved TV presenter in the country, was shot to death outside a restaurant in the capital on June 7, 1999. Some twenty-five years later, his murder, which remains unsolved, was reexamined in two different formats and on two different streaming platforms: a documentary shown on TelevisaUnivision's ViX (accessible of course in both Mexico and the United States) and a bio series on the globally distributed Amazon Prime Video. The two series are, respectively, *El show: crónica de un asesinato* (The Show: Chronicle of a Murder, 2023) and *¿Quién lo mató?* (Who Killed Him?, 2024), which became the highest rated title in Spanish in the history of its platform, in spite of its subject being unknown outside Mexico. Both series examine the contradictions in the unique public figure of Stanley (a beloved and gifted presenter who was also a brutal bully of his crew and audience) and in the true crime mystery to which his murder gave rise (a story that remains incomplete and unresolved in spite of its multiple multimedia reiterations).

15 Produ, "Actriz Kate del Castillo: Estamos trabajando en la posibilidad de hacer la versión de *Las Aparicio de Argos*," November 30, 2015.

16 Aliana González, "Trabajar con Peter Blake y ser bisagra entre el modo de funcionar de EE UU y México fue un reto y un aprendizaje: Productores ejecutivos de *El candidato*," *Produ*, July 31, 2020.

The death itself and the depictions of that death occurred at different but parallel periods of radical change in the Mexican media landscape. Stanley, long a key celebrity on monopoly national network Televisa, had recently transferred the latest iteration of his long-running talk-cum-variety show to its newly created rival Azteca. A quarter of a century later, the first episode of the documentary was broadcast on Televisa's heritage free-to-air channel and benefited in traditional style from cross-promotion on that channel, such as discussion on talk shows (coincidentally, Stanley's son Paul is a regular presenter on Televisa's veteran morning show *Hoy* [Today]), before being corralled to the broadcaster's new premium streamer ViX+, beyond the reach of the national network's more modest audiences. Meanwhile the bio series, which bowed on Amazon's pay platform just months later and was released in weekly installments, revealed the unusual interest of a deep-pocketed global streamer in a domestic-themed real-life drama.

On June 6, 2023, *Produ* reported on the world premiere screening of *El show*, which took place significantly at the International Film Festival of Guadalajara, Mexico's largest showcase for local cinema.[17] In this prestigious setting, the series was presented to the public and press by producer Carlos Sosa and "director and journalist" Diego Enrique Osorno, with Sosa stressing the sheer labor involved in the production, which he says took some fourteen months.

Negotiating the new world of simultaneous broadcasting and streaming, the first episode would be premiered free to air on Televisa and the latter episodes posted on the network's platform ViX+, starting the next day. Sosa said the show will be an "exercise in nostalgia" for some viewers and of "discovery" for others, no doubt younger and unfamiliar with this "relevant debate on justice." The "exemplary case" of the murder, he continued, occurred at a key political moment of "alternation" (when the ruling PRI party was about to be voted out of government) but also suggested a "revision of entertainment," of how "we escape or survive reality." Going into more detail, we learn that the team interviewed two hundred people and filmed sixty, including all those "principally involved." Finally, Sosa offers an industrial motive for the existence of his series: there is now "more demand than supply." "People," he continues, "want to see many documentaries and [streaming] platforms are booming." It is therefore "logical," he says, that this kind of content should be made "showing a part of the reality of [our] country."

[17] Vanessa Maldonado, "Carlos Sosa de Viento del Norte: nos tomó 14 meses realizar la serie documental *El show. Crónica de un asesinato* para N+," *Produ*, June 6, 2023.

Meanwhile, as previously announced by *Produ* on January 11, 2023, Amazon Prime Video was preparing its own as-yet-unnamed miniseries on the assassination.[18] Here Alonso Aguilar, Amazon's recently appointed "Head of Local Originals," is quoted as saying that the murder marked a "turning point in the history of Mexico" and that his company is "proud" of the "talented team and cast" it has assembled. This quality cast includes celebrity Diego Boneta (who had recently starred as singer Luis Miguel in rival Netflix's bio series), longtime real-life singing star Belinda, and Luis Gerardo Méndez, a respected actor known both for a Netflix comedy and for more challenging roles in mainstream and indie feature films. Related news reports at the time add that Amazon was investing $300 million in Mexican original content, including a development deal with Boneta's production company.

Multimedia Bio Series

The third trend I have identified in recent Mexican production is one related to my previous case, the multimedia bio series. A relatively early example is of the masked wrestler and movie star Blue Demon (real name Alejandro Muñoz Moreno, 1922–2000), whose artistic afterlife stretches across several media. An eponymous bio series made by Sony Pictures TV in collaboration with national broadcaster Televisa in 2016–2017 cast as the wrestler Tenoch Huerta, the actor and activist most identified with the fledging Afro-Mexican pride movement (later to take on a Mayan-tinged role in the Marvel cinematic universe). It preceded a novel published in 2022 that focused on a key moment when Blue Demon clashed with eternal rival Santo over a stolen mask (a scene also re-created in the series)[19] and a comprehensive exhibit that was inaugurated also in 2022 in Mexico City's Museo del Estanquillo dedicated to the legacy of Carlos Monsiváis, the essayist who was one of the most respected literary figures in twentieth-century Mexico. Combining cinema, television, literature, and the fine arts, then, the figure of Blue Demon has, via a quality television that proclaims itself "cinematic," become a nexus for multimedia creativity.

The *Blue Demon* TV series was made, like my other titles, at a time of rapid transition in modes of audiovisual production and distribution. Thus, on

[18] Andrés Briceño Izarra, "Amazon Prime Video prepara miniseries sobre el presentador de TV mexicano Paco Stanley," *Produ*, January 11, 2023.

[19] Iván Farías, *El misterio de la máscara perdida* (Mexico City: Dark & Glow Press, 2022).

January 18, 2016, Sony Pictures Television (SPT) announced *Blue Demon* as "a new co-production with Televisa, to be shot on location in and around Mexico City and distributed by Televisa in Latin America and SPT around the world."[20] SPT's spokeswoman said there was a "growing demand in the region for stories about real people," a trend that had begun not in Mexico but in Colombia. In an interview the following month (February 18, 2016) she confirmed that "over half" SPT's content was now in the new genre of bio series and that the company wanted to show "Mexican, Latin American, and global" audiences the "fantastic figure" of the wrestler.[21] Once more, however, she stressed that the series would be shot entirely on location in Mexico, apparently a competitive advantage even for this transnational TV drama. Like other series I treat in this article, then, *Blue Demon* serves as a site of negotiation between the national and the global.

Later that year (September 2, 2016) *Produ* made a set visit to the series.[22] Coincidentally, the crew were shooting a scene re-creating not a wrestling bout but the glamorous premiere of one of Blue Demon's feature films with an array of period-clad extras. In the accompanying video the episode director stressed the importance of the unique locations, little known even in the capital itself. This one was the plush Teatro Fru Fru in the Historic Center, which has long been dark. The producer also stressed the authenticity of the decor and wardrobe, carefully crafted to fit the period. Clearly, then, this is an example of quality, even scholarly, TV, which fills the gap left in the national and international schedules by the now unfashionable genre of the classic literary adaptation. A few days earlier (August 23, 2016) the significantly titled "VP of Formats and New Contents" at Televisa had also stressed the cinematic prestige of these new series, implicitly distinguishing them from the network's old telenovelas. He had chosen to go "outside Televisa" itself for talent and hire prestigious film directors for this project.[23]

A very different public figure, but one also gifted with a bio series, is Mexico's greatest movie star, a figure re-created in *María Félix, la doña* (2022). This series was first shown as a flagship title on TelevisaUnivision's then new premium

[20] Maribel Ramos-Weiner, "NATPE-Angélica Guerra de SPT: *Blue Demon* será la próxima coproducción con Televisa," *Produ*, January 18, 2016.

[21] Maribel Ramos-Weiner, "Sony Pictures Television: The Bio Series Represent More than Half of our Production in the Region this Year," *Produ*, February 18, 2016.

[22] Produ, "Produ visitó el set de *Blue Demon*," *Produ*, September 2, 2016.

[23] Produ, "Televisa: 8 Series Ready and 10 in Preparation, with 20 Episodes per Season and a New Narration," *Produ*, August 23, 2016.

streaming service, explicitly intended to attract paying subscribers to ViX+. It was created by storied producer Carmen Armendáriz, herself the daughter of a movie star of Félix's era, whose account of the lengthy and expensive production process of this quality title is given in the specialist press. While the series was promoted as a homage to the "woman with the heart of a man," its textual contradictions, which circle around feminist and LGBTQ issues (Félix's conflicted attitudes to her lesbian personal assistant and gay son and best friend), correspond to unresolved tensions in postauthoritarian Mexican media. Indeed, Félix is presented here more as a (serial) wife and mother than as an independent agent who crafted, as she did, a unique persona and career.

Vital to current scholarly discussion here are Niamh Thornton's recent book on the star, most especially on her posthumous multimedia legacy in such varied areas as jewelry design and drag performance,[24] and Martín Echevarría and Ruben Arnoldo González's account of the continuing struggle for democracy in Mexican media since 2000, when, as noted earlier, there was a transfer of power from the PRI party that had ruled the nation for some seventy years.[25] While there is now no government censorship of television, still the industry may wish to curry favor with more conservative audiences, even as it ventures into controversial social topics.

On June 30, 2022, *Variety* announced the launch date of the premium streaming service ViX+, costing a modest $6.99 per month in the United States and MX$119 in Mexico.[26] And *María Félix, la doña* was cited first in the list of "premium series" available at launch. The company's chief transformation officer stated to *Variety* that "Spanish-language viewers" (said to be some 600 million strong) "will finally have a subscription streaming service with content that authentically reflects our culture." The bio series stands out once more here as a unique period title amid the contemporary content in different genres. And the huge free tier of Televisa's beloved back catalog serves to discourage "churn" (the cancelation of subscriptions that is the bane of all streaming platforms). The aim is to retain viewers who have gone, in the resonant industry phrase, "into the wild" (that is, no longer subscribing) within the familiar ecosphere of the

[24] Niamh Thornton, *María Félix: A Mexican Film Star and Her Legacy* (Woodbridge, UK: Tamesis, 2023).

[25] Martin Echevarria and Ruben Arnoldo Gonzalez, eds., *Media and Politics in Post-Authoritarian Mexico: The Continuing Struggle for Democracy* (Cham, Switzerland: Palgrave Macmillan, 2024).

[26] Todd Spangler, "TelevisaUnivision Sets ViX+ Premium Service Pricing, Launch Date," *Variety*, June 30, 2022, https://variety.com/2022/digital/news/vix-plus-pricing-launch-date-televisaunivision-1235306723/.

company as a whole.[27] Given the special characteristics of the huge, diverse, yet relatively financially undeveloped Spanish-speaking market, TelevisaUnivision and ViX(+) thus dare not turn their backs on ordinary television even as they pride themselves on select titles of premium, quality programming.

Let us look once more at the trade press to trace within this general context the specific production process of the limited series, which coincided with the twentieth anniversary of the star's death. Focus throughout was on the executive producer as creative force behind the show. On July 12, *Produ* interviewed Carmen Armendáriz, stressing once more the fact that her project is one of only two original titles with which the new premium streamer was launched.[28] Armendáriz said that she had been trying to make the project for five years, an odyssey that ended only at the perfect time with the "birth" of ViX+. In the course of that lengthy development period the number of episodes was reduced from forty-two to just eight (heritage telenovelas can reach some two hundred episodes). Armendáriz said that she realizes that this is insufficient, but she has concentrated on what she says is "important" in the biography: Félix's childhood, her four marriages, her emotional and physical losses, and the relationship with her family (these priorities fail to align with the series' publicity, which promotes female independence and rebellion as the key factors).

While Armendáriz said she consulted with an ethnographer to document accurately the mid-twentieth-century period, the series director María Fernanda Suárez stressed the role of "fiction" in period series and the importance of "evocative beauty": in the authentic locations and art design, of course, but also in the costuming and "bearing" of the actors. A lengthy seven months after preproduction began, shooting was coming to an end. When the series was finally aired, generalist press coverage was ecstatic. Álvaro Cueva, Mexico's sole serious TV critic, praised it as "masterful": now, he wrote with satisfaction, Mexicans need no longer be envious of foreign biographical titles like *The Crown*.[29] A prestige bio series could thus serve as a focus for national pride even at a time of internationalization of production.

From the account above, the key features and virtues of my corpus are clear. The titles constitute a wide range of recent TV drama, from telenovela to true crime

[27] Jennifer Maas, "TelevisaUnivision CEO Wayde Davis Says ViX's Free Tier Is Pivotal against Churn," *Variety*, 11 May 2022.

[28] Vanessa Maldonado, "Carmen Armendáriz de Televisa: La miniserie *María Félix, la doña* resume 50 décadas [*sic*] de la actriz en ocho episodios," *Produ*, July 12, 2022.

[29] Álvaro Cueva, "*María Félix* de ViX+," *Milenio*, July 21, 2022.

and celebrity biography, and are variably available across the Spanish-speaking world and in the United States and UK. Trade journals and general press sources enable us to sketch a portrait of the creative processes of showrunners, who are often powerful women. Beyond this specific Mexican context, however, my corpus addresses issues of general interest to media studies scholars interested in other territories and in making connections between television and other media, especially cinema.

An Unstoppable Advance?

The most recent survey for the year 2024 by OBITEL offers the chance to compare production and consumption of TV serials and series in Mexico at the present time with nine other Spanish- and Portuguese-language territories during what the scholarly collective calls "the unstoppable advance of streaming."[30] Its initial premise is one we have observed in such cases as the remake of classic *novela El extraño retorno de Diana Salazar*: while free-to-air or linear television remains wedded to the heritage telenovela, albeit with fewer episodes than before, the new genre of the short-form series is dominant on streaming platforms.[31]

According to OBITEL, although the percentage of time devoted on the legacy networks' weekly schedules to "fiction" (i.e., drama and comedy) in Mexico has risen markedly in recent years (from 15.9 percent in 2019 to 27.2 percent in 2023), it still lags behind Spain and the US Hispanophone channels (41.0 percent and 40.7 percent, respectively).[32] Yet Mexico leads the way in hours of first-run, locally produced shows in its own market, beating its closest competitor, Brazil, by a total of 1,523 to 1,476 hours.[33] Mexico also comes a resounding first in hours exported to other markets, besting its Brazilian rival by 2,999 to 1,242 hours.[34] The growing convergence with the United States is also shown by the fact that Mexico's northern neighbor showed no fewer than 1,157 hours of Mexican content while exporting only 255 of its own in Spanish.

As in the days of the golden age of Mexican cinema, which was exhibited throughout Latin America in the 1940s and 1950s, and, more recently, of globally

[30] OBITEL, *Observatorio Iberoamericano de Ficción Televisiva 2024* (Porto Alegre: Globo, 2024).
[31] Ibid., 21.
[32] Ibid., 23.
[33] Ibid., 25.
[34] Ibid., 27.

exported telenovelas, Mexico remains a regional leader in legacy media. Indeed, measured in terms of content hours, its external or export market amounts to almost twice its internal presence at home: at 183 percent, that foreign footprint is rivaled only by the smaller territory of Colombia at 176 percent.[35] And contradicting OBITEL's initial premise on the rise of streaming series, the hours of exhibition of local telenovelas on linear channels, both in the ten Latin American territories as a whole and in Mexico itself, have risen to reach the prepandemic levels of 2019.[36] This trend was no doubt aided by innovations in form and content that attracted the interest of attentive viewers, such as the revival of the relatively rare didactic and historical telenovela subgenres I identified as the first trend in current production. As we have seen in this article, then, the legacy genre is not yet dead. Indeed, the total hours of shorter-form series screened in Mexico (much lower of course than the figure for telenovelas, with their still lengthy runs of episodes) did not increase from 2019 to 2023. My focus in this article on TelevisaUnivision and its digital twin ViX is thus justified by the network's heavy load of production and consumption.

Conversely, although OBITEL laments the lack of reliable public data for SVOD services, still it claims that Netflix is the leading such platform in nine countries, including Mexico, by number of subscribers.[37] Yet the US-based streamers Netflix and Amazon Prime Video (both treated in this article) premiered only ten and five titles, respectively, in Mexico in 2023, while TelevisaUnivision's ViX offered twenty-four new locally produced shows, plus two additional coproductions.[38] The disproportionate focus of foreign scholars on Netflix, then, due no doubt to the ease of access around the world, does not correspond to the experience of Mexican viewers in their home country, even those relatively wealthy consumers who enjoy the luxury of broadband and are thus not limited to the still popular free-to-air service. Moreover, one recent article has suggested that Netflix's "Top Ten" list systematically undercounts and distorts audience behavior in the Ibero-American region by failing to account for genre-related differences, such as the continuing preference for long-run telenovelas.[39]

[35] Ibid., 28.
[36] Ibid., 30.
[37] Ibid., 31.
[38] Ibid., 33.
[39] Michael L. Wayne and Nahuel Ribke, "Loved Everywhere? Netflix's Top 10 and the Popularity of Geographically Diverse Content," *Convergence: The International Journal of Research into New Media Technologies* 0, no. 0 (2024): 1–17.

As for the format of local streaming premieres in Mexico, they are indeed overwhelmingly "series" (like ViX's *Extraño retorno* remake once more), with a lesser but significant slot given up to even shorter "miniseries" (like Prime Video's *¿Quién lo mató?*).[40] My case studies here do not seem to bear out OBITEL's observation that Mexican titles appeal to young protagonists in an attempt to capture the attention of the elusive youth demographic (although the more youthful supporting characters of didactic *novela Vencer el pasado* did indeed inspire participation from their peers on social media). But my article does bear witness to OBITEL's further observations on the content of current titles: many are anchored by female protagonists who boast "complexity and depth" of psychology.[41] These trends are confirmed in this article by examples as diverse as the troubled, time-traveling Diana Salazar, the president's unwilling but resourceful widow in *Ingobernable*, and the barely fictionalized and emotionally traumatized María Félix in *La doña*. It is no accident, then, that OBITEL's "highlights of the year" are free-to-air telenovelas and VOD series focused on the empowerment of women of all kinds.[42] We have once more seen this trend in several titles here: from the *Vencer* franchise that focuses on finally triumphant female friendship across generations via the newly professional protagonist of the remade *Diana Salazar* to María Félix, acclaimed as the "woman with the heart of a man" even as she submits to successive and sometimes abusive spouses.

My own corpus and OBITEL's latest survey thus both bear out the general intuition that Televisa's linear and streaming services and their heritage genre of telenovela remain central to the Mexican viewing experience. And although I have barely mentioned audience reception here, very recent research based on informant interviews and focus groups in Mexico and Spain does suggest some distinguishing features of young viewers (Millennials and Generation Z) in each territory. For example, when asked for their perspectives on cruelty and historical truthfulness in current TV drama (a question posed most urgently by the sometimes gruesome *Diana Salazar* format), more Mexicans agree that audiovisual fiction should not be "a mere escape from reality" but should "confront the viewer with the cruelty of human nature."[43] Likewise, Mexicans

[40] OBITEL, *Observatorio Iberoamericano de Ficción Televisiva 2024*, 34.

[41] Ibid., 35.

[42] Ibid., 44.

[43] Carlos Fernández-Rodríguez, Luis Miguel Romero-Rodríguez, and Belén Puebla-Martínez, "Audiovisual in the Streaming Era: Millennials' and Centennials' Perspectives on Cruelty and Historical Truthfulness," *Online Journal of Communication and Media Technologies* 13, no. 4 (2023): 9.

were more in agreement than Spaniards that global problems should be shown "without concessions" and "aimed at raising awareness" and, also, that the "dystopian themes" in TV shows arise mainly from "media exaggeration" (this sounds like a warning for hyperbolic thrillers based on conspiracy theories like *Ingobernable* and *El candidato*). It is also notable that the documentary and drama series based on the murder of Paco Stanley stake a claim to both uncovering a disturbing truth (raising awareness) and warning against manipulation (media exaggeration), especially and ironically that performed by television itself. It seems younger Mexican audiences are primed to welcome such shows, as indeed they did.

Mexicans and Spaniards agree that "series based on historical events should show historical rigor and fidelity" (surely a factor to be borne in mind when making or watching bio series set in the twentieth century or historical *novelas* that claim to represent the brutalities of the colonial period). But it is Mexicans who are more likely than Spaniards to say that "the consumption of films and television today is a depressing experience." This observation is somewhat undermined by the continuing and evident loyalty of Mexican audiences to Mexican content, a phenomenon noted throughout Latin America by Joe Straubhaar and colleagues in *From Telenovelas to Netflix*.[44]

Interestingly, another article by the same research team found that the young Mexican streaming viewers in the focus group had a stronger sense of "third person effect" than Spaniards.[45] That is, "they believed the content consumed… affected others more than themselves, especially when dealing with 'sensitive' topics like abortion, politics, or immigration."[46] While, as we have noted in this article, graphic scenes of sex and violence are indeed more common in streaming series than in legacy serials, it seems Mexican viewers trust their own independent response to such controversial material and are more concerned by its supposed effect on their less savvy fellow citizens. Moreover, as we have also seen, free-to-air broadcasters have indeed incorporated such modern topics as gender violence into their *novelas* without abandoning traditional romance or

[44] Joe Straubhaar, Melissa Santillana, Vanessa de Macedo Higgins Joyce, and Luiz Guilherme, *From Telenovelas to Netflix: Transnational, Transverse Television in Latin America* (Cham, Switzerland: Palgrave Macmillan, 2021), 49–84.

[45] Carlos Fernández-Rodríguez, Luis M. Romero-Rodríguez, and Belén Puebla-Martínez, "Millennials' and Centennials' Perspectives on Screening Narratives in Spain and Mexico: Spiral of Silence, Bandwagon Affect, and Third Person Effect," *Research Journal in Advanced Humanities* 4, no. 3 (2023), https://doi.org/10.58256/rjah.v4i3.1217.

[46] Ibid., 50.

frightening off the older, poorer, and more conservative target audience, which remains loyal to the daily rhythms of linear TV. Indeed, that is the avowed aim of Rosy Ocampo's hugely successful *Vencer* franchise.

The merger with Univision by legacy network Televisa and the infusion of foreign capital on the new streaming platforms mean that the industrial base of production in Mexico is currently by no means in crisis. And we have seen that the trend for the centrality of female protagonists has smoothly transitioned from legacy linear media (the *Vencer* franchise once more) to the new genres of streaming series, even in apparently inhospitable genres such as the political thriller, which was rebooted for the "uncontrollable" star Kate del Castillo in *Ingobernable*, and the bio series, initially male focused in Mexico but now redirected from Blue Demon and his fellow male celebrities to María Félix as the quintessential protagonist, indeed *la doña*, of the nation's entertainment industry. The immediate prospect, industrially and aesthetically, remains unpredictable, when even giants like Televisa and Netflix have stumbled in recent times. But television drama in Mexico, both broadcast and streaming, surely has its future assured, drawing as it does on a lengthy and invaluable heritage of intellectual property and experienced as it is in exporting a content that is so prized by its local viewers to a wider Spanish-speaking world and beyond.

Bibliography

Briceño Izarra, Andrés. "Amazon Prime Video prepara miniserie sobre el presentador de TV mexicano Paco Stanley." *Produ*, January 11, 2023.

Cueva, Álvaro. "*María Félix* de ViX+." *Milenio*, July 21, 2022.

Echevarria, Martin, and Ruben Arnoldo Gonzalez, eds., *Media and Politics in Post-Authoritarian Mexico: The Continuing Struggle for Democracy*. Cham, Switzerland: Palgrave Macmillan, 2024.

Farías, Iván. *El misterio de la máscara perdida*. Mexico City: Dark & Glow Press, 2022.

Fernández-Rodríguez, Carlos,, Luis Miguel Romero-Rodríguez, and Belén Puebla-Martínez. "Audiovisual in the Streaming Era: Millennials' and Centennials' Perspectives on Cruelty and Historical Truthfulness." *Online Journal of Communication and Media Technologies* 13, no. 4 (2023): e202350.

Fernández-Rodríguez, Carlos, Luis M. Romero-Rodríguez, and Belén Puebla-Martínez. "Millennials' and Centennials' Perspectives on Screening Narratives in Spain and Mexico: Spiral of Silence, Bandwagon Effect, and Third Person Effect." *Research Journal in Advanced Humanities* 4, no. 3 (2023). https://doi.org/10.58256/rjah.v4i3.1217.

González, Aliana. "Trabajar con Peter Blake y ser bisagra entre el modo de funcionar de EE UU y México fue un reto y un aprendizaje: Productores ejecutivos de *El candidato*." *Produ*, July 31, 2020.

Izarra, Richard. "Rosy Ocampo de Televisa en #PRODUprimetime: Franquicia *Vencer* en preparación de cuarta y quinta temporadas y todas disponibles al mercado mundial." *Produ*, July 14, 2021.

Laugier, Sandra. "Taking TV Series Seriously." *Open Philosophy* 5 (2022): 250–3.

Laugier, Sandra, ed. "Une politique de la culture populaire." In *Les Séries: laboratoires d'éveil politique*, 9–28. Paris: CNRS Editions, 2023.

Maas, Jennifer. "TelevisaUnivision CEO Wayde Davis Says ViX's Free Tier Is Pivotal against Churn." *Variety*, May 11, 2022.

Maldonado, Vanessa. "Carmen Armendáriz de Televisa: La miniserie *María Félix, la doña* resume 50 décadas [*sic*] de la actriz en ocho episodios." *Produ*, July 12, 2022.

Maldonado, Vanessa. "Carlos Sosa de Viento del Norte: nos tomó 14 meses realizar la serie documental *El show. Crónica de un asesinato* para N+." *Produ*, June 6, 2023.

Maldonado, Vanessa. "*El extraño retorno de Diana Salazar* y la serie animada de comedia *Circo Gómez* entre los contenidos de ficción de TelevisaUnivision." *Produ*, May 16, 2023.

Maldonado, Vanessa. "Epigmenio Ibarra de Argos: México debe tener más confianza en las series de TV." *Produ*, July 24, 2015.

Maldonado, Vanessa. "Patricio Wills de W Studios: Trabajando en *Juana la virgen* y *El extraño retorno de Diana Salazar*." *Produ*, January 24, 2024.

OBITEL. *Observatorio Iberoamericano de Ficción Televisiva*. Porto Alegre: Globo, 2022.

OBITEL. *Observatorio Iberoamericano de Ficción Televisiva*. Porto Alegre: Globo, 2024.

Plohn, Cynthia. "Patricio Wills y Carlos Bardasano de W Studios comenzaron producción de *El extraño retorno de Diana Salazar* para ViX." *Produ*, August 22, 2023.

Produ. "Actriz Kate del Castillo: Estamos trabajando en la posibilidad de hacer la versión de *Las Aparicio* de Argos." November 30, 2015.

Produ. "Netflix: Kate del Castillo Will Be the Star of Our New Original Mexican Series *Ingobernable* Produced by Argos." July 24, 2015.

Produ. "Produ visitó el set de *Blue Demon*." September 2, 2016.

Produ. "Televisa: 8 Series Ready and 10 in Preparation, with 20 Episodes per Season and a New Narration." August 13, 2016.

Ramos-Weiner, Maribel. "NATPE-Angélica Guerra de SPT: *Blue Demon* será la próxima coproducción con Televisa." *Produ*, January 18, 2016.

Ramos-Weiner, Maribel. "Sony Pictures Television: The Bio Series Represent More Than Half of Our Production in the Region This Year." *Produ*, February 18, 2016.

Spangler, Todd. "TelevisaUnivision Sets ViX+ Premium Service Pricing, Launch Date." *Variety*, June 30, 2022. https://variety.com/2022/digital/news/vix-plus-pricing-launch-date-televisaunivision-1235306723/.

Straubhaar, Joe, Melissa Santillana, Vanessa de Macedo Higgins Joyce, and Luiz Guilherme Duarte. *From Telenovelas to Netflix: Transnational, Transverse Television in Latin America*. Cham, Switzerland: Palgrave Macmillan, 2021.

Thornton, Niamh. *María Félix: A Mexican Film Star and Her Legacy*. Woodbridge, UK: Tamesis, 2023.

Wayne, Michael L., and Nahuel Ribke. "Loved Everywhere? Netflix's Top 10 and the Popularity of Geographically Diverse Content." *Convergence: The International Journal of Research into New Media Technologies* 0, no. 0 (2024): 1–17.

Wright, Amy E. *Serial Mexico: Storytelling across Media from Nationhood to Now*. Nashville: Vanderbilt University Press, 2023.

7

Watching Old Egyptian TV Shows Online: YouTube and Nostalgia for Television's Past

Egor Korneev

When exploring the rich history of Arab television, one is immediately struck by the abundance of old Egyptian TV shows available on YouTube. Produced between the early 1960s and 2000s, these shows are organized into numerous playlists and garner up to millions of views online, provoking vivid reactions from Arab users. Mostly uploaded to the official YouTube channels of the National Media Authority (NMA)—the Egyptian public service broadcaster—they seem to frequently generate a longing for the past remembered or imagined as a time of higher cultural and moral standards and greater communal cohesion. "The best days and the best TV series.... In them [you find] benefit and the respected generation," says one of the Arabic comments under the iconic Egyptian TV series *Al-Ḥilmīya Nights* (*Layālī al-Ḥilmīya*, 1987–95, dir. Ismāʿīl ʿAbd al-Ḥāfiẓ).[1] While the exact formulation may vary, a similar sentiment frequently echoes across the comments sections of this and hundreds of other Egyptian TV shows uploaded to YouTube.

The present chapter attempts to define the role that watching old serials on YouTube plays as a part of the diverse televisual practices in contemporary Egypt and in relation to recent developments in national, regional, and global TV. In particular, it asks why the legacies of Egyptian state-controlled television might be of interest to Arabic-speaking audiences at a time characterized by the proliferation of pan-Arab satellite channels, transnationalization of broadcasting flows, and, most importantly, the growing popularity of streaming video services.[2] Finding the answer to this question gives way to a broader analysis

[1] The original comment, as well as the vast majority of other comments under Egyptian TV shows on YouTube, was written in Arabic: "*Aḥlā ayyām wa-aḥlā musalsalāt fīhā fāyda wa-jīl muḥtaram*." See "Musalsal Layālī Al-Ḥilmīya."

[2] Tourya Guaaybess, ed., *National Broadcasting and State Policy in Arab Countries* (Houndmills, UK: Palgrave Macmillan, 2013), 52.

of TV's past and present in Egypt, which has always functioned as one of the most important production and broadcasting centers of the Middle East and North Africa, with its programs and serials being in high demand both within the country and in other parts of the Arab world.

Since the emergence of Arab TV in the late 1950s to early 1960s, Egypt has remained one of the main players in the regional television industry.[3] Although its dominance in the region is mostly associated with the era of state-controlled network television (1960s–90s), even today Egypt continues to be the most prolific producer of televised Arabic drama and provides the infrastructure for the operation of many regional satellite channels.[4] Being the most populated Arab country with around 110 million people, it also constitutes a huge market for television programs and serials created in other parts of the Middle East and North Africa (MENA) and beyond. Today, around 98 percent of Egyptians have a TV set at home, which helps explain the high demand within the nation for content broadcast through traditional terrestrial and satellite channels.[5]

In spite of traditional TV's continued popularity, there is, recent research demonstrates, growing interest among Egyptians in internet-based video platforms.[6] The number of Egyptian internet users is rising rapidly every year, with around 86 million users (78 percent of the population) noted in May 2023.[7] The increasingly important digital sphere deeply influences local spectatorship habits, as it allows Egyptians to watch TV programs and serials on demand via popular global and regional streaming video services such as YouTube, Netflix, Shahid, and Watch IT. These emerging tendencies challenge the status of satellite and terrestrial broadcasting as the main means of consuming TV content by the local population, especially in urban centers where access to a high-speed internet connection is more prevalent.[8]

The growing popularity of video streaming services has changed Egyptians' relationship with national television's past. At the same time, I argue, the streaming television era ushered in by Netflix and other internet-based video

3 Marwan M. Kraidy and Joe F. Khalil, *Arab Television Industries* (Houndmills, UK: Palgrave Macmillan, 2009), 125–8.

4 Rasha Allam, *Entertainment Media Industry in Egypt* (London: Routledge, 2019), 127.

5 Ibid., 125, 127.

6 Rasha Allam and Hesham Dinana, "The Future of TV and Online Video Platforms: A Study on Predictors of Use and Interaction with Content in the Egyptian Evolving Telecomm, Media & Entertainment Industries," *SAGE Open* 11, no. 3 (2021): 9, https://doi.org/10.1177/21582440211040804.

7 Ibid., 2; Alaa El-DIn, "Egypt's Internet Users Increase by 1.4 Million in May," *Daily News Egypt*, August 23, 2023, https://www.dailynewsegypt.com/2023/08/23/egypts-internet-users-increase-by-1-4-million-in-may/.

8 Allam and Dinana, "Future of TV and Online Video Platforms."

platforms has not led to a rupture with the legacies of Egyptian state-controlled TV but rather has reshaped this older form of television into an object of nostalgia. My analysis adds a new dimension to the growing body of research dedicated to the role of streaming television in the country, which tends to emphasize either the business models video streaming platforms and TV channels adopt to become economically profitable in the local market, or questions of mass media ownership and how internet-based video platforms change the balance between private and state-controlled players.[9] While building on this research, the following analysis will attend to the nostalgic function of watching old Egyptian TV content online. To better contextualize this practice and its social meaning, we should start with the existing theorizations of the relationship between nostalgia and mass media.

Studying Mass-Mediated Nostalgia

The wide appeal of nostalgic content is not a new phenomenon exclusive to the mass media industries of the present. More broadly tied to modernity and its concept of an irreversible time and a teleology of progress, nostalgia temporarily gains prominence in certain cultural contexts, for instance, in late nineteenth-century Europe, the United States of the mid-1970s, or the globalizing West of the late twentieth century.[10] However, the in-depth study of nostalgia's role in popular culture is a relatively new trend tied to postmodernist developments of the late 1980s to the 1990s. Founding scholars of memory studies of that period, such as David Lowenthal and Andreas Huyssen, were interested in how the unstable conditions of the present encourage mass media to nostalgically represent or monumentalize the past, which anchors the audiences and provides

[9] Rasha Allam and Sylvia Chan-Olmsted, "The Development of Video Streaming Industry in Egypt: Examining Its Market Environment and Business Model," *Journal of Media Business Studies* 18, no. 4 (2021): 285–303; Rasha Allam, "From Flabby to Fit: Restructuring the Public Broadcasting System in Egypt"; Nagy, "Advancing the Egyptian Public Broadcaster to Compete in the Digital Era," *Global Media and Communication* 17, no. 3 (2021): 297–320, https://doi.org/10.1177/1742766521102397; Rasha Allam, "Egyptian Media Ownership and Pluralism: Overview, Performance, and Challenges," in *Media Ownership in Africa in the Digital Age: Challenges, Continuity and Change*, ed. Winston Mano and Loubna El Mkaouar (London: Routledge, 2023).

[10] Svetlana Boym, *The Future of Nostalgia* (New York: Basic Books, 2001), 10, 11, 13; Anna M. Ciasullo, "Afterword: In Love with the Past," in *Netflix Nostalgia: Streaming the Past on Demand*, ed. Kathryn Pallister (Lanham, MD: Lexington Books, 2019), 235; Linda Hutcheon and Mario Valdés, "Irony, Nostalgia, and the Postmodern: A Dialogue," *Nuevas Poligrafías* 3 (2000): 18, https://doi.org/10.22201/ffyl.poligrafias.2000.3.1610.

ontological security.[11] In their works, nostalgia was typically colored as regressive, conservative, or leading to oversimplified perceptions of history.[12] The later generation of memory studies scholars, whom Alex Bevan associates with the "poststructuralist revisions" in the field, tried to provide a less escapism-centered account of contemporary nostalgia, attaching it to a variety of audiences and functions, as well as providing a more nuanced cultural and economic context for its proliferation.[13] Among other things, specialists of this wave started paying attention to the role of on-demand video streaming platforms—especially Netflix—in using nostalgia as an indispensable part of their digital identity in the United States and other Western countries.[14]

Focused on nostalgic uses of streaming video services beyond the West, my research benefits from an array of nostalgia conceptualizations developed in the 2000s to 2010s. First, it borrows from Linda Hutcheon's idea that nostalgia depends on the dialectical relationship between the present and the past: it is impossible without a present perceived as imperfect, from which it exiles us by temporarily bringing the past near.[15] Second, the Egyptian longing for television's past can be better explained through the dichotomy of restorative and reflective nostalgia developed by Svetlana Boym in her renowned *The Future of Nostalgia* (2001).

Whereas restorative nostalgia rarely calls itself nostalgia and "attempts a transhistorical reconstruction of the lost home" in the present by reviving the "pristine tradition" hidden in the past, reflective nostalgia doubts the possibility of restoring it and "thrives in longing itself."[16] It seems that the Egyptian case provides us with both types of nostalgia, which I will try to demonstrate below.[17]

Finally, Paul Grainge's differentiation of nostalgia as a *mode* (intentionally encoded in mass-mediated texts) and a *mood* (an audience response to

[11] David Lowenthal, *The Past Is a Foreign Country* (Cambridge: Cambridge University Press, 1985); Andreas Huyssen, *Present Pasts: Urban Palimpsests and the Politics of Memory*, Cultural Memory in the Present (Stanford, CA: Stanford University Press, 2003).

[12] Carney, "A Dizi-Ying Past: 'Magnificent Century' ('Muhtesem Yuzyil') and the Motivated Uses of History in Contemporary Turkey," PhD diss., Indiana University, 2015, 43–75; Ciasullo, "Afterword," 237.

[13] Alex Bevan, *The Aesthetics of Nostalgia TV: Production Design and the Boomer Era* (New York: Bloomsbury Academic, 2019), 14–15.

[14] Kathryn Pallister, ed., *Netflix Nostalgia: Streaming the Past on Demand* (Lanham, MD: Lexington Books, 2019).

[15] Hutcheon and Valdés, "Irony, Nostalgia, and the Postmodern," 20–1.

[16] Boym, *Future of Nostalgia*, xvii–xix.

[17] When studying Turkish historical drama of the 2000s–2010s, Joshua Carney came to the same conclusions, stating that while the serials dealing with the early Republic of Turkey evoked reflective nostalgia, shows about the Ottoman Empire created restorative nostalgia. See Carney, "A Dizi-Ying Past"; Josh Carney, "Resur(e)Recting a Spectacular Hero: Diriliş Ertuğrul, Necropolitics, and Popular Culture in Turkey," *Review of Middle East Studies* 52, no. 1 (2018): 93–114, https://doi.org/10.1017/rms.2018.6.

mass-mediated texts) will be instructive in my analysis of the Egyptian state-controlled broadcaster's content on YouTube.[18] Stefanie Armbruster has noted that a nostalgic mode can be signified through both newly created audiovisual texts and the policy of series revivals and reboots used by streaming video services.[19] In turn, Giulia Taurino has analyzed how, in addition to the previously mentioned means, Netflix actively employs the clustering of numerous older television texts in its catalog into subcategories ("Nostalgic '90s," "Totally Awesome '80s," etc.), thus making nostalgia "a genre itself" instantaneously available to every subscriber.[20] Taurino's description of nostalgia as a separate genre is somewhat akin to how Alex Bevan has characterized many American quality television shows of the 2000s–2010s as "nostalgia TV," which intentionally benefited from the White longing for the boomer era of the 1950s to 1960s as the "American golden age."[21]

Although the Egyptian case analyzed in this chapter does not include the production of new shows evoking historical longing, Bevan's "nostalgia TV" seems to be a useful term to describe a sector of the country's emerging digital environment capitalizing on its televisual legacy, some of which was itself in the nostalgic mode. Thus, insights into how the nostalgic mode is encoded will be applied to the Egyptian TV environment and to the digital marketing of older televisual texts through the interface and capabilities of streaming platforms. To achieve this, the analysis will now turn to the primary vehicle for eliciting nostalgia in Egypt, that is, television shows of the late 1980s to 1990s.

Egyptian "Nostalgic '90s" and Their Aftermath

Across the Arab world, the period between the late 1980s and early 2000s is often remembered as the heyday of Egyptian television dramas.[22] Known in Arabic as *musalsalāt*, they gained public acclaim domestically and abroad partly due to

[18] Paul Grainge, *Memory and Popular Film* (Manchester: Manchester University Press, 2003).

[19] Stefanie Armbruster, *Watching Nostalgia: An Analysis of Nostalgic Television Fiction and Its Reception*, Cultural Studies 48 (Bielefeld, Germany: Transcript Verlag, 2016), 12.

[20] Giulia Taurino, "Crossing Eras: Exploring Nostalgic Reconfigurations in Media Franchises," in Pallister, *Netflix Nostalgia*, 11.

[21] Bevan, *Aesthetics of Nostalgia TV*, 13–19.

[22] Yāsmīn ʿĀdil, "Liʾ ʿUshshāq Drāmā al-Tisʿīnīyāt: Musalsalāt Ramaḍānīya Qadīma Lā Tunsā," Al-Jazīra Online, March 17, 2023; Yasmeen Salah, "Maʿa Al-Iḥtifāl Bi-60 ʿāman Min al-Drāmā: Daʿūnā Nastadhkir Ajmal Musalsalāt al-Tisʿīnāt Miṣrī," *Harper's Bazaar Arabia*, August 23, 2023; Zahrāʾ Abū al-ʿInayn, "21 Musalsalan Min Afḍal Musalsalāt Miṣrīya Qadīma Lā Tazāl Taḥẓā Bi-Shaʿbīya Ḥattā al-Yawm," *Al-Mashhad*, August 1, 2023.

the participation of prominent Egyptian directors, scriptwriters, and actors.[23] Egyptian TV series of this time belonged to various genres and dealt with diverse themes. Together with drama series and comedies shedding light on the everyday problems and class struggles of Egyptians (*musalsalāt drāmīya wa- kūmīdīya*), the audiences enjoyed historical-religious productions about the remote Arab-Islamic past (*musalsalāt dīnīya tārīkhīya*) featuring classical Arabic instead of the commonly used Egyptian dialect.[24] Another popular category comprised serials focused on the southern part of the country (Upper Egypt, or *al-Ṣaʿīd*), its more traditional and tumultuous social environment that, nevertheless, was depicted as being home to honest, brave, respectful, and patriotic people (*banū al-balad*).[25] *Musalsalāt* of the period were typically shot annually for Ramadan and consisted of around thirty episodes to be viewed every night of the Muslim holy month.

This wide variety of television dramas of the 1990s was almost solely produced by the Egyptian state-controlled media conglomerate known at the time as the Egyptian Radio and Television Union (ERTU) and also commonly referred to as Maspero.[26] Created in 1971 as a state body to manage all the national broadcasting services and production, ERTU continued to play this role until 2001 when private channels were officially allowed into the country.[27] Thus, for thirty years, it functioned as a state broadcaster monopolizing the production and airing of all Egyptian TV content through a limited number of terrestrial channels.[28]

Within this media context, the Egyptian state deemed television a powerful instrument of social engineering and nation-building and, during the turbulent 1990s, actively invested in it to fight the growing influence of local Islamist organizations, thus, according to the official rhetoric, "confronting terrorism

[23] Lila Abu-Lughod, *Dramas of Nationhood: The Politics of Television in Egypt* (Chicago: University of Chicago Press, 2005), 130.

[24] Jakob Skovgaard-Petersen, "The Khuṭba Scene in Arab Religious Films and TV Dramas," in *Muslim Preaching in the Middle East and Beyond: Historical and Contemporary Case Studies*, ed. Simon Stjernholm and Elisabeth Özdalga (Edinburgh: Edinburgh University Press, 2020), 38.

[25] Abu-Lughod, *Dramas of Nationhood*, 57–70.

[26] This title is borrowed from the name of the organization's headquarters in Cairo. See "Maspero Television Building," *Wikipedia*, n.d., https://en.wikipedia.org/wiki/Maspero_television_building.

[27] Before 1971, the Egyptian state-controlled media conglomerate was called the Egyptian General Corporation for Radio and Television (*al-Muʾassasa al-Miṣrīya al-ʿĀmma lil-Idhāʿa wa-l-Tilīfizyūn*), established in 1960. Maha Sherif Hussein Nagy, "Advancing the Egyptian Public Broadcaster to Compete in the Digital Era" (Master's thesis, American University in Cairo, 2019), 18–22; Allam, "Egyptian Media Ownership," 225.

[28] Kraidy and Khalil, *Arab Television Industries*, 15–17.

with media."[29] This was not the Egyptian state's first attempt to minimize the mass media presence of local Islamist movements. On the contrary, this process can be traced back to its opposition to the Muslim Brotherhood in the 1940s to 1950s and other ideologically related organizations in the subsequent decades.[30] Importantly, this new strategy of "confronting terrorism with media" was a state response to a gradual Islamization of the mediascapes that had been taking place in Egypt and the broader Middle East since the 1970s as a part of the Islamic Revival movement (*al-Ṣaḥwa al-Islāmīya*).[31]

Amid the growing popularity of Islamist organizations, the state investments in television sought to assert the Egyptian national culture and "correct" interpretations of Islam that did not endanger the secular nation-state's legitimacy.[32] Probably, these investments allowed ERTU to hire top film casts and crews, which, at the time, could not find jobs in the local cinematic industry, which was experiencing a deep economic crisis.[33] In 1992–4, state funding also materialized in a huge new media complex near the capital—the Egyptian Media Production City (EMPC)—containing higher-quality shooting sets and television equipment.[34] All these factors contributed to creating a new wave of TV shows, including historical dramas.

One of the most iconic series of this epoch is *Al-Ḥilmīya Nights* (*Layālī al-Ḥilmīya*, 1987–95). This multiyear television series written by Usāma Anwar ʿUkāsha (1941–2010) and directed by Ismāʿīl ʿAbd al-Ḥāfiẓ (1941–2012) addresses the transformation of Cairo from the early 1940s to early 1990s.[35] It centers on two prominent families residing in the al-Ḥilmīya neighborhood and,

[29] Abu-Lughod, *Dramas of Nationhood*, 167.

[30] Talha İ Duman, "The Evolution of a Social Movement: The Muslim Brotherhood's Path Between Political Ambition and the Rejection of Armed Struggle," *Religions* 15, no. 11 (2024): 1352, https://doi.org/10.3390/rel15111352.

[31] The global religious trend of the 1970s–2000s known as the Islamic Revival is usually associated with the Arab states' defeat in the Six-Day War (1967) and the Islamic Revolution in Iran (1979). It describes the formation of numerous Muslim organizations and communities questioning the legitimacy of the existing political structures and Westernized lifestyle in the Islamic world. Aaron Rock-Singer, *Practicing Islam in Egypt: Print Media and Islamic Revival* (Cambridge: Cambridge University Press, 2020), 1–21; Ira M. Lapidus, *History of Islamic Societies* (Cambridge: Cambridge University Press, 2014), 521–3.

[32] Karin van Nieuwkerk, *Performing Piety: Singers and Actors in Egypt's Islamic Revival* (Austin: University of Texas Press, 2013), 115.

[33] Shafik, *Popular Egyptian Cinema: Gender, Class, and Nation* (Cairo: American University in Cairo Press, 2007), 5.

[34] State Information Service, "Media Production City," May 19, 2016, https://www.sis.gov.eg/Story/67988/Media-Production-City?lang=en-us.

[35] Amjad Miṣbāḥ, "'Layālī al-Ḥilmiīya': Aḥdāth Wa-Shakhṣīyāt Maḥfūra Fī al-ʿAql al-ʿArabī," *ʿArab Journal*, April 12, 2022, https://arab-j.net/4594; "Musalsal Layālī Al-Ḥilmīya."

through their rivalry, traces the way it changed from an aristocratic into a more traditional district. Already famous by the late 1980s, ʿUkāsha and ʿAbd al-Ḥāfiẓ were able to dive into a social history of modern Egypt through a masterful plot acted out by some of the most prominent Egyptian cinema and television stars of the time, such as Yaḥyā al-Fakharānī (b. 1945), Ṣalāḥ al-Saʿdanī (1943–2024), Ṣafīya al-ʿUmrī (b. 1949), and many others. It paid attention to the main political and economic changes in the country, such as the overthrow of the monarchy (1952), the subsequent socialist rule of Gamal Abdel Nasser (Jamāl ʿAbd al-Nāṣir) (1956–70), and the economic "opening" (*infitāḥ*) initiated by the next president, Anwar al- Sādāt (1970–81). *Al-Ḥilmīya Nights* was marketed as a drama series (*musalsal drāmī*) and focused on the way these radical political and economic transformations shaped the lives of the show's protagonists, with some of them moving from the village to Cairo, emigrating or being imprisoned for their political views. Thus, the serial offered a reflection on the nation's recent history and the vicissitudes experienced by many Egyptians. It modeled and reenacted the past using costumes and sets, showing more old-fashioned and transforming morals and modes of social interaction through skillful acting and engaging plotlines. For a country where the late 1980s to 1990s marked a general dissatisfaction with the linear modernization and Westernization model, the growth of social inequality, and the deterioration of public services, such a portrayal provided an opportunity to relive the past and meditate on how it shaped the national present.[36] In Svetlana Boym's terms, such a historical emotion can be described as reflective nostalgia built on the sentiment of desiring and yet doubting the possibility of restoring the past, thus stressing the longing itself.[37]

Another historical television series enjoyed by Egyptian audiences at the time was *The Unknown Prince* (*Al-Amīr al-Majhūl*, 1997).[38] It was shot by Aḥmad Tawfīq (1930–2005), an actor and then television director specializing in religious-historical drama (*musalsalāt dīnīya tārīkhīya*) from the 1960s. While many of his previous works focused on widely revered Islamic personalities, this show was dedicated to the more politically charged figure of Hārūn al-Rashīd (766–809)—one of the most recognizable rulers of the ʿAbbāsid Caliphate (750–1258), who, nevertheless, does not possess a special religious status in

[36] Armbrust, *Mass Culture and Modernism in Egypt*, 6–8, 16; Īmān al-Ṭālibī, "Ayna Ikhtafā Zaman Layālī Al-Ḥilmīya?," Al-Jazīra Online, November 13, 2017.

[37] Boym, *Future of Nostalgia*, xix.

[38] "Musalsal Al-Amīr al-Majhūl Hārūn al-Rashīd."

the traditional Islamic historiography.[39] However, in popular culture, his reign is commonly associated with the Arab-Islamic golden age, the peak of the caliphate's political and economic might, and the time when traditional Islamic sciences and arts flourished.

Focused on palace intrigues and political affairs, *The Unknown Prince* emphasizes the glory and prosperity of the Arab-Islamic empire through vivid and shimmering decorations, costumes, and sets built in the new EMPC. The atmosphere of might and grandeur becomes prominent from the very beginning of the show's first episode, dedicated to the transition of power from Hārūn al-Rashīd's father, Caliph al-Mahdī (744–85), to his older son and successor, Mūsā al-Hādī (761–86). The latter is presented to us as a powerful commander who hears about his father's death in a military camp while planning a new campaign over a detailed map of the region (Figure 7.1). In turn, his brother Hārūn al-Rashīd shares the same news with the court in a lavishly adorned royal palace with golden walls and glossy, ornate decorations (Figure 7.2). In both scenes,

Figure 7.1 Mūsā al-Hādī, the caliph's successor, holds a military council. *Al-Amīr al-Majhūl: Harūn al-Rashid* (The Unknown Prince) © Radio & Television Union of Arab Republic of Egypt 1997.

[39] As Hārūn al-Rashīd is not commonly associated with a special religious status, some television critics suggested treating Aḥmad Tawfīq's show about him not as a religious-historical series (*musalsal dīnī*) but rather as a purely historical text (*musalsal tārīkhī*) not concerned with Islamic values and practices. Muḥammad Ṣāliḥ, "Al-Drāmā Wa-Iḥtirām al-Tārīkh?," *Al-Ahrām*, February 24, 1997.

Figure 7.2 Hārūn al-Rashīd addresses the court. *Al-Amīr al-Majhūl: Harūn al-Rashid* (The Unknown Prince) © Radio & Television Union of Arab Republic of Egypt 1997.

the sense of power and grandeur is also conveyed through the triumphal soundtrack, whose style and rhythm resemble modern-day military marches. Thus, using sound and production design, the series initially presents the rich imperial center as its crucial setting, which will remain the main location for the unfolding storylines.

In addition to lavish decorations, the series attracted many viewers due to the performance of Nūr al-Sharīf (1946–2015), an Egyptian film star already famous by the 1990s who popularized the historical-religious drama during the decade.[40] Although provoking controversy around the historical integrity of the

[40] Muḥammad Ṣāliḥ, "Jā'izat Aḥsan Mumaththil Kamān Wa-Kamān: Nūr al-Sharīf Mumaththilan Ma'a Sabq al-Iṣdār Wa-l-Taraṣṣud," *Al-Ahrām*, July 25, 1996.

caliph's representation among historians and some newspaper columnists, the series was praised for showing Hārūn al-Rashīd as a truly Islamic leader and a lover of art. For many critics, this depiction was necessary amid the negative representations of Islam promoted in the West during the 1990s. Some public commentators also attended to the caliph's framing as a fighter against the Persian influence brought to the court by the influential Iranian vizier family of the Barmakids.[41] Emphasizing the caliph's Arab identity as opposed to Iranian culture favored nationalist sentiments important for the Egyptian state propaganda of this period, which prompted the famous Egyptian historian ʿAbd al-ʿAẓīm Ramaḍān (1925–2007) to suspect the show's scriptwriter of glorifying Saddam Hussein's (Ṣaddām Ḥusayn's) war against Iran (1980–8) through the portrayal of Hārūn al-Rashīd's conflict with the Barmakids.[42]

Overall, while *Al-Ḥilmīya Nights* reflected on the dramatic history of modern Egypt to better understand the national present, *The Unknown Prince* provided the audience with an idealized snapshot of the premodern Arab-Islamic glory. The latter portrayal seems to evoke what Boym calls a restorative nostalgia, a desire to return to the better times associated with the pristine Islamic tradition, cultural flourishment, political might, and Arab national pride.[43] Both historical melodramas were watched by many viewers in Egypt as well as other Arab countries.

The highly successful Egyptian television productions of the 1990s also include dozens of other shows, such as *Mountain Wolves* (*Dhiʾāb al-Jabal*, 1992, dir. Majdī Abū ʿUmayra) dedicated to Upper Egypt (*musalsal ṣaʿīdī*), the critically acclaimed social drama *I Will Not Live in the Shadow of My Father* (*Lan Aʿīsh Fī Jilbāb Abī*, 1995, dir. Aḥmad Tawfīq), and the comedy *Wanīs's Diary* (*Yawmīyāt Wanīs*, 1994, dir. Aḥmad Badr al-Dīn and Ṣubḥī Muḥammad).[44] However, by the early 2000s, the heyday of ERTU came to an end as Egypt witnessed a gradual growth in the presence of private satellite TV channels that quickly succeeded in attracting a larger market share, owing partly to higher content quality and new television formats. In this new media environment, the state broadcaster's role

41 Muḥammad Ṣāliḥ, "Ḥikāya," *Al-Ahrām*, February 2, 1997; Muṣṭafā al-Ḍamarānī, "Al-Amīr al-Majhūl Wa al-Ḥaqīqa al-Ghāʾiba," *Al-Ahrām*, February 26, 1997; Āmāl Bakīr, "Kayfa Kana Al-Amīr al-Majhūl Malikan Li' al-Aʿmāl al-Fanīya?!," *Al-Ahrām*, February 21, 1997.

42 ʿAbd al-ʿAẓīm Ramaḍān, "Al-Musalsalāt al-Tārīkhīya Bayn al-Jadd Wa al-Hazal," *Al-Ahrām*, March 1, 1997.

43 Boym, *Future of Nostalgia*, xvii–xviii.

44 "Musalsal Dhiʾāb Al-Jabal"; "Musalsal Lan Aʿīsh Fī Jilbāb Abī"; "Musalsal Yawmīyāt Wanīs."

was further diminished after ERTU's one-sided and progovernment coverage of the 2011 revolution.[45]

Seeking to salvage the media conglomerate, successive Egyptian governments endeavored to transform it into a public service broadcaster, an entity designed to be financially and politically autonomous from the state and an impartial source of information for the public.[46] In December 2016 ERTU was rebranded as the National Media Authority (NMA), while retaining the control of twenty-three national TV channels (both terrestrial and satellite) and seventeen radio stations.[47] However, it did not help the organization solve its main problem—an increasing lack of funding. Since the late 2000s, the broadcaster's revenues have been significantly lower than expenditures, which has led to growing debt (reaching 2.5 billion dollars in 2017) and an acute financial deficit.[48] To keep the allegedly independent public service broadcaster operating, today the state continues to fund it and thus fully controls the aired content. In the meanwhile, NMA functions in a "firefighting mode," trying to raise its revenues without much investment in content quality.[49] It is within this context of financial crisis that the organization tries to take maximum advantage of ERTU/Maspero's rich back catalog of drama and other types of old content by making it available on YouTube.

Maspero YouTube Channels and Nostalgia

Whether YouTube functions as television has been hotly debated by mass media scholars since the late 2000s. Being a top-rated free streaming service financed by advertising revenues, this digital platform combines the TV-inspired logic of the channel and branding with greater agency and control on the part of users, who can personalize the content through self-selection.[50] However, for analyzing Egyptian nostalgic serials online, it is important to emphasize the features

45 Nagy, "Advancing the Egyptian Public Broadcaster," 23.
46 Allam, "From Flabby to Fit," 297–8.
47 Allam, "Egyptian Media Ownership," 225. Some of the NMA-controlled TV channels operate nationwide, while others are designed and aired for audiences in particular regions of Egypt, such as Upper Egypt, Delta, or Alexandria.
48 Ibid., 229; Allam, "From Flabby to Fit," 307–8; Nagy, "Advancing the Egyptian Public Broadcaster," 20–1, 25, 27–8.
49 Nagy, "Advancing the Egyptian Public Broadcaster," 89.
50 Lori Kido Lopez, "ISAtv: YouTube and the Branding of Asian America," in *From Networks to Netflix: A Guide to Changing Channels*, ed. Derek Johnson (New York: Routledge, 2018), 255–6.

YouTube (like other streaming platforms described later) shares with television, particularly its programming logic of the *flow*.

Theorized as a distinctly televisual experience in the 1970s by Raymond Williams, the flow describes a pattern of continuous viewing that transcends the boundaries of discreet audiovisual texts and implies a certain passivity on the part of the viewer.[51] Christopher Cox demonstrates how the flow remains a persistent feature of YouTube and Netflix. As the platforms prioritize the *bundled consumption* of videos enhanced by the autoplay function, they "foster an experience of flow emanating from a bundling of dispersed elements."[52] This allows viewers to "switch on" different flows, but at the same time the interface "organizes the pathways we are enticed to follow."[53] Following this logic, it seems that YouTube channels moderated by NMA in Egypt provide an experience "beyond mere encounters with individual television programs"[54] by giving instantaneous access to ERTU's large back catalog and prompting users to navigate its flow.

The digitized collection of TV shows is one of NMA's main treasures, as the state-controlled broadcaster (known for its otherwise weak online presence) possesses the largest media library in the Middle East and North Africa.[55] The main body sharing the NMA content on YouTube is called the Maspero Channels Network (*Shabakat Qanawāt Māsbīrū*), which was registered in November 2015. It is divided into several branches that are in charge of distinct content: Maspero Drama (2.9 million subscribers)[56] uploads dramatic serials (the 1980s–2000s), Maspero Zaman (1.1 million)[57] focuses on a variety of the earliest digitized TV shows (the 1960s–80s), Maspero Kids (820,000)[58] specializes in children's content, the Egyptian Radio (412,000)[59] deals with radio dramas of the past,

51 Raymond Williams, *Television: Technology and Cultural Form* (New York: Schocken Books, 1975).

52 Christopher M. Cox, "Programming—Flow in the Convergence of Digital Media Platforms and Television," *Critical Studies in Television* 13, no. 4 (2018): 444, https://doi.org/10.1177/1749602018796681.

53 Ibid., 441–4.

54 Ibid., 444.

55 NMA was criticized for its inability to create a customized digital platform for the live broadcasting and media library modeled after European, Japanese, and Turkish counterparts. Instead, NMA has a website that does not provide access to its back catalog and contains only a low-quality "live demo" (*bathth tajrībī*) of some state-owned TV channels and radio stations powered by the French online video-sharing platform Dailymotion. Allam, "From Flabby to Fit," 308–9; Nagy, "Advancing the Egyptian Public Broadcaster," 4, 5, 86.

56 YouTube, "Māsbīrū Drāmā" (all channel pages listed in bibliography).

57 YouTube, "Māsbīrū Zamān."

58 YouTube, "Māsbīrū Aṭfāl."

59 YouTube, "Al-Idhāʿa Al-Miṣrīya."

Maspero's Memory (253,000)[60] uploads the historically significant speeches of Egyptian presidents and high officials, Maspero's Religious Recordings (156,000)[61] prioritizes Islamic content mostly produced in the twentieth century, Maspero News (113,000)[62] gives access to the latest news bulletins, and the Voice of People (15,000)[63] covers sessions of the Egyptian Parliament. The state-controlled broadcaster has also created an official community on Facebook called Watch Maspero (*Shūf Māsbīrū*), which has around three million followers and regularly posts shorter videos from old shows with brief descriptions.[64]

Although NMA owns at least five other active YouTube channels, the biggest share of followers and views goes to Maspero Drama, where some popular old serials may gain up to 1.5–17 million views on each episode.[65] The administrators upload new videos every week (sometimes duplicating already existing files), whereas the statistics show a steady growth of followers, with two to four hundred thousand new participants each year (2019–23).[66] All the other Maspero branches with old content demonstrate similar growth patterns, although on a smaller scale than Maspero Drama.[67] The available numbers indicate that Egyptian TV shows produced by ERTU in the 1960s to 2000s remain a particularly attractive category of videos on YouTube and have a steadily increasing audience. Their popularity can be explained by several interrelated factors.

First, the important aspect praised in the comment sections of the Maspero Channels Network on YouTube is the quality of old Egyptian TV shows. Many viewers describe the plots and acting in old serials as being superior to contemporary productions—a fact that can be explained by the participation of Egypt's then leading cinema stars and scriptwriters. Naturally, the domination of Egyptian television over other Arab national TV industries in the past serves as proof that it possessed some unique qualities, which were lost by the subsequent generation of creators. According to numerous comments on YouTube, early Egyptian movies, *musalsalāt*, and radio dramas are frequently associated with

60 YouTube, "Dhākirat Māsbīrū."
61 YouTube, "Tasjīlāt Māsbīrū Al-Dīnīya."
62 YouTube, "Akhbār Māsbīrū."
63 YouTube, "Ṣawt Al-Shaʿb."
64 Facebook, "Shūf Māsbīrū."
65 For instance, take a look at the number of views on the popular TV series *Al-Ḍawʾ al-Shārid* (Stray Light, 1998, dir. Majdī Abū ʿUmayra): "Musalsal Al-Ḍawʾ al-Shārid."
66 Youtubers.me, "Māsbīrū Drāmā Youtube Channel Statistics," n.d., https://us.youtubers.me/d4e01655-9806-43eb-842c-6897aec140bc/youtuber-stats.
67 For example, see the available statistics for Maspero Kids: Youtubers.me, "Māsbīrū Aṭfāl Youtube Channel Statistics," n.d., https://np.youtubers.me/c69b98a5-8548-4011-89bd-17eb83965ddf/youtuber-stats/en.

the heyday of Egyptian national culture, whereas NMA's present poor condition might be read as its decline.[68]

In addition, ERTU's media library on YouTube can be attractive because it contains some older Egyptian drama genres that are almost extinct today. Probably, the most notable example is the already mentioned category of historical-religious TV series (*musalsalāt dīnīya tārīkhīya*), which was pioneered by Egyptian TV creators in the early 1960s and remained prolific in the Arab world until the early 2000s.[69] The mass production of Egyptian historical-religious TV series almost stopped in the 2000s due to ERTU's financial problems, and since then the genre has been largely associated with the Syrian TV industry and funding from the Gulf region.[70] Struggling to find the newer Egyptian serials about Islamic and ancient history, viewers resort to older productions, some of which garner millions of views on YouTube.[71] As becomes evident from series such as *The Unknown Prince* (1997), Egyptian *musalsalāt dīnīya tārīkhīya* provide an idealized and nostalgic image of the glorious Islamic past, which usually responds with the melodramatic imagination of premodern Arab-Islamic glory to the postcolonial fragmentation of the modern-day Arab world. Thus, viewing these serials on YouTube today may generate an interesting phenomenon of nostalgia within nostalgia, when an older televisual text watched out of yearning for the times when it was produced may in turn evoke longing for an even more distant historical period depicted on screen. This is not unique to *musalsalāt dīnīya* and also applies to Egyptian drama series of the 1990s reflecting on the nation's past—such as the already discussed *Al-Ḥilmīya Nights*. Using Paul Grainge's terms, we can describe such a complex structure of nostalgia as comprising both a nostalgic *mode*, intentionally encoded in the televisual text depicting a past, and a nostalgic *mood* felt by the audience when watching it online.[72] Also contributing to this structure is the Maspero Channels Network that intentionally stimulates a bundled consumption of uploaded old serials. As Giulia Taurino demonstrates in her study of Netflix,

[68] Muḥammad Ṣubḥī, "Māsbīrū Zamān: Al-Iflās al-Fannī Bi-Ṣīghat al-Ḥanīn," *Ultraṣawt*, November 25, 2016.

[69] According to my estimate, between 1962 and 2004, around 130 locally produced *musalsalāt dīnīya* were broadcast on Egyptian television.

[70] *Al-Ahrām*, "Al-Drāmā al-Dīnīya Matā Taʿūd Ilā Makānatiha?," July 25, 2011.

[71] Among the most popular Egyptian historical-religious TV series, we find *Imām al-Duʿā* (2003, dir. Muṣṭafā Al-Shāl), *al-Fursān* (1994, dir. Ḥusām al-Dīn Muṣṭafā), *ʿUmar ibn ʿAbd al- ʿAzīz* (1994, dir. Aḥmad Tawfīq), *Hārūn al-Rashīd* (1997), *Abū Ḥanīfa al-Nuʿmān* (1997, dir. Ḥusām al-Dīn Muṣṭafā), and *ʿAṣr al-Aimma* (1998, dir. Ḥusām al-Dīn Muṣṭafā). See "Musalsal Imām Al-Duʿā"; "Musalsal Al-Fursān"; "Musalsal ʿUmar Ibn ʿAbd Al-ʿAzīz"; "Musalsal Al-Amīr al-Majhūl Hārūn al-Rashīd"; "Musalsal Abū Ḥanīfa Al-Nuʿmān"; "Musalsal ʿAṣr Al-Aʾimma."

[72] Grainge, *Memory and Popular Film*.

organizing nostalgic content into a digital archive with numerous subcategories and thus marketing it as a separate genre can generate more demand for it.[73] What Taurino calls "nostalgia-by-indexing" describes well how the Maspero Channels Network clusters its old catalog into thematic branches on YouTube, creating a "hyper-mediated conglomerate of the past" with a distinctive nostalgic mode.[74]

A key factor in cultivating the nostalgic mode is free on-demand access to full TV shows. Thus, it stands out among the Egyptian private channels that either lock their content on subscription-based premium streaming services or upload only promotional excerpts to YouTube.[75] According to the Egyptian press, the decision to make ERTU's legacy available for free stemmed from the intention to protect it from piracy—as some of its materials were spread online illegally on third-party websites. Given the acute economic crisis the organization is going through, this step was also meant to get revenue from YouTube advertisements.[76] Although this measure has not had a noticeable effect on NMA's financial situation, it is nonetheless a passive source of income that accumulates more money each year thanks to the online audiences' constant growth.

Sharing old ERTU materials on YouTube for free has some apparent advantages, especially for audiences that cannot pay for subscription-based streaming platforms but desire unlimited on-demand access to Arabic televisual content. Indeed, this policy aligns with how the Egyptian state-controlled broadcaster has been traditionally associated with free services that all socioeconomic classes can access.[77] In the given circumstances, creating an

[73] Taurino, "Crossing Eras."

[74] The scholar of Turkish TV Joshua Carney uses the term "hyper-mediated conglomerate of the past" when analyzing Lynn Spigel's study of how US TV channels situated old reruns among other types of content in 1990s programming. I believe that the same term can be applied to the TV content made available on video streaming platforms. See Carney, "A Dizi-Ying Past," 56; Lynn Spigel, "From the Dark Ages to the Golden Age: Women's Memories and Television Reruns," *Screen* 36, no. 1 (1995): 16–33, https://doi.org/10.1093/screen/36.1.16.

[75] The most popular private TV channels functioning in Egypt and prominent on YouTube are MBC Miṣr (12.6 million subscribers), Al Hayah (9.15 million), Al Nahar TV (8.8 million), ON TV (8 million), dmc (7.15 million), CBC (4.2 million), Sada Elbalad (4.1 million), Al-Qāhira wa-l-Nās (4.1 million), and TeN TV (3.1 million). They usually make available short excerpts (two to ten minutes) designed to encourage users to watch full shows somewhere else—either on subscription-based streaming platforms or using satellite TV.

[76] Khālid Ibrāhīm, "Al-Ḍawʾ al-Shārid Wa-Ūbrā ʿĀyida Wa al-Marāghī Aḥdath al-Munḍammīn Li' 'Māsbīrū Drāmā,'" *Al-Yawm al-Sābiʿ*, March 10, 2016.

[77] Since the launch of local television in 1960, the state has never required its citizens to pay a separately defined license fee for the broadcaster's operations. Instead, it resorted to adding 0.2 piasters (~0.007 US cents) to the electricity bills of each household—a small amount set in 1968 and never changed since then. In the 2010s, some Egyptian officials suggested increasing the license fee and designating it as a separate bill, but the government has always refused to discuss this reform, fearing the public backlash it may provoke. Some local media specialists think that Egyptians will completely stop watching the NMA channels if a new tax is imposed. See Nagy, "Advancing the Egyptian Public Broadcaster," 20–1; Allam, "From Flabby to Fit," 307.

entire media catalog on YouTube complies with the historical mission of ERTU that provided Egyptians with free televisual broadcasting, thus addressing them less as customers and more as citizens.[78] Such a strategy might evoke nostalgia not only for the old content per se but also for the mode of engagement that characterized the Egyptian state TV of the past.

This specific mode of engagement is also related to the appeal of collective viewership. For instance, the popular program *Ramadan Riddles* (*Fawāzīr Ramaḍān*), broadcast during the Muslim holy month between the 1960s and 2000, still reminds many Arabs of spending time with family after *ifṭār*, when the fast-breaking meal was accompanied by a tradition of television watching and riddles solving.[79] Understandably, this older practice of collective spectatorship may generate nostalgic moods at a time characterized by the fragmentation of television audiences caused by niche satellite TV channels and streaming services. Unable to re-create these viewership patterns, the Maspero Channels Network on YouTube nevertheless provides its followers with audiovisual texts associated with creating these patterns and fosters an experience of television flow by bundling these television texts together.

Streaming Video Services in Egypt and Their Cultural Impact

The analysis of Egyptian state television's turn to YouTube would be incomplete without appreciating its relation to the global and regional streaming services, such as Netflix and Shahid, that have been increasingly penetrating the Arab entertainment media markets since the 2010s. This relation demonstrates how old content is increasingly being used in an attempt to capitalize on nostalgia TV in the digital space. And yet the economic problems that the country is permanently experiencing may be listed as one of the major reasons why the SVOD model remains unfavorable for many.[80]

[78] Kraidy and Khalil, *Arab Television Industries*, 15–17.

[79] One of a few text posts in the community section made on the Maspero Drama YouTube channel is dedicated to the collective watching of *Fawāzīr* after *ifṭār*. See Hind Mezaina, "Fawazeer Ramadan," *Culturist*, August 6, 2013, https://www.theculturist.com/home/fawazeer-ramadan.html; "Fākirīn Zamān Fī-Ramaḍān," communiy post, Māsbīrū Drāmā YouTube Channel, n.d., https://www.youtube.com/channel/UCrIy5XTmRPwZPOlilXDWacA/community?lb=UgkxWVk41gKhDNwySz2qERb3Pg4uK5tSdKIB.

[80] Allam, "From Flabby to Fit," 299.

The advent of privately owned streaming services in Egypt took place in 2011 when the Saudi-owned media conglomerate MBC introduced its pan-Arab video platform, Shahid.[81] Since then, it has become one of the major players in the local video streaming market by offering both a free service financed by advertising revenues and a premium platform available to customers paying for subscriptions. The latter model, analogous to Netflix, was launched in 2014 under the name Shahid Plus, but then was significantly rebranded in 2019 as Shahid VIP in an attempt to rival the US giant entering the MENA region.[82] In turn, Netflix started penetrating the local market in January 2016 by offering access to its global library, but this strategy has not achieved great success.[83] The next significant step was taken in 2019 when Netflix produced its first Arabic originals, *Jinn* (2019, dir. Mir-Jean Bou Chaaya) and *Dollar* (2019, dir. Samer Al Barkawi), and by so doing forced its competitors in the region to follow suit and adjust to the innovative models of TV content production and distribution. It was in this context that Shahid started creating its own original series, such as *El Diva* (The Diva, 2019, dir. Randa Alam) and *Fī Kull Usbūʿ Yawm Jumʿa* (Every Week Has a Friday, 2020, dir. Mohamed Shaker), and achieved success in promoting them within the region.[84]

As the statistics for 2022–3 indicate, Shahid VIP's subscription base in the Middle East and North Africa is growing even faster than that of Netflix.[85] Nowadays, Shahid VIP is the "largest premium VOD service in the world outside China, the USA, and India," having a bigger total number of Arab users than Netflix or any alternative video platform.[86] However, in Egypt the number of Netflix subscribers seems to still exceed that of Shahid VIP. According to the statistics for 2021, Netflix and Shahid VIP had 700,000 and 150,000 subscribers respectively in Egypt.[87] While these are significant numbers, they are small relative to the overall population and suggest that the majority of Egyptians still cannot afford to pay for these services.

81 Allam and Chan-Olmsted, "Development of Video Streaming," 286–7.

82 Nagy, "Advancing the Egyptian Public Broadcaster," 45; Fadi G. Haddad and Alexander Dhoest, "Netflix Speaks Arabic, Arabs Speak Netflix: How SVOD Is Transforming Arabic Series Screenwriting," *Journal of Arab & Muslim Media Research* 14, no. 2 (2021): 261–80. https://doi.org/10.1386/jammr_00034_1.

83 Nagy, "Advancing the Egyptian Public Broadcaster," 48.

84 Haddad and Dhoest, "Netflix Speaks Arabic," 262.

85 *Arab News*, "Shahid VIP to Dominate MENA's Streaming Market for Next 5 Years: Report," March 24, 2022, https://arab.news/57vt7.

86 Allam and Chan-Olmsted, "Development of Video Streaming," 286; *Al Arabiya*, "'Shāhid VIP' Tuṣayṭir ʿAlā Sūq al-Sharq al-Awsaṭ Bi-Naḥw 3 Malāyīn Mushtarik," March 25, 2022.

87 Allam and Chan-Olmsted, "Development of Video Streaming," 287.

Catering to privileged sections of society, Netflix and Shahid VIP's Arabic originals are mostly created by international teams and are pan-Arab in nature, meaning that they try to avoid associations with distinct Arab nationalities on the levels of both production and storytelling. By so doing, the creators aim to attract subscribers from different parts of the Arab world.[88] Despite such an attitude, both Shahid VIP and Netflix originals tend to frequently include Egyptian actors, characters, and stories because of how "relatable and accepted" they are due to the long history of Egyptian film and TV's domination in the region and because Egypt remains the most populated Arab country. Therefore, appealing to local tastes in Egypt has a greater likelihood of expanding the subscriber base in the region as a whole.[89]

Netflix and Shahid share a lot in common in terms of the kind of television experience they promote in the Arab world. Their Arabic originals are usually positioned as a superior alternative to the traditional popular format of *musalsalāt* associated with both old Egyptian TV series and newer productions for satellite channels. The Arab filmmakers and writers working for the streaming services consider the older genre of *musalsalāt* a televisual form prone to "repetition, prolongation, and simplified plotlines, primarily due to the broad target audience of people watching the series during other activities like socializing and eating."[90] Netflix and Shahid VIP present their Arabic originals as a high-end alternative to this conventional format, as they encourage a more critical mode of viewership and provide subscribers with unconventional plotlines.[91] Therefore, the present activity of these leading streaming services is not so much focused on Egyptian nostalgia TV and instead seems to position itself against longing for older forms and experiences of Arab television.

Although still inaccessible for many, Netflix and Shahid have already significantly transformed the emergent digital television industry in Egypt and the broader Arab world.[92] Some smaller regional projects, such as Icflix, launched in 2013, could not stand the competition and were eventually closed.[93] Other MENA streaming services, although unable to compete with the US and

[88] Haddad and Dhoest, "Netflix Speaks Arabic," 276.
[89] Ibid., 270.
[90] Ibid., 273.
[91] Simultaneously, Shahid VIP provides access to numerous *musalsalāt* of a more traditional format to cater to the tastes of its diverse audience.
[92] Haddad and Dhoest, "Netflix Speaks Arabic," 263.
[93] Nagy, "Advancing the Egyptian Public Broadcaster," 49.

Saudi leaders in terms of scale and subscription rates, have nonetheless found their niche audiences and demonstrate gradual growth.[94]

Of particular note with regard to the rise of nostalgia TV in Egypt is Watch IT, a platform established in 2019 by a large Egyptian private media group, United Media Services (*al- Muttaḥida li' l-Khidmāt al-I'lāmīya*), that has informal ties to the Egyptian state and the General Intelligence.[95] Mostly targeting Egyptian citizens with locally specific content, it initially suffered from recurrent technical difficulties and therefore lost many subscribers after a free trial.[96] At that time, Watch IT also tried to capitalize on the demand for older television programs on YouTube and was advertised as "bringing back original Egyptian dramas, TV shows, and movies from the 1960s and beyond" and thus "educating Egypt's next generation."[97] To achieve that, in 2019, it bought the exclusive rights to have a noticeable part of ERTU's productions on its premium digital platform, which led to the materials' deletion from the Maspero YouTube channels. This action provoked an extremely negative reaction from viewers, who accused NMA of selling the common national legacy instead of keeping it freely available for all citizens.[98] Coupled with the streaming platform's technical failures, the focus on the old content soon proved unsuccessful and resulted in the service's rebranding in 2021. Since then, Watch IT focused on its own originals, Egyptian "family-friendly entertainment content," and "a commitment to preserve the authenticity and the Egyptian and Arab identity."[99] Although ERTU's media library remained one of the platform's assets, it was also returned to the Maspero YouTube channels, to the pleasure and delight of

[94] For example, the Dubai-based services Starz On (previously known as Starz Play) and OSN+ (OSN Streaming) demonstrate a gradual growth in audiences.

[95] The Egyptian private media group United Media Services (*al-Muttaḥida li' l-Khidmāt al-I'lāmīya*) is known for owning some of the most influential Egyptian private channels established between 2001 and 2017, such as ON TV (2016), Al Hayah TV (2008), dmc (2017), and CBC (2011). Comprehensive information about media ownership in Egypt is provided on the website of Media Ownership Monitor Egypt., "Egypt 2018," n.d., https://egypt.mom-rsf.org/en/.

[96] Allam and Chan-Olmsted, "Development of Video Streaming," 294–6.

[97] "How Watch iT! Engaged 400 Million People."

[98] This critique can be found in the reviews section of the Facebook group *Shūf Māsbīrū* (Watch Maspero) belonging to NMA. See "Shūf Māsbīrū."

[99] As a part of this strategy, in 2022–3 the Egyptian private conglomerate United Media Services started to invest in the local historical TV series production. Using its network of satellite TV channels and the streaming service Watch IT, United Media Services attempts to popularize its newly produced content dealing with Ancient Egyptian and Islamic history. Among the recent works of this kind, we should mention the multiseason documentary *Umm al-Dunyā* (2022–, dir. Maḥmūd Rashshād), about the Ancient Egyptian past, and the historical drama *Risālat al-Imām* (2023, dir. al-Layth Ḥajw), about the famous Islamic theologian Muḥammad al-Shāfi'i (767–820). *CMOs*, "WATCH IT Unveils New Brand Identity," 2021. https://www.cmosmagazine.com/en/watch-it-unveils-new-brand-identity/; "Musalsal Umm Al-Dunyā"; "Musalsal Risālat Al-Imām."

their devoted followers.[100] Thus, free access to the entire back catalog proved to be a significant factor that conditioned the popularity of the Maspero Channels Network on YouTube.

Conclusion

When studying the influence of Netflix in the Western context, mass media scholars frequently stress its role in effectively capitalizing on nostalgia as a mode of engagement with audiences, which has proven especially effective when supported by a rich back catalog of shows.[101] In a similar vein, within the media environments of Egypt and the broader Arab world, attempts have been recently made to capitalize on locally produced digitized television catalogs of the twentieth century—an endeavor in which television nostalgia also played a crucial role. This is the context within which the current practice of watching old Egyptian TV content on YouTube needs to be situated.

To cope with an acute financial deficit, the Egyptian state-controlled broadcaster (NMA) created the Maspero Channels Network in 2015 and started uploading old Egyptian TV shows to YouTube. Its popularity became possible not only because of the abundance of old content available on the platform but also due to the modes of spectatorship that YouTube offers as a streaming video service that effectively combines the features of free televisual flow typical of the Egyptian state broadcaster with the advantages of the video-on-demand model. With free and unlimited access to the back catalog of Egyptian drama and other types of old TV shows, it thus deepens the viewers' nostalgic experiences. The Egyptian streaming service Watch IT's unsuccessful attempt to provide subscription-based access to local old content demonstrated the significance of keeping ERTU's digital library free. Only then can it remain available for audiences unable or unwilling to pay for the national broadcaster's legacy. This conclusion once again reminds us that television nostalgia is a complex experience that depends on both offered content and modes of engagement provided by digital platforms.

[100] There are some threads on X (Twitter) discussing this matter. For instance, see Sāra ʿUthmān, "Fīh Khabar Laṭīf…, " X (Twitter), March 7, 2021, https://x.com/Sarah_Othmann/status/1368601343145951241.

[101] Pallister, *Netflix Nostalgia*.

The streaming television era ushered in by Netflix and other internet-based video platforms did not lead to a rupture with the legacies of Egyptian state-controlled TV but reshaped it as an object of nostalgia. However, we should not forget that this attraction to entertainment like older *musalsalāt* is not shared by everybody, especially when it comes to younger online viewers who constitute the majority of internet users in the region.[102] Having been brought up after the "golden age" of ERTU, many of them cannot relate nostalgically to Egyptian traditional television and, instead, notice its technological inferiority in comparison with contemporary productions.[103] Thus, for a considerable part of the younger audience, old Egyptian TV might function as the Other of the contemporary higher-quality television manifested in satellite channels' productions and even more sophisticated Arabic originals created by Netflix and Shahid VIP.[104] This reminds us of the complexity of Egyptian television audiences that would inevitably have different responses to nostalgic content popularized by the NMA and other regional players.

Nevertheless, the nostalgic moods provoked by ERTU's televisual legacy cannot be ignored or devalued. The YouTube activity of the Maspero Channels Network is an important sign that Egyptian nostalgia TV is commodifiable and demanded by certain segments of Egyptian and broader Arab audiences.[105] According to the comment sections of numerous uploaded shows, many middle-aged Arab viewers associate Egyptian serials with the better times of their childhood, higher cultural and moral standards, and greater communal

[102] Allam and Chan-Olmsted, "Development of Video Streaming," 296.

[103] Ṣubḥī, "Māsbīrū Zamān."

[104] Michael Newman and Elana Levine analyzed how the advent of the masculinized image of US TV in the 2000s led to the further denigration of traditional TV genres, such as soap opera and sitcom, and positioned them as the Other of the new high-quality television. It seems that a similar tendency can be noticed in the contemporary Egyptian television context. See Michael Z. Newman and Elena Levine, *Legitimating Television: Media Convergence and Cultural Status* (London: RoutledgeLondon, 2012).

[105] It would be unfair to claim that the Egyptian televisual legacy remains important only within the country's borders, as many of the sentimental comments about it on YouTube and other platforms are posted by viewers from different parts of the Arab world. Concurrently, old Egyptian TV content is being spread on private digital platforms not directly related to Egypt. For instance, one of the most popular Lebanese YouTube channels, Shoof Drama, which possesses an impressive library of Arabic serials and has more than eight million followers, seems to have recently bought the rights to add ERTU's TV shows to its catalog. Despite being simultaneously available on Maspero Drama and Shoof Drama, famous Egyptian serials continue to gather hundreds of thousands of views on both YouTube channels, with the Lebanese web page having more non-Egyptian Arab viewers. The pan-Arab YouTube channel Shoof Drama is owned by a Lebanese media group that also created a separate free streaming service, Shoof Max. See Shūf Max, "Dhikrayāt Lā Tamūt," n.d., https://shoofmax.com/; YouTube, "Shoof Drama."

cohesion. These older television texts serve as means of temporarily exiling audiences from the imperfect present associated with political and ideological crises, as well as deep economic instability in Egypt and other parts of the contemporary Arab world.

Bibliography

Abū al-ʿInayn, Zahrāʾ. "21 Musalsalan Min Afḍal Musalsalāt Miṣrīya Qadīma Lā Tazāl Taḥẓā Bi-Shaʿbīya Ḥattā al-Yawm." *Al-Mashhad*, August 1, 2023. https://almashhad.com/article/459217420284850-entertainment/956848526799019-18-/مسلسلا-من-أفضل-مسلسلات-مصرية-قديمة-لا-تزال-تحظى-بشعبية-حتى-اليوم

Abu-Lughod, Lila. *Dramas of Nationhood: The Politics of Television in Egypt*. Chicago: University of Chicago Press, 2005.

ʿĀdil, Yāsmīn. "Liʾ ʿUshshāq Drāmā al-Tisʿīnīyāt: Musalsalāt Ramaḍānīya Qadīma Lā Tunsā."Al-Jazīra Online, March 17, 2023. https://www.aljazeera.net/arts/2023/3/17/-مسلسلات-التسعينات.لعشاق-دراما.

Alaa El-DIn, Mohamed. "Egypt's Internet Users Increase by 1.4 Million in May." *Daily News Egypt*, August 23, 2023. https://www.dailynewsegypt.com/2023/08/23/egypts-internet-users-increase-by-1-4-million-in-may/.

Al-Ahrām. "Al-Drāmā al-Dīnīya Matā Taʿūd Ilā Makānatiha?" July 25, 2011.

Al Arabiya. "'Shāhid VIP' Tuṣayṭir ʿAlā Sūq al-Sharq al-Awsaṭ Bi-Naḥw 3 Malāyīn Mushtarik." March 25, 2022. https://www.alarabiya.net/last-page/2022/03/25/-شاهد-VIP–%20،%20ل20%وفقًا20%وذلك-مليون-من-بأكثر-الأوسط-الشرق-في-لسوق-تسيطر-على
Datax مشترك20%=text:~#وكانت20%قاعدة20%مشتركي20%22%شاهد VIP,2022

Allam, Rasha. "Egyptian Media Ownership and Pluralism: Overview, Performance, and Challenges." In *Media Ownership in Africa in the Digital Age: Challenges, Continuity and Change*, edited by Winston Mano and Loubna El Mkaouar, 223–34. London: Routledge, 2023.

Allam, Rasha. *Entertainment Media Industry in Egypt*. London: Routledge, 2019.

Allam, Rasha. "From Flabby to Fit: Restructuring the Public Broadcasting System in Egypt." *Global Media and Communication* 17, no. 3 (2021): 297–320. https://doi.org/10.1177/17427665211023978.

Allam, Rasha, and Sylvia Chan-Olmsted. "The Development of Video Streaming Industry in Egypt: Examining Its Market Environment and Business Model." *Journal of Media Business Studies* 18, no. 4 (2021): 285–303.

Allam, Rasha, and Hesham Dinana. "The Future of TV and Online Video Platforms: A Study on Predictors of Use and Interaction with Content in the Egyptian Evolving Telecomm, Media & Entertainment Industries." *SAGE Open* 11, no. 3 (2021). https://doi.org/10.1177/21582440211040804.

Arab News. "Shahid VIP to Dominate MENA's Streaming Market for Next 5 Years: Report." March 24, 2022. https://arab.news/57vt7.

Armbrust, Walter. *Mass Culture and Modernism in Egypt*. Cambridge: Cambridge University Press, 1996.

Armbruster, Stefanie. *Watching Nostalgia: An Analysis of Nostalgic Television Fiction and Its Reception*. Cultural Studies 48. Bielefeld, Germany: Transcript Verlag, 2016.

Bakīr, Āmāl. "Kayfa Kana Al-Amīr al-Majhūl Malikan Li' al-Aʿmāl al-Fanīya?!" *Al-Ahrām*, February 21, 1997.

Bevan, Alex. *The Aesthetics of Nostalgia TV: Production Design and the Boomer Era*. New York: Bloomsbury Academic, 2019.

Boym, Svetlana. *The Future of Nostalgia*. New York: Basic Books, 2001.

Brightcove. "How Watch iT! Engaged 400 Million People." n.d. https://www.brightcove.com/en/customers/watchit/.

Carney, Josh. "Resur(e)Recting a Spectacular Hero: Diriliş Ertuğrul, Necropolitics, and Popular Culture in Turkey." *Review of Middle East Studies* 52, no. 1 (2018): 93–114. https://doi.org/10.1017/rms.2018.6.

Carney, Joshua. "A Dizi-Ying Past: 'Magnificent Century' ('Muhtesem Yuzyil') and the Motivated Uses of History in Contemporary Turkey." PhD diss., Indiana University, 2015.

Ciasullo, Anna M. "Afterword: In Love with the Past." In *Netflix Nostalgia: Streaming the Past on Demand*, edited by Kathryn Pallister, 235–42. Lanham, MD: Lexington Books, 2019.

CMOs. "WATCH IT Unveils New Brand Identity." 2021. https://www.cmosmagazine.com/en/watch-it-unveils-new-brand-identity/.

Cox, Christopher M. "Programming—Flow in the Convergence of Digital Media Platforms and Television." *Critical Studies in Television* 13, no. 4 (2018): 438–54. https://doi.org/10.1177/1749602018796681.

Ḍamarānī, Muṣṭafā al-. "Al-Amīr al-Majhūl Wa al-Ḥaqīqa al-Ghāʾiba." *Al-Ahrām*, February 26, 1997.

Duman, Talha İ. "The Evolution of a Social Movement: The Muslim Brotherhood's Path between Political Ambition and the Rejection of Armed Struggle." *Religions* 15, no. 11 (2024): 1352. https://doi.org/10.3390/rel15111352.

Facebook. "Shūf Māsbīrū." Channel page, n.d. https://www.facebook.com/ShoofMaspero/?locale=ar_AR.

"Fākirīn Zamān Fī-Ramaḍān." Communiy post. Māsbīrū Drāmā YouTube Channel, n.d. https://www.youtube.com/channel/UCrIy5XTmRPwZPOlilXDWacA/community?lb=UgkxWVk41gKhDNwySz2qERb3Pg4uK5tSdKIB.

Grainge, Paul. *Memory and Popular Film*. Manchester: Manchester University Press, 2003.

Guaaybess, Tourya, ed. *National Broadcasting and State Policy in Arab Countries*. Houndmills, UK: Palgrave Macmillan, 2013.

Haddad, Fadi G., and Alexander Dhoest. "Netflix Speaks Arabic, Arabs Speak Netflix: How SVOD Is Transforming Arabic Series Screenwriting." *Journal of Arab & Muslim Media Research* 14, no. 2 (2021): 261–80. https://doi.org/10.1386/jammr_00034_1.

Hutcheon, Linda, and Mario Valdés. "Irony, Nostalgia, and the Postmodern: A Dialogue." *Nuevas Poligrafías* 3 (2000): 29–54. https://doi.org/10.22201/ffyl.poligrafias.2000.3.1610.

Huyssen, Andreas. *Present Pasts: Urban Palimpsests and the Politics of Memory*. Cultural Memory in the Present. Stanford, CA: Stanford University Press, 2003.

Ibrāhīm, Khālid. "Al-Ḍawʾ al-Shārid Wa-Ūbrā ʿĀyida Wa al-Marāghī Aḥdath al-Munḍammīn Li' 'Māsbīrū Drāmā.'" *Al-Yawm al-Sābiʿ*, March 10, 2016. https://www.youm7.com/story/2016/3/10/2622235/الضوء-الشارد-وأوبرا-عايدة-والمراغى-أحدث-المنضمين-لماسبيرو-دراما.

Kraidy, Marwan M., and Joe F. Khalil. *Arab Television Industries*. International Screen Industries. Houndmills, UKUK: Palgrave Macmillan, 2009.

Lapidus, Ira M. *History of Islamic Societies*. Cambridge: Cambridge University Press, 2014.

Lopez, Lori Kido. "ISAtv: YouTube and the Branding of Asian America." In *From Networks to Netflix: A Guide to Changing Channels*, edited by Derek Johnson, 255–63. New York: Routledge, 2018.

Lowenthal, David. *The Past Is a Foreign Country*. Cambridge: Cambridge University Press, 1985.

"Maspero Television Building." *Wikipedia*, n.d. https://en.wikipedia.org/wiki/Maspero_television_building.

Media Ownership Monitor. "Egypt 2018." n.d. https://egypt.mom-rsf.org/en/.

Mezaina, Hind. "Fawazeer Ramadan." *Culturist*, August 6, 2013. https://www.theculturist.com/home/fawazeer-ramadan.html.

Miṣbāḥ, Amjad. "'Layālī al-Ḥilmiīya': Aḥdāth Wa-Shakhṣīyāt Maḥfūra Fī al-ʿAql al-ʿArabī." *ʿArab Journal*, April 12, 2022. https://arab-j.net/4594.

"Musalsal Abū Ḥanīfa Al-Nuʿmān." 1997. https://www.youtube.com/watch?v=Qrik7n1JKFg&list=PL9KkecclNUBTwDB9Y8ie6s2DUBr6O1Nip.

"Musalsal Al-Amīr al-Majhūl Hārūn al-Rashīd." 1997. https://www.youtube.com/watch?v=GkALH2JHF9A&list=PLhN_FqGqSxOe5DzfmzD5xdN0PbdO2xkuX.

"Musalsal Al-Ḍawʾ al-Shārid." 1998. https://www.youtube.com/watch?v=2m1HxRRYxFo&list=PL9KkecclNUBTWq1c0x34DbqbaDF7Xb_&index=32.

"Musalsal Al-Fursān." 1994. https://www.youtube.com/watch?v=AcHmAS13q5w&list=PL9KkecclNUBR1zmFh_IyYCr7qTupdEkJd&index=3.

"Musalsal ʿAṣr Al-Aʾimma." 1998. https://www.youtube.com/watch?v=Lo5wJYZfnaw&list=PL9KkecclNUBSodpfOkysCzYue-1HGoMaJ&index=3.

"Musalsal Dhiʾāb Al-Jabal." 1992. https://www.youtube.com/watch?v=ykbmXxzSHCg&list=PL9KkecclNUBRR_NDfzTa6LN0zVnkpt2st&index=2.

"Musalsal Imām Al-Duʿā." 2003. https://www.youtube.com/watch?v=jnp917I0EVs&list=PL9KkecclNUBSQTgZaNQFzMNFjEWyc2rZv&index=3.
"Musalsal Lan Aʿīsh Fī Jilbāb Abī." 1995. https://www.youtube.com/watch?v=Vmj6oKG3xMg&list=PL9KkecclNUBR1nC2sX8O3s2nJgjkk-Eiz&index=2.
"Musalsal Layālī Al-Ḥilmīya." 1987. https://www.youtube.com/watch?v=8FWAn3M381Q.
"Musalsal Risālat Al-Imām." 2023. https://elcinema.com/work/2077524/.
"Musalsal ʿUmar Ibn ʿAbd Al-ʿAzīz." 1994. https://www.youtube.com/watch?v=Cd8_f7TskzI&list=PLtNk7av9R5XKiwK28Ss2QdQxWTlXp-waW.
"Musalsal Umm Al-Dunyā." 2022–4. https://www.imdb.com/title/tt26422326/?ref_=tturv_ov_i.
"Musalsal Yawmīyāt Wanīs." 1994. https://www.youtube.com/watch?v=M16801ZhgUA&list=PLFNMg0MdiESZdVpzSsTbL7DlhMwgdWo-i.
Nagy, Maha Sherif Hussein. "Advancing the Egyptian Public Broadcaster to Compete in the Digital Era." Master's thesis, American University in Cairo, 2019.
Newman, Michael Z., and Elana Levine. *Legitimating Television: Media Convergence and Cultural Status*. London: Routledge, 2012.
Nieuwkerk, Karin van. *Performing Piety: Singers and Actors in Egypt's Islamic Revival*. Austin: University of Texas Press, 2013.
Pallister, Kathryn, ed. *Netflix Nostalgia: Streaming the Past on Demand*. Lanham, MD: Lexington Books, 2019.
Ramaḍān, ʿAbd al-ʿAẓīm. "Al-Musalsalāt al-Tārīkhīya Bayn al-Jadd Wa al-Hazal." *Al-Ahrām*, March 1, 1997.
Rock-Singer, Aaron. *Practicing Islam in Egypt: Print Media and Islamic Revival*. Cambridge: Cambridge University Press, 2020.
Salah, Yasmeen. "Maʿa Al-Iḥtifāl Bi-60 ʿāman Min al-Drāmā: Daʿūnā Nastadhkir Ajmal Musalsalāt al-Tisʿīnāt Miṣrī." *Harper's Bazaar Arabia*, August 23, 2023. https://ar.harpersbazaararabia.com/الفن/مسلسلات-التسعينات-مصري
Ṣāliḥ, Muḥammad. "Al-Drāmā Wa-Iḥtirām al-Tārīkh?" *Al-Ahrām*, February 24, 1997.
Ṣāliḥ, Muḥammad. "Ḥikāya." *Al-Ahrām*, February 2, 1997.
Ṣāliḥ, Muḥammad. "Jāʾizat Aḥsan Mumaththil Kamān Wa-Kamān: Nūr al-Sharīf Mumaththilan Maʿa Sabqal-Iṣdār Wa-l-Taraṣṣud." *Al-Ahrām*, July 25, 1996.
Shafik, Viola. *Popular Egyptian Cinema: Gender, Class, and Nation*. Cairo: American University in Cairo Press, 2007.
Shūf Max. "Dhikrayāt Lā Tamūt." n.d. https://shoofmax.com/.
Skovgaard-Petersen, Jakob. "The Khuṭba Scene in Arab Religious Films and TV Dramas." In *Muslim Preaching in the Middle East and Beyond: Historical and Contemporary Case Studies*, edited by Simon Stjernholm and Elisabeth Özdalga, 30–47. Edinburgh: Edinburgh University Press, 2020. https://doi.org/10.3366/edinburgh/9781474467476.003.0003.

Spigel, Lynn. "From the Dark Ages to the Golden Age: Women's Memories and Television Reruns." *Screen* 36, no. 1 (1995): 16–33. https://doi.org/10.1093/screen/36.1.16.

State Information Service. "Media Production City." May 19, 2016. https://www.sis.gov.eg/Story/67988/Media-Production-City?lang=en-us.

Ṣubḥī, Muḥammad. "Māsbīrū Zamān: Al-Iflās al-Fannī Bi-Ṣīghat al-Ḥanīn." *Ultraṣawt*, November 25, 2016. https://www.ultrasawt.com-/-ماسبيرو-زمان-الإفلاس-الفني-بصيغة-الحنين/محمد.صبحي/راصد/عشوائيات

Ṭālibī, Īmān al-. "Ayna Ikhtafā Zaman Layālī Al-Ḥilmīya?" Al-Jazīra Online, November 13, 2017. https://www.aljazeera.net/blogs/2017/11/13/أين-اختفى-زمن-ليالي-الحلمية

Taurino, Giulia. "Crossing Eras: Exploring Nostalgic Reconfigurations in Media Franchises." In *Netflix Nostalgia: Streaming the Past on Demand*, edited by Kathryn Pallister, 9–23. Lanham, MD: Lexington Books, 2019.

ʿUthmān, Sāra. "Fīh Khabar Laṭīf" X (Twitter), March 7, 2021. https://x.com/Sarah_Othmann/status/1368601343145951241.

Williams, Raymond. *Television: Technology and Cultural Form*. New York: Schocken Books, 1975.

YouTube. "Akhbār Māsbīrū." Channel page, n.d. https://www.youtube.com/©MasperoNews.

YouTube. "Al-Idhāʿa Al-Miṣrīya." Channel page, n.d. https://www.youtube.com/©EgyRadio.

YouTube. "Dhākirat Māsbīrū." Channel page, n.d. https://www.youtube.com/©MasperoMemory.

YouTube. "Māsbīrū Aṭfāl." Channel page, n.d. https://www.youtube.com/©MasperoKids.

YouTube. "Māsbīrū Drāmā." Channel page, n.d. https://www.youtube.com/©MasperoDrama.

YouTube. "Māsbīrū Zamān." Channel page, n.d. https://www.youtube.com/©MasperoZaman.

YouTube. "Ṣawt Al-Shaʿb." Channel page, n.d. https://www.youtube.com/©SawtElSha3b.

YouTube. "Shoof Drama." Channel page, n.d. https://www.youtube.com/©shoofdramaone.

YouTube. "Tasjīlāt Māsbīrū Al-Dīnīya." Channel page, n.d. https://www.youtube.com/©MasperoEldenia.

Youtubers.me. "Māsbīrū Aṭfāl Youtube Channel Statistics." n.d. https://np.youtubers.me/c69b98a5-8548-4011-89bd-17eb83965ddf/youtuber-stats/en.

Youtubers.me. "Māsbīrū Drāmā Youtube Channel Statistics." n.d. https://us.youtubers.me/d4e01655-9806-43eb-842c-6897aec140bc/youtuber-stats.

8

Streaming the Nation: Salvaging the National Filipino Audience

Daniel Rudin

Mikhail Red's psychological horror film *Deleter* (2022) uses a mixed-genre approach, visually compelling art direction, and R-rated depictions of rape and suicide to highlight the psychological trauma experienced by online content moderators in the Philippines. The film unexpectedly swept the Forty-Eighth Metro Manila Film Festival (MMFF), a government-run event promoting the local film industry. While MMFF entries are seldom seen by non-Filipinos, Red has blazed a more cosmopolitan path, guaranteeing sales through premieres in the United States and the United Arab Emirates, and via streaming platforms like Amazon Prime Video.[1] In effect, he has, with transnational capital and audiences, undercut what MMFF was designed to protect—the economic autonomy of Filipino movies. This essay takes *Deleter* as a case study in how subscription video on demand (SVOD) exposes the inner contradictions of national cinema in the Philippines. It begins by situating the MMFF and "Philippine cinema" within postcolonialism's turn from economic protectionism to cultural nationalism. It then assesses the contradictions inherent in national cinema, which are heightened by the commodity conditions under which streaming operates. Finally, it compares *Deleter*'s aesthetic approach to Red's previous work, which,

1 *Deleter* began streaming on Amazon Prime Video on March 16. Said Red, "If we tap the global market, and we have that international buffer to sustain yung profit ng isang film [the profit of a film], we can do more daring, challenging stories that don't have to cater to just the local taste. It should work with the local audiences, but it should have legs internationally, by doing that we can support more ambitious budgets and projects." See "Rappler Talk Entertainment: Mikhail Red on 'Neomanila" and the Future of PH Cinema," *Rappler*, March 7, 2019, https://www.youtube.com/watch?v=B6W2k8l0Guw; One Carlo Diaz, "Filipino Film *Deleter* to Hit UAE Cinemas after Huge Local Success," *National News*, January 8, 2023, https://www.thenationalnews.com/arts-culture/film-tv/2023/01/08/filipino-film-deleter-to-show-in-uae-cinemas-after-huge-local-success/; Leah C. Salterio, "*Deleter* Not Originally Meant for the MMFF," *Philstar Global*, January 14, 2023. https://www.philstar.com/entertainment/2023/01/14/2237416/deleter-not-originally-meant-mmff.

contrary to his claims, is less concerned with the mass Filipino audience than with isolated individuals living under an increasingly technologically mediated globalization.

Nationalism and the Metro Manila Film Festival

The MMFF is one of the oldest festivals in Southeast Asia. In the late 1950s, the movie industry entered into a crisis attributed to the unionization of its theaters, the Paramount Decree, and competition from foreign films.[2] In response, the industry lobbied for quotas on film imports.[3] In 1966 the government established the Manila Film Festival, which prohibited the screening of non-Filipino films during the festival.[4] In 1975, Imelda Marcos expanded the festival to the broader Metro Manila region. Initially run by a private foundation, the MMFF was eventually brought entirely within the purview of the government bureaucracy.[5]

Restricting foreign films within the greater metro region for the MMFF's two-week duration ensures local studios full market capture—making it *the* major Philippine festival.[6] Unlike the Cinemalaya and Q-Cinema festivals that attract a middle-class cineaste viewer, the MMFF cultivates a mass audience—which even votes on Best Picture.[7] It begins at an auspicious time to get moviegoers into theaters—Christmas day—and features public pageantry that "weaves together national history, mainstream movies, and notions of the cinema as a spectacle

[2] The Paramount Decree was American antitrust legislation that prohibited vertically integrated production and distribution. US jurisprudence is often followed in the Philippines, as occurred in this case. See Bienvenido Lumbera, "Philippine Problems in Film History," in *Readings in Philippine Cinema*, ed. Rafael Ma. Guerrero (Manila: Experimental Center of the Philippines, 1983), 74–5; also Joel David, *Wages of Cinema: Film in Philippine Perspective* (Quezon City: University of Philippines Press, 1998), 70.

[3] Mauro Tumbocon Jr. and Lena Pareja, "Philippine Motion Picture Producers Association," in *CCP Encyclopedia of Philippine Art, Volume VII: Philippine Film*, ed. Nicanor Tiongson (Manila: Cultural Center of the Philippines, 1994), 292.

[4] Bienvenido Lumbera, "Philippine Film: 1961–1992," in *CCP Encyclopedia of Philippine Art, Volume VII: Philippine Film*, ed. Nicanor Tiongson (Manila: Cultural Center of the Philippines, 1994), 46.

[5] Initially, the festival was run by the Movie Workers Welfare Foundation (MOWELFUND), an organization founded by the actor and right-populist president Joseph Estrada. The festival profits went toward supporting workers in the film industry. Estrada proposed to Imelda Marcos the idea of expanding the Manila Film Festival, which lead to the MMFF and MOWELFUND's role managing it.

[6] Supreme Court E-Library, "Presidential Decree No. 824," November 7, 1975, https://elibrary.judiciary.gov.ph/thebookshelf/showdocs/26/17327.

[7] Jasmine Trice, *City of Screens: Imagining Audiences in Manila's Alternative Film Culture* (Durham, NC: Duke University Press, 2021), 140.

for the people."[8] Jurors of the competition have included working-class members of the general viewing public.[9]

Nevertheless, for Jasmine Trice the festival's "utopic, nationalist vision of local film culture" seldom lives up to its promise.[10] The mass audience that MMFF seeks to cultivate consists largely of workers who, subsisting on ten dollars a day, justify moviegoing only once a year during the holidays with friends and family.[11] Lacking an audience with disposable income, some studios are able to produce only one commercial success out of ten.[12] It therefore falls on the MMFF—which, one must recall, shields local films from international competition—for studios to break even. And since studios keep box office numbers secret, MMFF receipts are the surest indicator of success on the national level.

Despite the festival's protectionist agenda, a wide discrepancy remains between the commercial success of Hollywood films and those that are locally produced. Case in point is *Deleter*, which after an exceptional performance grossed four million dollars at the MMFF, or half of the overall box office profits.[13] By comparison, Marvel's *Doctor Strange in the Multiverse of Madness*, at twelve million dollars, earned more than all MMFF films combined.[14] It is in this context that Red uses SVOD as a "buffer" to leverage globalization to his advantage. In doing so, he mirrors the strategy of major studios in the Philippines such as Star Cinema and Viva Films.[15] After decades of declining viewership, these studios often send their films directly to proprietary streaming apps to engage the diaspora of nearly two million Filipinos working abroad. In keeping with this trend toward internationalization, last year's MMFF held an additional run in Los Angeles—a city with a Filipino American population of half a million.[16]

Given studios' dependence on the global market logic of SVOD, the MMFF's goal of developing a national audience through government protectionism

8 Ibid., 139–40.

9 Agimat, "Something New in 2010 MMFF," http://www.agimat.net/film/n101122b.php.

10 Trice, *City of Screens*, 140.

11 Trice, *City of Screens*, 64.

12 Mikhail Red in discussion with the author, October 4, 2023.

13 Granted, these numbers were likely reduced by the country's severe Covid restrictions. See Leah C. Salterio, "Deleter Not Originally Meant for the MMFF," *Philstar Global*, January 14, 2023, https://www.philstar.com/entertainment/2023/01/14/2237416/deleter-not-originally-meant-mmff.

14 Statista, "Highest Grossing International Movies in the Philippines in 2022, by Box Office Revenue," https://www.statista.com/statistics/655233/philippines-top-10-movies-by-revenue/.

15 Both studios were established during a peak in the production of Tagalog films in the 1980s and early 1990s.

16 Pew Research, "Top 10 U.S. Metropolitan Areas by Filipino Population, 2019," https://www.pewresearch.org/social-trends/fact-sheet/asian-americans-filipinos-in-the-u-s/#top-10-u-s-metropolitan-areas-by-filipino-population-2019.

seems no longer viable. Nevertheless, as Patrick Campos notes, the task of creating a national cinema has persisted into the present era, only now as a cultural politics "linked with the logic of globalization [which makes] the idea of the 'national'... easily deployable in service of marketing and branding."[17] SVOD exploits nationalism's marketability. But at the same time it preserves contradictions inherent to national cinema by, as film historian Andrew Higson puts it, "repress[ing] internal differences [and] tensions."[18]

National cinema is a by-product of nationalism in the global South, which proliferated in the postwar period, usually under developmental regimes with strong-man leadership. The Philippine postwar, postcolonial period was dominated by "elite democracy"—that is, a patronage system of landed elites who received favorable exchange rates and state subsidies to grow native industries. Under the second Marcos presidency, particularly after the 1972 declaration of martial law, the economy shifted to what has been called "crony capitalism." This new system attempted to effect Fordist welfare statism through authoritarian politics. Like many other industries taken under the state's wings, cinema was given preferential support. It particularly aligned with Imelda Marcos's cultural patronage—hence her tie-up with the MMFF—leading to what Tolentino calls the "apparatus of cinema."[19]

With the fall of Marcos, this apparatus was largely disassembled. Nevertheless, the premise of the nation-state and its nationalist legitimating ideology was retained by the incoming neoliberal regime. But this new nationalism was more spectral and less anchored in mass politics. Eventually, as the nation-building aspirations of postcolonialism receded in the rearview mirror, the contradictions of national cinema were subsumed in the machinery of globalization.[20] Today, national cinema reasserts itself as an ideology or reified discourse.[21] The origins of this reification lie in national cinema's constitution during two periods: economic protectionism during the 1950s, and the cultural turn of the 1960s to 1970s.

17 Patrick Campos, *The End of National Cinema: Filipino Film at the Turn of the Century* (Quezon City: University of the Philippines Press, 2016), 11.

18 Andrew Higson, "The Concept of National Cinema," *Screen* 30, no. 4 (Autumn 1989): 62.

19 Roland Tolentino, *Contestable Nation State: Cinema, Cultural Politics, and Transnationalism in the Marcos-Brocka Philippines* (Quezon City: UP Diliman Press, 2014), 56–60.

20 For an elaboration of this theme, see Lisandro Claudio's compelling use of cultural memory studies to assess post-Marcos nationalism in *Taming People's Power: The EDSA Revolutions and Their Contradictions* (Quezon City: Ateneo University Press, 2013).

21 Campos, *End of National Cinema*, x. For Campos, film studies must account for the continued presence of Filipino nationalism "in making sense of the now in cinema."

National Cinema and the "Cultural Turn"

Tagalog cinema has from the outset had to compete with Hollywood. Early Tagalog films eked out an existence on the margins of the film industry by playing in second-run and provincial theaters.[22] While upper-class patrons frequented the first-run foreign theaters, the lower classes—many of them recent immigrants to an industrializing Manila—preferred Tagalog films. These viewers were called the "*bakya* crowd" (a deprecating term referring to the wooden slippers worn by the poor living in backward barrios), whose standards were "unsophisticated, undiscriminating, uncouth even."[23]

The nationalist movement arising in the 1950s postcolonial period blamed industrial underdevelopment on continued American dominance. The movement's spokesman, Claro Recto, called for the "strengthening of Filipino capitalism."[24] Recto's "economic nationalism" was the ideological precondition for national cinema, which was considered part of the industrial capacity needed for national self-sufficiency within global Fordist capitalism.[25] On the policy level, state intervention sought to bring about national production during more than a decade of import substitution industrialization.[26] In effect, this meant using state subsidies and tariffs to replace American imports with Filipino-made commodities.

After the first major flowering of Tagalog films in the 1940s to 1950s, the local film market was replaced in the 1960s by independent actor-producers making Hollywood knockoffs. A new cinema emerged in the 1970s that was oriented toward building a national audience, pivoting from considerations of national

[22] Lumbera, "Philippine Problems in Film History," 69.

[23] Pete Lacaba, "Notes on Bakya: Being an Apologia of Sorts for Filipino Masscult," in *Readings in Philippine Cinema*, ed. Rafael Ma. Guerrero (Manila: Experimental Center of the Philippines, 1983), 117; Bienvenido Lumbera, "Approaches to Philippine Film," in *The Urian Anthology: 1970–1979*, ed. Nicanor G. Tiongson (Manila: M. L. Morato, 1983), 100.

[24] Renato Constantino, *The Philippines: The Continuing Past* (Quezon City: Foundation for Nationalist Studies, 1978), 292.

[25] While there is not space to discuss the meaning of Fordism in this essay, it may simply be taken as the period from 1930 to 1960 when the American industrial model became economically hegemonic and when the New Deal coalition with its social welfare state became the center of the global political order. See David Harvey, *The Condition of Postmodernity: An Inquiry into the Origins of Cultural Change* (Malden, MA: Blackwell, 1990).

[26] Paul Hutchcroft, "Situating the Philippine Political Economy: Import-Substitution Industrialization in Comparative Perspective," in *Kasarinlan* 4, no. 4 (1989): 39–56; Frank H. Golay, *The Philippines: Public Policy and National Economic Development* (Ithaca, NY: Cornell University Press, 1961), 412, quoted in Paul Hutchcroft, "Predatory Oligarchy, Patrimonial State: The Politics of Private Domestic Commercial Banking in the Philippines," PhD diss., Yale University Press, 1993, 186.

economy toward national culture. In what has been called the "cultural turn," these young nationalists used the theater and the cinema to construct a "critical definition of the role of the Philippines within the neocolonial framework."[27]

For Recto's protégé, Renato Constantino, American cinema reinforced counteridentification with Filipino culture. Hollywood's neo-imperialist values were also an "ideological battering ram" on a national consciousness repressed since the American defeat of the Filipino Revolution at the turn of the twentieth century.[28] Whereas Recto's economic nationalism did not fundamentally push against the American-dominated global order, Constantino opposed all things American—including Hollywood—and advocated commandeering the media (and cinema) to spread a new order of nationalist values. Constantino's program was echoed by the Filipino Maoists, who advocated a "second propaganda movement" modeled after nineteenth-century nationalism. Thereafter, national cinema was defined in terms both pedagogical and anti-imperialist.

For Higson, the concept of national cinema is always prescriptive, "citing what ought to be... rather than describing the actual cinematic experience of popular audiences." As such, histories of national cinema have the function of "asserting national autonomy [against] Hollywood's international domination."[29] Nationalist critic Bienvenido Lumbera argued the local consumption of Western cinema was a form of escapism that led to political quiescence.[30] Along similar lines, Filipino filmmaker Lino Brocka claimed that the cultural products of Hollywood blocked the Filipino imagination.[31] In 1974 he issued a manifesto calling for the use of cinema to develop a national audience "by gathering experience that is not alien to the majority of Filipinos at a particular time; by compressing and systematizing this experience for them; and by giving back this now crystallized experience to them in films they would enjoy and be moved by and take as their own."[32]

27 Constantino, *Philippines*, 272.

28 Renato Constantino, *Insight and Foresight* (Quezon City: Foundation for Nationalist Studies, 1977), 131–2; Renato Constantino, *Identity and Consciousness: The Philippine Experience* (Quezon City: Renato Constantino, 1974), 49. Constantino states that cinema was propaganda that "seep[ed] down to the barrios through media which should [have otherwise] been vehicles for the propagation of a Filipino identity."

29 Higson, "Concept of National Cinema," 54.

30 Lumbera, "Approaches to Philippine Film," 99; Higson, "Concept of National Cinema," 56.

31 Marlina Gonzales-Tamrong, "LA-Based Pinoy Newspaper Features Brocka," *Manila Chronicle*, May 6, 1987.

32 Lino Brocka, "Philippine Movies: Some Problems and Prospects," in *Readings in Philippine Cinema*, ed. Rafael Ma. Guerrero (Manila: Experimental Center of the Philippines, 1983), 260–1.

It was in the service of liberating the nation that national cinema turned from despising the *bakya*—the masses—as a symptom of national poverty toward embracing its transformative potential. Only the *bakya* could overcome Hollywood's economic domination and the cultural imperialism that it engendered. Like Brocka, Lumbera believed a film should "reflect… the culture of Filipinos."[33] Lumbera and the critics' group to which he belonged resolved the contradiction between cinema's global political economy and the ideal of a national culture by separating "technique" from vernacular "sensibility… bear[ing] the stamp of the culture that molded the artist." The critics saw their role as setting "guideposts for [other] critics and directors," ultimately concluding "that it was valid to demand that a Filipino film be Filipino in sensibility."[34]

Film historian Roland Tolentino calls Brocka an "unconscious Maoist" because his program echoes Mao Zedong's edict that culture elevate the public's aesthetic standards while popularizing revolutionary themes.[35] Mao's belief in liberation of the nation through struggle at the level of cultural revolution played a significant role in the "cultural turn." Following Nikita Khrushchev's 1956 denunciations of Stalin, Mao accused the Soviet Union of economism, or of having overemphasized the struggle at the point of (factory) production. In part due to this debate, both during and after the Chinese Cultural Revolution there was a turn away from labor organizing on the factory floor toward a "new type" of democratic revolution that located the revolutionary subject in "the masses." National cinema in turn shifted its focus to fomenting a revolution that would cohere a diverse public around a common culture, creating the preconditions for national economic autonomy. Using art and literature, nationalists envisioned a collective suffering and remaking of the nation by its own subaltern—the *bakya*.

Rather than advocate for an economic or industrial transformation in Recto's sense, the new cultural nationalism expressed a "doubled subject": a free Filipino collectivity on the one hand, and an object of historical and economic circumstances on the other.[36] Whereas a postcolonial nation's people were oppressed by America, they were through the act of resistance seen as constituting

[33] Lumbera, "Nunal sa Tubig Revisited," in *The Urian Anthology: 1970–1979*, ed. Nicanor G. Tiongson (Manila: M. L. Morato, 1983), 242–3.

[34] Lumbera, "Approaches to Philippine Film," 99; Lumbera, "Nunal sa Tubig Revisited," 242–3.

[35] Roland Tolentino, *Contestable Nation Space: Cinema, Cultural Politics, and Transnationalism in the Marcos-Brocka Philippines* (Quezon City: UP Diliman Press, 2014), 144.

[36] Caroline Hau, *Necessary Fictions: Philippine Literature and the Nation, 1946–1980* (Quezon City: Ateneo de Manila University Press, 2018), 27.

a national identity. This "identity" was always, however, prescriptive rather than actual. In reality, the concept of "national cinema" repressed the cosmopolitan global forces within which cinema from the Philippines has always existed.

The End of National Cinema

For Caroline Hau, the politics of the cultural turn attempted to make good on a postcolonial experience of conflict and fragmentation. This experience, common to both rich and poor alike, was taken by the nationalist intelligentsia as a means of programmatic unity.[37] But in the course of time, economic and political failures attenuated the program—requiring national cinema, too, to double down on its idealist dimension in the face of cinema's ever-present international realities. National cinema was then absorbed in reverse order from its development—first, as a rarified cultural product on the festival circuit, then as an economic category slotted into the "existing geography" of national fare, which was reorganized and distributed internationally by SVOD.[38]

The decline of national cinema mirrors the decline of postcolonial nationalism more broadly. Internal differences within the nationalist movement began to emerge following the 1986 People Power uprising that deposed Ferdinand Marcos. Brocka's political drama *Orapronobis* (Fight for Us, 1989) stands as an important testament to this era, both for its content and for the debate it sparked. Working with Cannes stalwart Pierre Rissient and the French company Stephan Films, Brocka produced a strident indictment of the incoming Aquino regime. He was pilloried for depicting Filipino vigilantes as cannibals, a furor that had less to do with the film than where it was financed and exhibited: Paris and New York City.[39] Brocka responded to his nationalist critics by saying that his film was not "designed to represent everything, not the cosmos, not the Philippines."[40] The scenes of cannibalism were removed in the US videotape release, and the movie was never exhibited in the Philippines in Brocka's lifetime.

Brocka's comeuppance in 1989 can be taken to embody the end of national cinema—his engagement with Rissient and Cannes expressed the limits of

[37] Ibid., 12.

[38] Ramon Lobato, *Netflix Nations: The Geography of Digital Distribution* (New York: NYU Press, 2019), 5.

[39] Daniel Rudin, "Counter/Public: The Politics of Committed Film in the Philippines," PhD diss., University of California, Santa Cruz, 2021.

[40] Lino Brocka, "Brocka on 'Orapronobis,'" *Philippine Daily Inquirer*, July 15, 1989.

developing a national audience *as an expression of national consciousness*. Yet the cultural-nationalist discourse his film embodied continues even under changed circumstances. More recently, Lav Diaz, beloved by an elite group within the Philippines and by cinephiles internationally, has claimed that the slow pace of his films reflects how Filipinos (in his words, Malays) experience time.[41] However, anyone who has spent a day in frenetic Manila may well conclude the opposite, that Filipinos live life at high velocity. In fact, Diaz's claim reflects a programmatic idea about Philippine culture rather than reality.

If a cultural-nationalist sentiment prevails among today's indie filmmakers, it is even more vociferously endorsed by international festivals. According to Red, festivals—not only in the West but also in cities like Busan—expect "a certain flavor" (for instance, "socially engaged" content) from Philippine artists.[42] Even if these festival films were widely popular (they usually are not), that they are construed as allegories of their country of origin tends to reduce the filmmaker to an emblem of an abstract national collectivity.

Paradoxically, Red suggests that working with transnational streamers "allows us to still tell very Filipino, very local stories."[43] Yet, when these same local stories play out on a global stage, the contradiction between national and international is often heightened. In 2019, filmmaker Erick Matti suggested that the Netflix episode of *Street Food Asia*, which centered on the southern Philippine city of Cebu, was an example of poverty porn—that it exploited images of poverty for profit rather than representing "authentic" culture. Having unleashed a firestorm of criticism from other netizens, Matti clarified that the featured cuisine—*lumpia shanghai* and *eel bakasi*—were "not IDENTIFIABLY regarded as a quintessential Filipino street food that can represent our food to the world." In response, one commenter pointed out the obvious: "Who gauges filipinoness [*sic*] anyway?"[44]

[41] Guilio Bursi, "Manila in the Claws of Darkness: Lav Diaz in Conversation with Giulio Bursi," *Mousse Magazine*, January 10, 2014, https://www.moussemagazine.it/magazine/lav-diaz-giulio-bursi-2014/.

[42] Mikhail Red, in discussion with the author, October 4, 2023.

[43] Ibid.

[44] GMA News Online, "Erik Matti Slams Netflix Show for 'Bad Research': 'Is the Eel Bakasi the Most Important Street Food to Showcase from Our Country?,'" https://www.gmanetwork.com/news/lifestyle/food/692981/erik-matti-slams-netflix-show-for-bad-research-is-the-eel-bakasi-the-most-important-street-food-to-showcase-from-our-country/story/. Matti was forced to later clarify that, based on the show, global viewers might assume the rest of the country lived in poverty. See ABS-CBN News, "Erik Matti Presents 'Inclusive' Story about PH through Food in HBO Asia Show," October 23, 2019, https://news.abs-cbn.com/life/10/23/19/erik-matti-presents-inclusive-story-about-ph-through-food-in-hbo-asia-show. For more on the fallacy of how foreign films are judged based on their context, see Neepa Mujumdar, "From Neorealism to Melodrama," in *Film Analysis: A Norton Reader* (New York: W. W. Norton, 2005), 510–27.

No one food—or film—can account for a given culture, much less for the myriad ethnic foods and indie films of the Philippines.

Both Matti's and Red's comments point—though in different ways—back to an old dream of Filipino filmmakers, of achieving a national cinema that is not merely independent in the economic sense but also free from the cultural influences of Hollywood. Unlike Thai director Apichatpong Weerasethakul, who for political reasons "would rather not represent Thailand," Matti and Red not only presume a Tagalog-speaking audience but also express how, as media scholar Ramon Lobato observes, "familiar anxieties about national sovereignty are returning in a different guise through internet distribution."[45] Streaming platforms do not simply replace the already-existing cinema or television ecologies but complexify "the existing geography of distribution."[46] That is to say, streaming builds off of but also reorganizes the previously established media viewerships, tearing down national barriers in its advance. For this reason, Lobato notes the insufficiency of spatial metaphors used to define cinema and television, suggesting we may at best "simply note the contradictions" involved in such definitions.[47]

Case in point is *Deleter*, which, despite having won the MMFF, does not fit preconceived notions of national cinema. Rather, *Deleter* is both a global media product and a function of vintage protectionism. If the ghost of cultural nationalism returns with streaming, it leaves unresolved contradictions between "reified discourses" of national cinema and international production and consumption.[48] Looked at from another angle, the nationalist movement had already created the preconditions for what Hans Magnus Enzensberger called the consciousness industry—a shift in social relations where individual consciousness becomes both last refuge and site of conformity.[49] The postcolonial Filipino intelligentsia that leveraged government power to institutionalize national cinema were also crucial to the development of an indigenous consciousness industry that is but one step from advocating for the state's domination over the individual. From today's perspective, national cinema has shed its utopian promise because it is confronted with a rather distressed version

[45] Campos, *End of National Cinema*, 12; Lobato, *Netflix Nations*, 16.
[46] Lobato, *Netflix Nations*, 5.
[47] Ibid., 70.
[48] Campos, *End of National Cinema*, x.
[49] Hans Magnus Enzensberger, *The Culture Industry: On Literature, Politics and the Media* (New York: Seabury Press, 1974), 14. Enzensberger's definition of the culture industry as "the industrialization of the human mind" is helpful in differentiating the concept from traditional notions of cinema or mass media.

of the postcolonial nation, one that *Deleter* exposes as a paradigmatic site of globalization, a sovereign accomplice in a system of world states that equally prey on their subjects.

Salvaging a National Audience

Mikhail Red comes from a family of filmmakers. His father, Raymond Red, a veteran of the independent movement, shoots commercials and makes an experimental film every five years for the indie festivals. The younger Red has chosen a different path. He sees the indie festival circuit, such as Cinemalaya, which he headlined with *Birdshot* (2016), as a launchpad rather than final destination. Red rejects mere survival in the festival circuit with the "same faces, programmers, and crowd." He contrasts this "echo chamber" to "a box office run of [two million dollars]" and a Filipino audience of a half million people.[50]

Red's approach is also a departure from that of Khavn De La Cruz, the most prolific independent filmmaker in the Philippines, who "makes films with [himself] as the primary audience." According to Khavn, "The hierarchy in Philippine theatres starts with the Hollywood mafia, then the local major studios. Next are the minor studios posing as independents.... Lastly, you have the indies; which have ultra-limited budgets or even none at all."[51]

Khavn's frustration with institutional hierarchy explains the motivations underlying the so-called indie filmmaking movement, which arose in the 2000s and has flourished ever since. The movement has, however, been criticized by filmmakers for failing to reach a broad national audience.[52] Matti's recent lament on Facebook captures this criticism: "Are we really getting our films to the audience it was actually made for? Or are we just bringing them to the small audience we embarrassingly deserve?"[53]

[50] Mikhail Red, in discussion with the author, October 4, 2023.

[51] Alexis Tioseco, "Family Meals, Family Values, and Philippine Cinema: An Interview with Independent Filmmaker Khavn De la Cruz," *Senses of Cinema*, February 2005, https://www.sensesofcinema.com/2005/filipino-cinema/khavn_de_la_cruz/.

[52] See, for instance, Roland Tolentino's critique, "Indie cinema bilang kultural na kapital" [Indie Cinema as Cultural Capital], *KPK Column, Bulatlat*), August 3, 2008, http://rolandotolentino.wordpress.com/2008/08/03/indie-cinema-bilang-kultural-na-kapital-kpk-column-bulatlat/. For Tolentino, indie cinema is a reduced political horizon that symptomizes the neoliberal economics of a middle-class intelligentsia. He argues that indie films flow from deindustrialization, in which the studio is displaced by the director's desire to accumulate enough cultural capital to go mainstream—avoiding national "audience development" of prior nationalist directors like Lino Brocka.

[53] Krissy Aguilar, "Erik Matti: Film Industry in 'Dire Situation' after Low Box Office Sales," Inquirer.net, February 8, 2019, https://entertainment.inquirer.net/317330/erik-matti-film-industry-in-dire-situation-after-low-box-office-sales. See also https://www.facebook.com/erikmatti/posts/10156167745526446.

Matti's sentiment not only recalls the nationalist critiques of Brocka and Lumbera; it also bespeaks a multidimensional crisis in the present. Responding to this crisis, Red argues that developing a national audience requires an international "buffer," which provides resources for the growth of a local market.[54] Yet there are not only fiscal limits but also aesthetic preferences of the Pinoy viewer to attend to. To this end, Red's formula is to entertain and engage the audience through a "subtext" of genre codes that "make[s] it easier for them to digest or enjoy." He characterizes this approach as a "trojan horse" or a manner of smuggling in either a political message, a formal innovation, or both, within familiar genres, "to start to develop a real sustainable audience locally."[55]

Deleter, Red's most successful attempt at this model, is a ghost story doubling as an allegory of global business process outsourcing (BPO) exploitation. The office building where the labor happens is haunted by a worker—a social media content moderator—who has committed suicide by plunging from the roof. By the end of the film, it is revealed through a flashback that this same worker was raped by her employer. Red avoids being too heavy-handed with his politics by "smuggling in" a critique of the global economy through a generic plotline of betrayal, death, and vengeance. Although the immediate villain is the Filipino boss—rather than some foreign firm—underneath the violence lies a sense of existential dread in the face of technology mediating how the characters interact with society.

On the aesthetic level, *Deleter*, like Red's earlier horror films, is in dialogue with the American independent studio A24's "elevated horror" genre. So called for bespoke art direction, auteur-like directorial control, and themes of psychological trauma, elevated horror attempts to push the genre beyond schlock and jump scares. That Red so closely links his own efforts to an American studio indicates the limits of the older generation's cultural nationalism. Not only is he on the forefront of bringing global trends and techniques into the star-obsessed Philippine market, Red also disrupts audience expectations by casting against type—by extracting the gorgeous Nadine Lustre from her typical romantic role and subjecting her to facial disfiguration, as in the closing scenes of *Deleter*—without losing them.

[54] "Rappler Talk Entertainment: Mikhail Red on 'Neomanila' and the Future of PH Cinema," *Rappler*, March 7, 2019, https://www.youtube.com/watch?v=B6W2k8l0Guw.

[55] Mikhail Red, in discussion with the author, October 4, 2023.

Red's ultimate aim of developing a national audience is a "balancing act" that involves juggling the need for genre with pushing the envelope.[56] To do so, he must combat audience preference for the dominant trend in the Philippines, the romantic comedy (rom-com). Star Cinema is the undisputed master of the rom-com, having developed it through the three highest-grossing local films.[57] In a recent twist of fate, the broadcast franchise of Star Cinema's parent company, the media conglomerate ABS-CBN, was allowed to expire after a messy and politicized showdown in Congress. Paradoxically, this created space for new genre experimentation by older outfits like Viva, which has since rebounded and even established two profitable online apps. Red has seized on the opportunities for genre mixing and experimentation opened within this context by partnering with Viva.

The key move Red made with *Deleter* was to tap Viva's top star, Lustre, and put her in an R-rated film.[58] Both the suicide theme and approach toward casting build on Red's previous film, *Eerie* (2018), which uses art photography, color grading, and crafted lighting to dramatize supernatural and Catholic themes common to Philippine horror—all in relation to a plot channeling adolescent crisis and trauma.[59] Just before *Eerie*'s launch, Red had released a neorealist political thriller called *Neomanila* (2017). At a public screening on *Neomanila*'s opening night, the theater was empty; but for Red and the cast, the film would not have even screened.[60] The following week *Eerie* launched and went on to gross five million dollars. The failure of the political thriller and the success of the elevated horror film was enough of a "wake-up call" for Red to take seriously the balancing act between his auteur approach and satisfying local viewers.[61]

Eerie was acquired by Netflix, while *Deleter* currently streams from Vivaone and Amazon Prime Video. By leveraging both theater and streaming, Red achieves a "50/50 buffer," with half the audience coming from abroad. Those who can afford streaming are a niche audience. In the Philippines, this means a middle-class inclined toward art films that legitimate the spectacle unfolding on SVOD platforms. By contrast, the mass cineplex audience with limited income

56 Ibid.

57 Ibid. The three films are *Hello Love, Goodbye* (2019), *The How of Us* (2018), and *Rewind* (2023), which earned thirteen, fifteen, and fifteen million, respectively.

58 *Deleter* was given an 18+ rating in the Philippines, and an X-rating on Amazon.

59 Prefiguring his choice of Lustre for *Deleter*, with *Eerie* Red cast Bea Alonzo, another rom-com superstar.

60 Mikhail Red, in discussion with the author, October 4, 2023.

61 Ibid.

values spectacle per se, without needing to indulge in prestige films. If he wants to capture this Filipino audience, Red cannot push his "socially relevant, politically charged" mixed-genre experiments too far beyond the threshold of what it will tolerate.[62] Having experienced success with the "buffer" model, Red now sets his target on discovering "what resonates with the mass audience in the [Philippine] cinemas." He wants to achieve an equal market share with Hollywood films by having "Filipinos support Filipinos on the big screen."[63]

From National Cinema to Streaming as Commodity

Although the SVOD market is growing rapidly and, per one recent estimate, there are three hundred thousand Filipino Netflix subscribers, significant economic and infrastructural problems remain if a broader audience is to be realized.[64] Filipinos are accustomed to free content, few own credit cards, piracy is rampant, and internet speeds lag far behind other regional countries.[65] In response, Netflix bundled their product with internet packages sold by major telecom companies, established an office in Manila, and has begun to produce more original content.[66]

Another problem is that the Philippine market remains organized, like in America, along oligopolistic lines. For Allen Scott, Hollywood's success is a function of "economies of scale" in both production and distribution.[67] This means corporate concentration, allowing the major studios to orchestrate multi-million-dollar blockbusters. In terms of distribution, the same principles obtain in the Philippines, with the shopping mall cineplexes owned by Shoemart Supermalls (SM) holding the largest market share. Michael Kho Lim narrates how beginning in 1985, SM was able to break the monopoly of the stand-alone theater cartels, leading eventually to the latter's obliteration.[68] The shift is significant for how cinema is consumed. In SM theaters, films draw customers in

62 "Rappler Talk Entertainment: Mikhail Red on 'Neomanila.'"
63 Mikhail Red, in discussion with the author, October 4, 2023.
64 Statista, "Estimated Number of Active Streaming Subscribers to Netflix in Philippines from 2017 to 2020," https://www.statista.com/statistics/607624/philippines-netflix-subscribers/.
65 Michael Kho Lim, "Netflix Philippines Report," 2018.
66 Ibid.
67 Allen J. Scott, "Hollywood and the World: The Geography of Motion-Picture Distribution and Marketing," *Review of International Political Economy* 11, no. 1 (2004): 34.
68 Michael Kho Lim, *Philippine Cinema and the Cultural Economy of Distribution* (London: Palgrave Macmillan, 2019), 114–21.

to patronize shopping mall vendors. Needless to say, these exhibition conditions discourage filmmaking risks and reinforce family friendly genres.

As Red has discovered, SVOD presents opportunities to overcome the shopping mall monopoly. However, if the transition from the theater to the cineplex radically transformed the relationship of the Filipino audience to the cinema, the same is true of SVOD—particularly since the lines between the overall experience and the individual product are diffuse as much as their technological advances are interdependent.[69] These mediations can be understood by analogizing the streaming consumer with the buyer at a department store, who believes they are purchasing a single good but who is actually purchasing the brand in its entirety. Oskar Negt and Alexander Kluge call this massing of products a macrocommodity. Much like the shopping mall, SVOD platforms sell a product that acquires value precisely because it is comprehensive and holistic.[70] While Amazon links online retail to streaming, Netflix effectively promotes 4K televisions, not to mention consumer goods via product placement and in their ad-supported plan. Moreover, a subscription facilitates water-cooler chatter at work, gives access to knowledge regarding vacation destinations, and provides an endless stream of bingeable serialized (and customized) content to prepare the psychic energies for the next day at work.

In short, the SVOD operates by aggregating commodities in the service of the consciousness industry, or "the needs of the consciousness of concrete individuals," who imagine themselves customers involved in a market exchange, but who are actually "producing the media cartel." An entire invisible apparatus—Netflix, Amazon—is set in motion with their purchase, suggesting a higher stage of commodity that could only find a conscious articulation under socialism.[71]

[69] Lobato, *Netflix Nations*, 37.

[70] Oskar Negt and Alexander Kluge, *Public Sphere of Experience: Analysis of the Bourgeois and Proletarian Public Sphere* (New York: Verso, 2016), 131.

[71] Ibid., 105, 130–3. For Negt and Kluge, just as the consciousness industry masses human needs, this concentration's potential is limited by the private nature of appropriation under which the media cartel necessarily masks itself. The experience bought and sold through SVOD potentially transcends the anarchy of commodity exchange. Yet the product takes on omnipotent characteristics as the viewers gives themselves over to its flow, deferring their own diffuse experience to consumption and the totalizing capitalistic world to which it belongs: a form of ownership without any rational relationship to that totality. A more lucid relation to totality would instead likely look something like Vertov's *Man with the Movie Camera*—distinctions between technology, the filmmaker, and the viewer would be sustained, but the fluid and dialectical interrelation would be widely available to all—the viewer would be capable of authorship, of becoming a filmmaker, in ways that SVOD facilitates in only the most obscure and mediated fashion.

In contrast to the massing of needs made possible by the media conglomerate—and by the SVOD corporations—independent artists must struggle to construct a cult following, much less a national audience. Just as SVOD platforms elicit the collaboration of viewers in rating films, thus guiding decisions on future products, the Filipino studios are able to amass information about the viewing desires of their target markets. Value is generated through not only the popular films but also the "non-earning films" that may flop yet can still produce "new stars, experimenting with a new genre, or testing the market, which may yield future economic value."[72] Could this same model, in tandem with an inexpensive SVOD launch, be used to develop quality independent films? Red's attempts will either show the way toward such a model or prove the exception. It is worth noting that during the 2023 MMFF, Star Cinema's *Rewind* (2023), a romantic drama, raked in approximately $6.8 million—far more than *Deleter*.[73] The return of the romance casts some doubt on Red's assertion that the market is shifting to mixed genres.

If anticipating aesthetic preferences is difficult, then how to account for more complex changes that SVOD is introducing to cinema? In their study of digital cinema, Elsaesser and Hagener suggest that the medium has fundamentally changed from "inside out… leaving outer appearances intact but, in the meantime, hollowing out the foundations… on which a certain medium or mode of representation was based."[74] By "medium" they mean film in its formal or technological sense and conventions received from twentieth-century cinema. They ask, what if digital cinema merely inherits the rituals of the past without managing to transform it commensurate to the needs of the present? For Lobato, this "theory of technological change" can also apply to SVOD.[75]

Lobato suggests that a history of television (of which he considers SVOD platforms an extension) reveals transformations changing it "from a mass medium to a niche one through technological and institutional developments" that atomize the mass viewership "into a series of personalized choices."[76] SVOD platforms, much like television, symptomize broader middle-class atomization.

72 Lim, *Philippine Cinema*, 122.

73 R. J. Mataro, "'Rewind' Reportedly Nears the P400 Million Mark as Highest-Grossing MMFF 2023 Film," https://www.lionheartv.net/2024/01/rewind-rp400-million/.

74 Thomas Elsaesser and Malte Hagener, *Film Theory: An Introduction through the Senses*, 2nd ed. (London: Routledge, 2015), 204–5. This very speculative proposition would appear rather utopian, whereas in fact digital cinema is mimetically reproducing most ontological premises of the cinema.

75 Lobato, *Netflix Nations*, 33–4.

76 Lobato, *Netflix Nations*, 23; James Bennett and Nikki Strange, *Television as Digital Media* (Durham, NC: Duke University Press, 2011), 2.

In this respect, they are a "domestic technology," one among many "machines for the home" that play a role in the reconstitution of bourgeois domesticity—by bringing the family, such as it is today, together in front of the television screen. Philippine SVOD services may perform a slightly different role, however, since part of their target market works abroad. In fact, for the oversees Filipino worker (OFW) the opportunity of class uplift often means an extended separation of spouses.[77]

To the extent that the Philippine middle class remains small, isolated, and internally fragmented, SVOD may perform its domestic function well, by bringing families together. But it seems unlikely that it will, a la Brocka and Red's aspirations, lead back to the formation of a national audience—save through the infusion of new genres or the consolidation of smaller indie operators.[78]

Streaming the Nation or Dramatizing Globalization?

If the question of whether Red's "trojan horse" and "international buffer" formulas will change the local film industry remains in doubt, the fact that *Deleter* resonated with a broad audience is without question. *Deleter* is a David-versus-Goliath story of the Pinoy worker in the clutches of global capitalism. It dramatizes international content moderation work that occupies nearly two million Filipinos averaging $325 a month.[79] With as many working abroad, the Philippines is organically entrenched in the global economy.[80] *Deleter* reveals this reality, shining a light on the clandestine occupation, the job's psychological trauma, and labor and sexual exploitation that implicitly lurk below the surface. Red strategically overlaps Lustre's star power or mass appeal with middle-class "social issues" (feminism, labor) in order to capture both mass and niche audiences. In a sense, the film has elements of a warped office drama—Philippine style.

77 Red keys in on this phenomenon during a scene in *Deleter* when Lustre, in the office bunk room, observes a young man sobbing into his phone "babe, babe." Presumably, he is bereft of his lover by virtue of being a migrant from the provinces himself, or due to her being somewhere abroad.

78 Lobato, *Netflix Nations*, 104–5.

79 Ralf Rivas, "Global Cost-Cutting to Help Boost Philippine BPOs' Growth in 2023," *Rappler*, March 1, 2023, https://www.rappler.com/business/global-cost-cutting-to-boost-philippines-process-outsourcing-growth-2023/; also Glassdoor, "BPO Salaries in Manila, Philippines?," https://www.glassdoor.com/Salaries/manila-bpo-salary-SRCH_IL.0,6_IM995_KO7,10.htm.

80 Statista, "Total Number of Overseas Filipino Workers (OFWs) from 2017 to 2022," https://www.statista.com/statistics/1287067/philippines-number-of-overseas-filipino-workers/.

Deleter is also, thematically and conceptually, situated within a contact zone where local control and international capital collide. Many BPO jobs are in export processing zones governed by the Philippine Economic Zone Authority (PEZA), which doles out perks such as tax holidays to attract foreign investment. *Deleter*'s ominous and repeated establishing shot of the modern glass tower where the story unfolds makes the building itself a character, faceless, cosmopolitan, impersonal (Figure 8.1). The polished (and deadly) stainless steel elevators, shots with the camera hovering just over the ledge of the rooftop overlooking metro Manila, scenes taking place inside the sterile convenience store at the tower's base, chain-vaping, stylish but frazzled employees—these are iconographic symbols to the average MMFF viewer who, even if not an office worker, will live in the shadow of such a building in Makati, Ortigas Center, San Juan, or Quezon City.

Several movie reviews—quite possibly written in buildings resembling the one depicted—took fault with *Deleter*'s lighting. Whenever the film's vengeful suicide spirit appears, the building's fluorescent lights flicker and then are extinguished.[81] Manila office buildings, as in any other global urban center, are

Figure 8.1 The title of the film *Deleter* appears over a low-angle shot of a generic office building, the location where most of the film takes place. The red stoplight in the lower right corner foreshadows the lighting color used in later scenes where the spirit takes her revenge. *Deleter*, directed by Mikhail Red © Viva Films & PelikulaRed 2022.

[81] Lé Baltar, "'Deleter' Review: Come for the Technohorror, Stay for Nadine Lustre," *Rappler*, December 28, 2022, https://www.rappler.com/entertainment/movies/deleter-review/.

occupied around the clock with lights that radiate an enveloping harshness. Global and local collide in the grinding activity that characterizes an agent's graveyard shift, servicing clients for whom it is day. In a convenience store adjacent to the office building, a white, almost radioactive mise-en-scène is used to dramatize middle-class life in the late neoliberal era. These scenes are rendered frightening not only by the harsh lighting but also by the spirit who appears on a CCTV screen above the cash register. By contrast, the office scenes in *Deleter* are exceedingly dim. In criticizing this point it would appear the reviewers expect verisimilitude—a Manila office building should be, like the convenience store, glaringly lit. Obviously, Red here takes poetic license to establish a grim and foreboding mood.

Other documentary details that become plastic in Red's hands are scenes with the fabled BPO bunk rooms made for resting but where call center workers are known to fornicate to relieve stress. The opposite is true in *Deleter*—the dimly lit sleeping quarters are like prison cells, replete with scratched messages on the bed that spell out "you won't be able to escape from here" and a worker sobbing while ghoulishly lit by his cell phone. Everyone seems trapped by the building, from the workers—more than likely migrants from the provinces who never seem to go home—to the security guard glued to the CCTV, to Lustre who is stuck in a reoccurring nightmare of an apparent act of bloody domestic violence.

Above all, everyone is trapped by technology. In this respect, the film mirrors the Covid pandemic lockdown conditions in 2020–2, which in the Philippines were exceedingly authoritarian, especially for the poorest.[82] *Deleter* is motivated, like Netflix's Korean series *Squid Game*, by both an existential threat and the authoritarian, every-man-for-himself response to it. If the visual principle of *Squid Game* is driven by a tension between red blood versus shiny surfaces of the game-show setting, *Deleter* harmonizes a blueish florescent glow with a saturated red like one might see in a photo lab. Blue is used for a silhouette of a body on the opaque door behind which, witnessed by Lustre, a coworker was raped. These colors come together as strobe lights in a club scene, where the supporting character's sexual advances seem to trigger some deep associations in Lustre (both guilt and childhood trauma). The scene ends with Lustre locking herself in a blue bathroom; the outside remains red; cutting back and forth between these two spaces produces a sense of bipolarity.

[82] Ditsi Carolina, "Manila Lockdown: One of the Longest COVID Lockdowns in the World," *Witness: Al Jazeera English*, October 19, 2021, https://www.youtube.com/watch?v=E5F3xA_zkFc&t=1s. Carolina documents how some, desperate for food during the lockdown, mounted bread riots.

The film has two endings—first, where the actual source of the haunting is discovered, and a second where Lustre has donned an informant's rubber mask and is being interviewed, ostensibly as part of an investigation into the recent events. Suddenly, however, she is made to view a video of herself, which resembles a clip from the beginning of the film where a woman has cut her face and is peeling off the flesh. Only this time the woman is Lustre. With blood pouring out of her mask, she fumbles for the key that will delete herself, thus ending the pain as well as the film.

If the establishing shot of the office building embodies the globalized outer world, another iconic shot from the film is a concrete staircase lit by a foreboding red light, punctuated by a flickering florescent tube. Lustre, bathed in red light, peers up into the stairwell, prefiguring her descent into the belly of the beast, or rather its Id (Figure 8.2). Like *Deleter*, Red's earlier work *Rekorder* (2013) showcased themes of repressed trauma. *Rekorder* is a film about a down-and-out cameraman, Maven (Ronnie Quizon), who uses his fading skills to drum up work by pirating films in the cinema with an outdated Handycam. Like *Deleter*, it is an allegorical tale with a moral dilemma but no resolution, and one that features multiple layers of mediation between different technologies.

But while *Deleter* uses cell phone and surveillance video camera footage to represent the point of view of the dead, it lacks the metatextual complexity and

Figure 8.2 The character Lustre stands at the threshold of the zone between the human world and digital netherworld of the spirits or of the psyche. There is a use of contrasting color tones of red and blue, which accentuates the passage. *Deleter*, directed by Mikhail Red © Viva Films & PelikulaRed 2022.

allegorical simplicity of *Rekorder*, which is just as much about the crisis in the film industry in its transition to digital as it is about violent crime and its digital mediation via the internet. In *Rekorder* Red seamlessly weaves together media shards from different devices in an engrossing story that plays out like a twisted remake of Vertov's *Man with the Movie Camera*. As we observe the vagaries of existence of a media artist living in the slums of the global South, we realize that instead of mediating, technology perverts, distorts, and advances an ethos of violence. Unlike *Rekorder*'s Maven, who is appropriately haggard, in *Deleter* Lustre's angelic face wedged into a cubicle where a worker would otherwise sit seems a bit odd. Hardly beaten down, she remains a graceful prima donna; not a hair is amiss even as she is stalked by a spirit. If in the last sequence Red subverts Lustre's beauty by covering it with a hideous mask, it is out of tune with the first false ending. As a result, the plot's structure lacks integrity.

Deleter implicitly strikes a note of moral caution against mindless doomscrolling. Yet unlike *Rekorder*, whose protagonist is vitally linked to the theme of content creation, *Deleter*'s video moderators don't produce but rather consume, and the ghost's videographic POV beggars belief—dead people don't film (Figure 8.3). Hence, these secondary layers of mediation and moralizing no longer quite plumb Red's capacity and penchant for psychological drama. *Rekorder*'s theme of the artist's cognition in relation to his instruments is replaced by the star's (video-mediated) facial mutilation. In the process, the film

Figure 8.3 Red uses a CCTV effect that eventually becomes the ghost's POV. *Deleter*, directed by Mikhail Red © Viva Films & PelikulaRed 2022.

is reduced to the political message on the one hand (for Red a "subtext") and a formal, albeit harmoniously arranged, mise-en-scène on the other.

As previously stated, Red justifies his genre-mixing approach through the idea of a "trojan horse," or as a means of surreptitiously inserting a social message while laying the groundwork for acceptance of new forms. In so doing, he runs the risk of on the one hand reducing the film to propaganda and on the other hand being undercut by the frivolousness of the genre.[83] There are also certain problems to *Deleter* that stem from nationalist assumptions and Red's own class position. Red expects that the film will inspire viewers to research BPO exploitation, yet the sheer scope of the industry means that many Filipinos are either employed by or know somebody in it. Finally, there are limitations to pursuing themes of "social relevance" for a national audience that may just not be interested. Why then not leverage SVOD's strengths to pursue a niche audience, rather than make a compromise film for the "masses"?

As this essay has argued, national cinema is a contradiction in terms. Further, in the context of globalization, history has proven mass culture a site of the consciousness industry rather than the reconciliation between the middle class and *bakya*. The consciousness industry's massing of human needs points toward a higher stage of commodity, a condition that the media cartels mask through private appropriation—and, in the Philippines, through consumption still indebted to the idea of national cinema. However, the national audience developed via SVOD faces the challenge of increasing fragmentation of the national community—both in class terms and as niche-market atomization. Related to this challenge is the question of whether Red's hiding the social critique in a trojan horse "makes it easier for [the mass audience] to digest or enjoy" or results in a vicarious identification with exploitation that prevents the viewer from reflecting on the work of art.[84] Are viewers simply seeing what is in front of them—a predigested and genre-coded realism—rather than having an

[83] The antinomy between form and content was precisely what nationalists like Lumbera encountered when, as mentioned above, he separated aesthetic "technique" from a vernacular "sensibility." See Lumbera, "Approaches to Philippine Film," 99, and "Nunal sa Tubig Revisited," 242–3. Instead of focusing on either form or content, Walter Benjamin argues for considering the author's position within the relations of production. For a filmmaker, this means precisely his or her relation to other filmmakers—that is to say, the way they change the tools of production, tools that are defined through aesthetic form as a function of practice in history. We must ask, is Red teaching other artists, and if so, what? See Walter Benjamin, "The Author as Producer," in *Walter Benjamin: Selected Writings: Volume 2, Part 2*, ed. Marcus Bullock, Howard Eiland, and Gary Smith (Cambridge, MA: Harvard University Press, 2005), 768–82.

[84] Mikhail Red, in discussion with the author, October 4, 2023.

aesthetic experience that interrogates the medium as a technology, an aspiration *Deleter* has taken up with themes of digital remediation that are perhaps better executed in the indie *Rekorder*? Even if it is seen only by a small audience, even if only by other artists, aesthetic judgment ought to be prioritized above aesthetic predigestion and pseudo-nationalist sympathies.

Bibliography

ABS-CBN News. "Erik Matti Presents 'Inclusive' Story about PH through Food in HBO Asia Show." October 23, 2019. https://www.abs-cbn.com/life/10/23/19/erik-matti-presents-inclusive-story-about-ph-through-food-in-hbo-asia-show.

Agimat. "Something New in 2010 MMFF." http://www.agimat.net/film/n101122b.php.

Aguilar, Krissy. "Erik Matti: Film Industry in 'Dire Situation' after Low Box Office Sales." Inquirer.net, February 8, 2019. https://entertainment.inquirer.net/317330/erik-matti-film-industry-in-dire-situation-after-low-box-office-sales.

Baltar, L é. "'Deleter' Review: Come for the Technohorror, Stay for Nadine Lustre." *Rappler*, December 28, 2022. https://www.rappler.com/entertainment/movies/deleter-review/.

Benjamin, Walter. "The Author as Producer." In *Walter Benjamin: Selected Writings: Volume* 2, *Part 2*, edited by Marcus Bullock, Howard Eiland, and Gary Smith, 768–82. Cambridge, MA: Harvard University Press, 2005.

Bennett, James, and Nikki Strange, eds. *Television as Digital Media*. Durham, NC: Duke University Press, 2011.

Brocka, Lino. "Brocka on 'Orapronobis.'" *Philippine Daily Inquirer*, July 15, 1989.

Brocka, Lino. "Philippine Movies: Some Problems and Prospects." In *Readings in Philippine Cinema*, edited by Rafael Ma. Guerrero, 259–62. Manila: Experimental Center of the Philippines, 1983.

Bursi, Giulio. "Manila in the Claws of Darkness: Lav Diaz Lav Diaz in Conversation with Giulio Bursi." *Mousse Magazine*, January 10, 2014. https://www.moussemagazine.it/magazine/lav-diaz-giulio-bursi-2014/.

Campos, Patrick. *The End of National Cinema: Filipino Film at the Turn of the Century*. Quezon City: University of the Philippines Press, 2016.

Carolina, Ditsi. "Manila Lockdown: One of the Longest COVID Lockdowns in the World." *Witness: Al Jazeera English*, October 19, 2021. https://www.youtube.com/watch?v=E5F3xA_zkFc&t=1s.

Claudio, Lisandro. *Taming People's Power: The EDSA Revolutions and Their Contradictions*. Quezon City: Ateneo University Press, 2013.

Constantino, Renato. *Identity and Consciousness: The Philippine Experience*. Quezon City: Renato Constantino, 1974.

Constantino, Renato. *Insight and Foresight*. Quezon City: Foundation for Nationalist Studies, 1977.

Constantino, Renato. *The Philippines: The Continuing Past*. Quezon City: Foundation for Nationalist Studies, 1978.

David, Joel. *Wages of Cinema: Film in Philippine Perspective*. Quezon City: University of Philippines Press, 1998.

Diaz, One Carlo. "Filipino Film *Deleter* to Hit UAE Cinemas after Huge Local Success." *National News*, January 8, 2023. https://www.thenationalnews.com/arts-culture/film-tv/2023/01/08/filipino-film-deleter-to-show-in-uae-cinemas-after-huge-local-success/.

Elsaesser, Thomas, and Malte Hagener. *Film Theory: An Introduction through the Senses*. 2nd ed. London: Routledge, 2015.

Enzensberger, Hans Magnus. *The Culture Industry: On Literature, Politics and the Media*. New York: Seabury Press, 1974.

Glassdoor. "BPO Salaries in Manila, Philippines?" https://www.glassdoor.com/Salaries/manila-bpo-salary-SRCH_IL.0,6_IM995_KO7,10.htm.

GMA News Online. "Erik Matti Slams Netflix Show for 'Bad Research': 'Is the Eel Bakasi the Most Important Street Food to Showcase from Our Country?'" https://www.gmanetwork.com/news/lifestyle/food/692981/erik-matti-slams-netflix-show-for-bad-research-is-the-eel-bakasi-the-most-important-street-food-to-showcase-from-our-country/story/.

Golay, Frank H. *The Philippines: Public Policy and National Economic Development*. Ithaca, NY: Cornell University Press, 1961.

Gonzales-Tamrong, Marlina. "LA-Based Pinoy Newspaper Features Brocka." *Manila Chronicle*, May 6, 1987.

Harvey, David. *The Condition of Postmodernity: An Inquiry into the Origins of Cultural Change*. Malden, MA: Blackwell, 1990.

Hau, Caroline. *Necessary Fictions: Philippine Literature and the Nation, 1946–1980*. Quezon City: Ateneo de Manila University Press, 2018.

Higson, Andrew. "The Concept of National Cinema." *Screen* 30, no. 4 (Autumn 1989): 36–47.

Hutchcroft, Paul. "Predatory Oligarchy, Patrimonial State: The Politics of Private Domestic Commercial Banking in the Philippines." PhD diss., Yale University Press, 1993.

Hutchcroft, Paul. "Situating the Philippine Political Economy: Import-Substitution Industrialization in Comparative Perspective." *Kasarinlan* 4, no. 4 (1989): 39–56.

Lacaba, Pete. "Notes on Bakya: Being an Apologia of Sorts for Filipino Masscult." In *Readings in Philippine Cinema*, edited by Rafael M a. Guerrero, 117–23. Manila: Experimental Center of the Philippines, 1983.

Lim, Michael Kho. "Netflix Philippines Report." 2018.

Lim, Michael Kho. *Philippine Cinema and the Cultural Economy of Distribution*. London: Palgrave Macmillan, 2019.

Lobato, Ramon. *Netflix Nations: The Geography of Digital Distribution*. New York: NYUNTYU Press, 2019.
Lumbera, Bienvenido. "Approaches to Philippine Film." In *The Urian Anthology: 1970–1979*, edited by Nicanor G. Tiongson, 94–101. Manila: M. L. Morato, 1983.
Lumbera, Bienvenido. "Nunal sa Tubig Revisited." In *The Urian Anthology: 1970–1979*, edited by Nicanor G. Tiongson, 240–5. Manila: M. L. Morato, 1983.
Lumbera, Bienvenido. "Philippine Film 1961–1992." In *CCP Encyclopedia of Philippine Art, Volume VII: Philippine Film*, edited by Nicanor G. Tiongson, 40–9. Manila: Cultural Center of the Philippines, 1994.
Lumbera, Bienvenido. "Philippine Problems in Film History." In *Readings in Philippine Cinema*, edited by Rafael Ma. Guerrero, 74–5. Manila: Experimental Center of the Philippines, 1983.
Mataro, R. J. "'Rewind' Reportedly Nears the P400 Million Mark as Highest-Grossing MMFF 2023 Film." https://www.lionheartv.net/2024/01/rewind-rp400-million/.
Mujumdar, Neepa. "From Neorealism to Melodrama." In *Film Analysis: A Norton Reader*, edited by Jeffrey Geiger and R. L. Rutsky, 510–27. New York: W. W. Norton, 2005.
Negt, Oskar, and Alexander Kluge. *Public Sphere of Experience: Analysis of the Bourgeois and Proletarian Public Sphere*. New York: Verso, 2016.
Olivares-Cunanan, Belinda. "Brocka Depicts Pinoys as Cannibals." *Philippine Daily Inquirer*, July 11, 1989.
Pew Research. "Top 10 U.S. Metropolitan Areas by Filipino Population, 2019." https://www.pewresearch.org/social-trends/fact-sheet/asian-americans-filipinos-in-the-u-s/#top-10-u-s-metropolitan-areas-by-filipino-population-2019.
"Rappler Talk Entertainment: Mikhail Red on 'Neomanila' and the Future of PH Cinema." *Rappler*, March 7, 2019. https://www.youtube.com/watch?v=B6W2k8l0Guw.
Rivas, Ralf. "Global Cost-Cutting to Help Boost Philippine BPOs' Growth in 2023." *Rappler*, March 1, 2023. https://www.rappler.com/business/global-cost-cutting-to-boost-philippines-process-outsourcing-growth-2023/.
Rudin, Daniel. "Counter/Public: The Politics of Committed Film in the Philippines." PhD diss., University of California, Santa Cruz, 2021.
Salterio, Leah C. "Deleter Not Originally Meant for the MMFF." *Philstar Global*, January 14, 2023. https://www.philstar.com/entertainment/2023/01/14/2237416/deleter-not-originally-meant-mmff.
Scott, Allen J. "Hollywood and the World: The Geography of Motion-Picture Distribution and Marketing." *Review of International Political Economy* 11, no. 1 (2004): 33–61.
Statista. "Estimated Number of Active Streaming Subscribers to Netflix in the Philippines from 2017 to 2020." https://www.statista.com/statistics/607624/philippines-netflix-subscribers/.

Statista. "Highest Grossing International Movies in the Philippines in 2022, by Box Office Revenue." https://www.statista.com/statistics/655233/philippines-top-10-movies-by-revenue/.

Statista. "Total Number of Overseas Filipino Workers (OFWs) from 2017 to 2022." https://www.statista.com/statistics/1287067/philippines-number-of-overseas-filipino-workers/.

Supreme Court E-Library. "Presidential Decree No. 824." November 7, 1975. https://elibrary.judiciary.gov.ph/thebookshelf/showdocs/26/17327.

Tioseco, Alexis. "Family Meals, Family Values, and Philippine Cinema: An Interview with Independent Filmmaker Khavn De la Cruz." *Senses of Cinema*, February 2005. https://www.sensesofcinema.com/2005/filipino-cinema/khavn_de_la_cruz/.

Tolentino, Roland. *Contestable Nation Space: Cinema, Cultural Politics, and Transnationalism in the Marcos-Brocka Philippines*. Quezon City: UP Diliman Press, 2014.

Tolentino, Roland. "Indie cinema bilang kultural na kapital." *KPK Column, Bulatlat*, August 3, 2008. http://rolandotolentino.wordpress.com/2008/08/03/indie-cinema-bilang-kultural-na-kapital-kpk-column-bulatlat/.

Trice, Jasmine. *City of Screens: Imagining Audiences in Manila's Alternative Film Culture*. Durham, NC: Duke University Press, 2021.

Tumbocon, Mauro, Jr., and Lena Pareja. "Philippine Motion Picture Producers Association." In *CCPCC Encyclopedia of Philippine Art, Volume VIII: Philippine Film*, edited by Nicanor G. Tiongson, 292–3. Manila: Cultural Center of the Philippines, 1994.

Zedong, Mao. *Talks at the Yenan Forum on Literature and Art*. N.p., 1942.

9

Rethinking English-Translation Subtitles in the Age of Streaming: *Decision to Leave* (2022) and the Korean Wave

Hiju Kim

A famous quote by Korean film *Parasite*'s director Bong Joon Ho reads, "Once you overcome the one-inch-tall barrier of subtitles, you will be introduced to so many more amazing films."[1] This quote has emerged in many articles and book chapters on Korean cinema and has been used to promote the idea of spotlighting international films in mainstream arenas.

Moreover, 2019, the year of *Parasite*'s release, has been heralded as a milestone for subtitled films in the age of *hallyu* or the Korean Wave of popular culture disseminating globally. Within this context, scholars across the realms of language and cultural studies have analyzed the subtitles of *Parasite* and other popular Korean works, such as the Netflix series *Squid Game*, to reveal the cultural intricacies as well as industrial dynamics hidden behind the seemingly uncomplicated "one-inch-tall barrier" evoked by Bong.

These intricacies and dynamics are often overlooked in popular discourse on subtitling. In the case of *Squid Game*, for instance, a debate surrounding the difference between Netflix's English subtitles and its closed captions produced a wealth of comments across social media communities and editorials, including headlines such as "You Could Be Watching Squid Game Wrong—All Because of a Netflix Error" and "Still Watching Squid Game on Netflix? Change This

[1] Justin Chang, "Commentary: 'The 1-Inch-Tall Barrier of Subtitles': Bong Joon Ho Rightly Calls out Hollywood Myopia," *Chicago Tribute*, January 7, 2020, https://www.chicagotribune.com/2020/01/07/commentary-the-1-inch-tall-barrier-of-subtitles-bong-joon-ho-rightly-calls-out-hollywood-myopia/.

Subtitle Setting Immediately."[2] However, treating subtitles as simply a "setting" on Netflix or any other platform obscures the role of subtitlers—whose names are often left out of these discussions—as well as of the culturally specific guidelines underlying the subtitling process.

These guidelines vary depending on the context. For example, pan-European screen translations have centralized practices outlined by organizations such as the European Association for Studies in Screen Translation, whereas the subtitling guidelines in Korea are less standardized and relate predominantly to honorifics and titles given to persons of certain social positions, dependent on conditions such as age and seniority.[3] Such context-specific variations impact significantly the translation from source text into target text.

This chapter closely analyzes *Decision to Leave*, a critically acclaimed 2022 film directed by the South Korean filmmaker Park Chan-wook, which deals with translation and mistranslation and thereby calls attention to the work of subtitled cinema in the present moment. At the time of writing, the film is widely available online through several international on-demand streaming services, including Mubi, its initial exclusive distributor and a platform for global arthouse and independent cinema. The film has dialogues that incorporate Korean, written and spoken Mandarin Chinese, as well as several English words. In China, the film premiered at the 2022 Hainan International Film Festival and was released in Chinese theaters the following year. It also premiered at the 2022 Cannes Film Festival, and shortly thereafter the film was released theatrically and distributed via streaming. The film's English subtitles were written by Darcy Paquet as part of

[2] Jack Shepherd, "You Could Be Watching Squid Game Wrong—All Because of a Netflix Error," GamesRadar+, October 4, 2021, https://www.gamesradar.com/squid-game-translation-english-korean-closed-caption/, and Jennifer Bisset, "Still Watching Squid Game on Netflix? Change This Subtitle Setting Immediately," *CNET*, October 26, 2021, https://www.cnet.com/culture/entertainment/watching-squid-game-on-netflix-change-this-subtitle-setting-asap/. These debates reveal that the English closed captions equate to a transcription of an English dubbed version of the audio for the hearing impaired. With reference to *Squid Game*, Jihyun Cho has contended that the English closed captions that include nonverbal descriptions are "more concise" than the standard English subtitles. See Cho, "Squid Game and the 'Untranslatable': The Debate around Subtitles Explained," *Conversation*, October 14, 2021, https://theconversation.com/squid-game-and-the-untranslatable-the-debate-around-subtitles-explained-169931, and Catalleya Storm, "Squid Game Exposes Issue with English vs. English CC," *Hearing Like Me*, December 22, 2021, https://www.hearinglikeme.com/squid-games-exposes-issue-with-english-vs-english-cc/.

[3] For examples of untranslatable Korean honorifics, titles, and appellations in *Squid Game* and *Parasite*, see Cho, "Squid Game and the 'Untranslatable'"; Jieun Kiaer and Loli Kim, "One-Inch-Tall Barrier of Subtitles: Translating Invisibility in *Parasite*," in *The Soft Power of the Korean Wave: Parasite, BTS and Drama*, ed. Youna Kim (London: Routledge, 2022), 90–103; and Evanthia Andrianaki, "An Analysis of the Translation of Culture-Specific Concepts in the Subtitles of *Parasite* and *Squid Game*," PhD diss., Ewha Womans University, 2021.

Mubi's service, and it has become Mubi's most streamed film in North America to date. Via Mubi, the film is also available in India, Turkey, and the UK, with subtitles available in French and Turkish as well. On peer-to-peer platforms such as YIFI Subtitles, one can download subtitles for the film in twenty-one additional languages, including Arabic, Danish, Dutch, Farsi/Persian, Greek, Hebrew, Indonesian, Italian, Japanese, Malay, Portuguese, Romanian, Spanish, Swedish, Thai, and Vietnamese. These subtitles are typically written and uploaded by various subtitlers working under their user ID pseudonyms. The subtitles I examine in this chapter are Paquet's official English subtitles on Mubi, which have subsequently been adopted by other major streaming platforms, including Kanopy.

Decision to Leave exemplifies the ways subtitled cinema can produce an immersive, multilingual, and multisensory experience for viewers, even as it might at times deflate aesthetic ambiguity or flatten out characters and cultural complexity. These aspects of subtitled cinema become particularly significant in the era of streaming, as platforms like Mubi habituate viewers to watching material in translation, thereby opening up at once new possibilities and limitations for engagement with film form as well as cultural specificities.

Theorists of subtitling have called attention to how, despite their known function as mediating agents, subtitles or closed captions can distort rather than neutrally transfer meaning. As scholar Jihyun Cho notes, the process of subtitling requires "not only bilingual competence" with an understanding of nuanced cultural elements but also "specific skills essential to deliver messages within a limited space on screens."[4] Indeed, the epistemological validity of on-screen translations goes beyond the question of semantic accuracy. In her study of English subtitles of Indian rap songs in Bollywood movies, Ritika Sinha notes the formal constraints of subtitling, where "generally a maximum of 2 lines and 35 characters" is recommended for translators to adhere to the "average reading speed of the viewers" in English. These constraints reinforce the fact that "translation is primarily contextual" and "cannot be explained through the mapping of linguistic correspondence between languages, or judged with respect to universal standards of quality and accuracy."[5] Jorge Díaz-Cintas even asserts

[4] Cho, "Squid Game and the 'Untranslatable.'"

[5] Ritika Sinha, "Lost in Translation: Culture-Bound Lexical Items in English Subtitles of the Rap Songs by Indian Rapper 'Badshah' in Bollywood Movies from 2016–2021," *Rupkatha Journal on Interdisciplinary Studies in Humanities* 13, no. 4 (2021): 2–3.

that the re-creations and substitutions for culture-bound terms "[raise] questions about the power of subtitling in the dissemination and entrenchment of certain concepts and realities in other cultural communities" and that "commercial forces and colonising practices cannot be excluded from this debate."[6] Building on these theories, and drawing on the example of *Decision to Leave*, I argue that subtitles ought to be seen as cultural texts in their own right that reveal complex layers of social and industrial contexts in which subtitlers operate.

Set in South Korea, *Decision to Leave* follows a Korean detective, Hae-jun, who finds himself attracted to a widow, formerly married to the victim in his murder case. The widow, Seo-rae, is a Chinese immigrant who has lived in Korea with her late husband and has learned to speak advanced Korean. As Hae-jun attempts to solve the mystery surrounding the late husband's death, he realizes that the murderer is the wife, Seo-rae, for whom he has by now developed romantic feelings. But with the case already closed as a suicide, Hae-jun leaves Seo-rae, telling her to throw her smartphone containing the evidence of her murder into the sea. After a series of intricate events, Seo-rae takes an action that leads to more deaths, including the death of her new husband, Ho-sin. At the end, she buries her own body in the sand near the sea, thus permanently leaving Hae-jun by taking her own life.

In the diegetic world of the film, wordplay and digital translation apps foster misunderstandings and lead to multilingual, multisensory humor. At times, paralingual forms of communication coupled with dialogue and audio mimicking AI-read texts intentionally evoke ambiguous moods and meanings that are not directly translatable. As a result, Hae-jun must often reassess the ways he uses his native language to express his interiority, which divulges his linguistic *limitations* regardless of his native abilities. Furthermore, Seo-rae's verbal expressions, as Hae-jun's "unresolved mystery," remain illegible throughout the film's plot, hence refusing to satisfy Hae-jun and, by extension, the non-Chinese-speaking audience in their desire to conquer the "unknown" epistemologically through translation.

Critic Moeko Fujii notes about the English subtitled version of the film that oftentimes when Seo-rae speaks into the AI translator in Chinese, "the non-Chinese-speaking audience sits through the beat of sound with Hae-jun," since

[6] Jorge Díaz-Cintas, "The Name and Nature of Subtitling," in *The Palgrave Handbook of Audiovisual Translation and Media Accessibility*, ed. Łukasz Bogucki and Mikołaj Deckert (Cham, Switzerland: Palgrave Macmillan, 2020), 168.

the subtitles do not offer translations of Seo-rae's spoken Chinese.[7] In other words, the English-speaking audience must wait for the subtitles as they are digitally translated into Korean for the protagonist Hae-jun, who is subject to the technological and linguistic algorithms that birth the translations. However, as Fujii adds, the same applies to Seo-rae in certain moments in the film. Although fluent in Korean, she must nevertheless look up unfamiliar words, which makes her unable to react or respond immediately, even when faced with Hae-jun's highly emotionally charged language that showcases his vulnerability. The film thus disrupts the notion of "translation as a highway." Translation emerges in the film as "not an innocuous agent of clarity wiping fog away from a window" but rather the fog itself, a two-way impasse that complicates the denominations of power, such as "colonized and colonizer, spectacle and voyeur, suspect and detective."[8] Any party in the audience who does not know either Chinese or Korean is subject to a narrative restriction through a spatiotemporal reliance on waiting, reading, or listening to the translations provided for the characters in the film. This raises the following question: how might a multilingual film like *Decision to Leave* prompt emotive and cognitive variations among audiences that know none, one, or all the languages involved? My textual analysis addresses this question and thereby reflects on how subtitled entertainment from the Korean Wave mediates the ideal of a transcultural and transnational universality in the age of streaming.

Subtitled Cinema: Constraints and Inventions

Scholar Niall Flynn expounds that "imperfections are part of the experience of subtitled cinema—assimilated or at least accepted by viewers as a necessary trade-off that highlights the inadequacy of translation."[9] *Decision to Leave* offers an exemplary case of the imperfections of subtitled cinema. First, the film itself thematizes the limitations of language, but the film's translation into multiple languages for global streaming adds another layer to its thematic commentary.

7 Moeko Fujii, "Refusing to Leave: Misread Language in Decision to Leave and the Noriko Trilogy," *Orion Magazine*, April 11, 2023, https://orionmagazine.org/article/late-spring-movie-yasujiro-ozu-reivew/.

8 Ibid.

9 Niall Flynn, "An Intimate Encounter: Negotiating Subtitled Cinema," *Open Library of Humanities* 2, no. 1 (2016): 1–25.

The English subtitles, for instance, incorporate substitutions that do not lexically correlate to the source text but convey additional information to deliver a thematic arc for viewers. At the same time, such subtitles subdue the subtleties of atmospheric or allegorical elements tied to the film's themes of mystery and misunderstandings that fuel a sense of desire. Paquet, as the subtitler, attempts to embody these different registers all within the text, by providing steppingstones over an unfathomable river of images and sounds hinged to the source language, in order to arrive safely at a destination in the target language spoken by viewers with varying degrees of linguistic and cultural competency. Given the limitations inherent within this translation process, subtitles exist as paratexts that often reveal *more* information rather than *less*, while nonetheless atmospherically reducing the poetic effects delivered through idiomatic expressions in the source language.[10]

In *Decision to Leave*, the subtitles often reveal information at intervals different from the original utterances because they might add what is hidden in the source text to maximize the reader's understanding of the film's central thematic arc. One example is found in the English subtitles for one of the film's diegetic songs, "Mist," a Korean song by Jung Hoon-hee first released in 1967. Here, the English subtitles modify the original lyrics to foreshadow an upcoming element in the dialogue. In the subtitles, the song's lyrics read, "I walk alone / on the street / and filled with mist / Is that your *silhouette*?" However, the lyrics that come after the phrase "and filled with mist" in the original song are "the single shadow of you who was once so kind." When the song plays in the film, the audio stops just after the Korean word for "once," which is the first word in the phrase in Korean and is articulated before "shadow." But the English subtitles complete the interrupted phrase with "Is that your *silhouette*"—in order to foreshadow the association of Seo-rae with a silhouette in a dialogue occurring fifty minutes into the film, when Hae-jun timidly talks about the beauty of Seo-rae's "silhouette," a word he dictates in English. The subtitles explicitly use the English word "silhouette" twice, first in alluding to the lyrics of the song and

[10] For a detailed history and theoretical implications of audiovisual subtitles, including silent movies, intertitles, early talkies, multilingual films, dubbing, accessibility issues, fansubbing, fandubbing, and crowdsourcing, see Elisa Perego and Ralph Pacinotti, "Audiovisual Translation through the Ages," in *The Palgrave Handbook of Audiovisual Translation and Media Accessibility*, ed. Łukasz Bogucki and Mikołaj Deckert (Cham, Switzerland: Palgrave Macmillan, 2020), 33–56, and Gian Maria Greco and Anna Jankowska, "Media Accessibility within and beyond Audiovisual Translation," in *The Palgrave Handbook of Audiovisual Translation and Media Accessibility*, ed. Łukasz Bogucki and Mikołaj Deckert (Cham, Switzerland: Palgrave Macmillan, 2020), 57–81.

second in a later scene involving Hae-jun's dialogue, perhaps for the purpose of maintaining narrative cohesion. The rearranged lyrics of the song in the subtitles reveal a clearer relevance of the soundtrack to the English-speaking audience, rather than to a Korean-speaking audience member who may be less clear about the song's narrative function since even the mention of "shadows" in the Korean lyrics gets cut off in the film.

The subtitles sometimes also invent alphabetized alliterations that imitate the effects of words through a different English consonant while maintaining similarity in meaning. When describing the reliability of nuclear reactors where his wife works as an operator, Hae-jun humorously utters a tagline from her workplace, "원자는 완전 안전 (*won-ja-neun wan-jeon an-jeon*)," meaning "Nuclear Is Completely Safe."[11] But to deliver the wordplay in the alliteration using the Korean equivalent of the sound *w* in English, Paquet creates a new alliteration using the letters "cl," hence changing the source text to the target text as "Clearly Cleaner Nuclear." In another case, to maintain an acronymic form in the original Korean, the subtitler creates a different acronym in English, even if it forsakes the meaning in the source text.

Such substitutions alter the experience of viewers watching the film with subtitles; while Korean-speaking viewers would be encouraged to decode or be left in the unknown regarding cultural elements such as referencing existing songs, audiences reading subtitles on streaming platforms are unambiguously provided with additional meaning, which restricts them from being active agents even when given the option to hit pause and reflect on the written phrases during or after the initial film viewing.

Another example of an inventive textual substitution in the film is when the subtitles attempt to deliver verbal humor. When Seo-rae's new husband, a stock market analyst, jokes that he is "not an *analyst* because [he] likes the anus," Mubi's subtitles translate the line as "not an *an a list* who likes rear ends."[12] In spoken Korean, the word used for "anus" in the sentence is "항문 (*hang-mun*)," a homonym of the Korean word for "학문 (*hak-mun*)," meaning "field of study." The ambiguity inherent in Ho-sin's joke has the potential to confuse Korean-speaking audiences, who may think the character is expressing his lack of interest in the scholarly realm of stocks. English subtitles, however, can accelerate legibility to solicit a reaction such as laughter the moment the words

[11] All Korean dialogue not mentioned as "subtitles" in this essay have been translated by the author.
[12] In this line, the character Ho-sin says the word "analyst" in English.

appear on the screen. By phonetically transcribing the letters "n," "a," and "l" for "analyst," and by writing "an a list," the subtitles not only break down the word "analyst" into three syllables to specify that he is referring to the three-syllabled English word "analyst" but also change the suffix from "-yst" to "-ist" to deliver the humor that he is referring to a type of person who has a fetish for anal sex. In the original Korean dialogue, there is a high possibility for this English-based joke to fall through, as is also made clear by the other characters listening to Ho-sin's joke in the film, who appear speechless, given the few seconds of confusion that his joke warrants.

In other instances, the nuances in the source text are elided to speedily arrive at informative substitutes. Some examples include writing "downtown Seoul" in lieu of the precise name of the street, "Euljiro," and changing "warm food" to "real food" when referring to the high quality of Hae-jun's warm, home-cooked meals for his wife. In one scene, the subtitles read "A Chinese name?" instead of the original line, "Is there a last name called Sa?" to inform the audience that there is no such last name "Sa" in Korea, hence foreshadowing that the character in question is undoubtedly a person of Chinese rather than Korean lineage. Although the English substitutes in the subtitles aim to alert viewers of the intended context potentially alluded to by the characters in the dialogue, they also end up modifying the quality of audience engagement with the film. If the subtitles were to keep the name of the street or translate the phrases more literally, non-Korean audiences may be prompted to either research or ponder the meaning of the original dialogue, not too differently from the way Korean audiences would watch the film. Equipped with additional context, readers of subtitles with no knowledge of the spoken language are made to rely more on given information rather than on actively decoding meaning in a form of exchange that merely hints at bits of information in the dialogue.

Apart from revealing pieces of information ahead of a scene, subtitles may change the modality of a word to describe a character's experience, often in clearer and less poetic terms than those used by the character. For example, a close-up shot reveals Seo-rae's smartphone as she looks up the meaning of "붕괴 (*boong-goe*)," a Korean word for "collapse" used by Hae-jun to describe his psycho-emotional state of being (Figure 9.1). For this shot, the subtitle shows the definition of the Korean word as "shattered" (Figure 9.2).

However, in Korean, "*boong-goe*" is used primarily to describe the physical collapse of large-scale architectures like buildings, and because such collapses

Figure 9.1 The subtitle translates the Korean word *boong-goe*, which Hae-jun uses to describe his psycho-emotional state of being, as "shattered." *Decision to Leave*, directed by Park Chan-wook © Moho Films 2009.

Figure 9.2 A close-up frame of Seo-rae's smartphone as she looks up the Korean word *boong-goe* using an online dictionary, with the English subtitle maintaining its translation of the word as "shattered." *Decision to Leave*, directed by Park Chan-wook © Moho Films 2009.

are often used in situations of unexpected accidents, an example phrase in the dictionary (that is revealed in the extreme close-up shot of her smartphone but omitted in the subtitles) reads, "붕괴 위험 (*boong-goe wi-heom*)," meaning "danger of collapse" (Figure 9.3). When Hae-jun first expresses his feelings to Seo-rae with the words "*Na-neun-yo wan-jeon boong-goe duet-seo-yo*" (I have completely collapsed), which prompts her to look up the definition of the word, the film subtitles this line as "I'm completely shattered," which fails to convey the monumental scale and physicality embedded in the word

Figure 9.3 An extreme close-up frame of Seo-rae's smartphone as she reads the dictionary definition of *boong-goe*, with the subtitle revealing its translation of the definition in the first line while omitting a contextual example written in the second line. *Decision to Leave*, directed by Park Chan-wook © Moho Films 2009.

boong-goe that would inspire a mental image of Hae-jun as a once-stable structure collapsing to the ground.[13] Additionally, the extreme close-up of Seo-rae's smartphone after she looks up the word reveals the full definition of *boong-goe* as a combination of two verb forms, "무너지고 (*moo-neo-ji-go*)" and "깨어짐 (*kkae-eo-jim*)," with *moo-neo-ji-go* being equivalent to a verb form of "collapse," combined with the Korean word for "and," and *kkae-eo-jim*, meaning "to BE broken." This results in the full dictionary definition of the word *boong-goe*, shown on the screen as "collapse *and* be broken," which poetically describes how Hae-jun sees himself in a certain trajectory of disintegration that involves a kinetic movement of his body collapsing to the ground *and* breaking. However, the English subtitles for this shot containing the full definition of the word read, "ruined or broken"—words that appear as adjectives or past participles of the words "ruin" and "break."

These translations connote an endpoint without a sense of movement or trajectory. Hence, a direct translation that could have worked is Hae-jun saying, "I have completely collapsed," which would comprise an intersemiotic mode of translating a dictionary definition on-screen to translate Hae-jun's internal state of being in a literal sense.

[13] It can also be translated as "I have been completely collapsed" or "I have become completely collapsed," as the Korean word *boong-gue*, when applied as a verb, is used passively. This aspect further describes Hae-jun as a passive entity, a victim of his falling for Seo-rae, and reveals the challenges of directly translating this expression into English, as the English word "collapse" is rarely used as an adjective.

Based on the subtitles, Hae-jun is a "shattered," "ruined *or* broken" character, which does describe his state of being in this scene but without the subtle affective elements embedded in the literal definition shown in the Korean dictionary. While the subtitles can speed up one's understanding of Hae-jun's state by having the audience accept his expression as figurative rather than literal, they also flatten out Hae-jun's linguistic imagery by offering a translator-subtitler's interpretation of Hae-jun's architectural "collapse" as being equivalent to a state of being internally "shattered."[14]

The film also thematizes translations as physical objects in the plot, which potentially reminds any audience reading subtitles about the mediated nature of their experience. In one scene, Hae-jun reads a printed transcript of Seo-rae's verbal monologues in Chinese recorded on her smartwatch as part of his case evidence. As Hae-jun reads out loud the translated texts in Korean, the printed texts fill up the film's entire frame, immersing the assumed Korean-speaking audience in the process of taking a set of translations without the means to verify its tonal reliability. Hae-jun reads the same transcript while imagining her voice playing out in his head—something like reading closed captions that indicate the inclusion of laughter, or like parenthetical directions in a screenplay—"웃음 (laughter)"—while the audience, as the extradiegetic spectator, hears and sees Seo-rae in the moments when she is laughing and speaking in Mandarin. Hae-jun, who does not know Chinese, reads each line in the transcript as a Korean reader would read the subtitles accompanying the images of Seo-rae on-screen. Meanwhile, readers and speakers of Mandarin Chinese who do not know Korean can only identify with Seo-rae, whose speech is being translated and read by Hae-jun and the Korean-speaking audience. As a result, character perspective, which is a key aspect of audience identification, is mediated through the fog of translations as well as subtitles.

On the whole, even if a viewer may not be able to identify with either or both Hae-jun and Seo-rae through linguistic affiliation, the sounds and scripts of distinct languages, coupled with the presence of subtitles on-screen, create an immersive, dreamy, and multisensorial aesthetic experience via

[14] Even some English-speaking audiences have on an intuitive level picked up on the limitations of the subtitles, especially since the protagonists' multilingual (mis)communication is the pivotal driving force behind the film's theme of misunderstandings and cultural gap that fuel romantic tension. On NPR, Glen Weldon highlights the English speakers' inevitable reliance on "the film's translators" to cue them in on the exact language gap between the protagonists and confirms that the function of the word "shattered" in the subtitles, for the scene featuring the online dictionary, is practically lost to English speakers. "Murder is Lost in Translation in 'Decision to Leave,'" NPR, October 24, 2022, https://www.npr.org/transcripts/1127208739.

multilingualism.[15] To draw on Flynn's description of subtitled cinema, the film offers multiple "registers of experience—aural, visual, graphic, aesthetic, textual," and the anticipation of subtitles at times functions "as if it were an aesthetic effect."[16] What goes beyond the representational parameters enters a space of negotiation in which readers of subtitles engage in multisensorial ways, in order to relate without fully grasping all the linguistic nuances.

To produce this multisensorial aesthetic experience, subtitles may at times veer toward deliberate mistranslation. In the film, Hae-jun and Seo-rae frequently rely on an AI translator app to intermedially communicate in translated Chinese and Korean, using speech-to-text and text-to-speech audio tools. The misunderstandings that come from the imperfections of AI translators lend themselves to humor, functioning as a catalyst that creates an intimate bond between the characters. To re-create this element of misunderstanding, the subtitles invent a new form of discrepancy in their indirect dialogue. When Hae-jun listens to an AI-translated text from Seo-rae's spoken Mandarin Chinese after secretly recording her verbally speaking to a cat, the machine dictates its translation as, "Another crow? To thank me for feeding you? No need for that. If you must give me a present, bring me the head of that kind detective. That is what I want." In the original Korean, the last line reads, "*please* bring me the *heart* of that kind detective. I kind of want that," which humorously shows the errors in AI translators that often misread whether the tone is intended to be extremely formal or informal. However, the English subtitle not only maintains an informal register throughout but also changes the word "heart" to "head." The reason for this significant change is that the Korean word for "heart" used in this scene, *simjang*, is an anatomical term to denote the physical heart rather than a figurative one to indicate her intimate feelings. In the film, Hae-jun's AI translator has incorrectly chosen the anatomical word to translate Seo-rae's Chinese word for "heart," "心 (*xīn*)," which is used in both anatomical and figurative contexts.

On Mubi's exclusive digital publication *Notebook*, Paquet assumes that the audience will understand the line "Bring me the head of that kind detective" as

[15] This emphasis on the characters' fluid use of their native and nonnative languages also differs from the multicultural films in which the characters speak English as a default language of communication. One example is *Late Autumn* (2010), another Korean film that depicts a China-Korea interethnic love story in which the protagonists use English as their mutual language. For more analysis on the issue of language in *Late Autumn*, see Dorothy Wai Sim Lau, "Tang Wei's Stardom in and beyond Sinophone Media Culture: Linguistic Versatility, Public Reception, Chinese-Korean Cultural Imagination," in *Reorienting Chinese Stars in Global Polyphonic Networks: Voice, Ethnicity, Power* (Singapore: Palgrave Macmillan, 2021), 89–109.

[16] Flynn, "Intimate Encounter," 3.

a reference to Sam Peckinpah's 1974 film *Bring Me the Head of Alfredo Garcia*. Paquet expressed his concerns about this translation to the director, and Park approved of the translational change.[17] The subtitler's assumption speaks to the type of curatorial connection Mubi imagines with its target audiences that primarily comprise cinephiles. To avoid what the translator-subtitler saw as the greater danger of a character's supposed misunderstanding not being transferred to the audiences at all, the English translation substitutes the word "heart" with "head," as the English word for "heart" refers to both the organ and one's feelings, similarly to the Chinese term as used by Seo-rae. A direct translation that could have worked here is, "Please bring me the cardiovascular organ of that kind detective," which could maintain the source text's uncannily violent yet erotically charged expression born from a mistranslation. This deliberate change of the word "heart" with "head" makes for an unambiguous sentence, yet it significantly reduces the intensity and erotically charged element of Hae-jun's (mis)understanding of Seo-rae's phrase.

Similar issues arise even for audiences from the Korean diaspora. When Hae-jun hears the word *simjang* through the AI translator, the word can be immediately interpreted by a Korean-speaking audience as either a downright digital mistranslation or an earnest expression intended by Seo-rae to metaphorically deliver her feelings. According to columnist Park Ki-chan, the word *simjang* is often used in North Korea when referring to a "deep, sincere feeling or mournful emotion," especially in literary works or romantic situations.[18] But because the word is not used in such contexts as frequently in South Korea as it is in North Korea, the AI translation of the word in the film for South Korean audiences could emerge as a purely anatomical reference, and yet one that suggests an ambiguous and comic expression with potentially erotic undertones. These discrepancies show that cultural variations exist even in the same language. Hence the subtitle's aim to deliver meaning "accurately" or without any ambiguity is not always possible, and subtitling should therefore not be thought of as a curative process that replicates the experience of the audiences watching the film in its original language.

[17] Darcy Paquet, "What I Learned Translating 'Decision to Leave,'" *Notebook*, December 19, 2022, https://mubi.com/en/notebook/posts/what-i-learned-translating-decision-to-leave.

[18] Park Ki-chan, "'Bring Me the Heart,' the Reason Why North Korean Expressions Come Through," *Columnist*, July 30, 2022, https://www.thecolumnist.kr/news/articleView.html?idxno=1192. Translations by the author of this chapter.

Subtitling the Korean Wave in the Age of Streaming

Language differences and the role of subtitles figure, though in a relatively auxiliary role, within mainstream discussions about the global popularity of Korean Wave entertainment. The online debate on the different subtitles settings on Netflix for *Squid Game* came about after the series had gained significant popularity for its provocative depiction of violence. Geon-Cheol Shin and Mark D. Whitaker detail the ways the Korean series would not have been possible without Netflix, as the director Hwang Dong-hyuk failed to find a domestic Korean producer to realize the show for ten years.[19] This phenomenon shows that even the "Korean wave" is not necessarily a centrifugal movement of cultural influence, spreading from South Korea to other places in the world, but rather a multidirectional flow often prescribed and promulgated by global streaming platforms as linguistic facilitators, with subtitles and online discussions occurring on local levels. Speaking of a "global Korean Wave," Shin and Whitaker further stress that what catches the interest of platforms like Netflix—which services content "to over 230 million subscribers world-wide"—is not the word "Korean" but rather "global."[20] Describing the visuals in *Squid Game* as "the color of bright red blood" speaking the "'original language' of humanity" to the audience at large, Shin and Whitaker also indicate that its "global persuasiveness in terms of genres and themes," such as the critique of class conflict, has been paired with an emphasis on visuals "to cross global language and cultural barriers."[21]

If Netflix has been crucial to the success of Korean Wave shows like *Squid Game*, then Mubi, which distributes and streams international films with subtitles, functions as a paratextual gateway for acclaimed works like *Decision to Leave* to assume the arthouse label as part of a curated, transnational canon. As Roderik Smits and E. W. Nikdel note, "MUBI launched in 2007… as an information-sharing platform for a global audience of online cinephiles,… developed to inform the public about the latest 'quality' auteur films from across the world."[22] Mubi's curation goes beyond English-language films, and

[19] Geon-Cheol Shin and Mark D. Whitaker, "Squid Game between Global OTT Services," in *The Korean Wave in a Post-Pandemic World: BTS, Cosmax and Squid Game* (Singapore: Springer, 2023), 527–59.

[20] Ibid., 547.

[21] Ibid., 537–8.

[22] Roderik Smits and E. W. Nikdel, "Beyond Netflix and Amazon: MUBI and the Curation of On-Demand Film," *Studies in European Cinema* 16, no. 1 (2019): 22–37.

the platform does not present subtitles as a barrier that one must overcome but rather as a niche point of connection for online cinephiles.

For its benchmarking of "quality" and "taste" based on user ratings, reviews, and personally curated lists of films from different parts of the world, Mubi earned Bloomberg's promotion of the platform as the capital for "independent, foreign, and classic film[s], not Hollywood blockbusters."[23]

Since its founding as The Auteurs in 2007, Mubi initially failed "as an aggregator of arthouse films," Matthias Frey notes, but later built "credibility in the industry... among users" based on "word-of-mouth" and the flow of activity in a, now-eliminated, anglophone social networking community called Mubi Social that functioned as a discussion board for passionate cinephiles.[24] Jennifer Hessler argues that "concepts of auteurism and transnationalism [are manifested]... in the site's distribution model,... platform design and in the social interactions," enabling the platform to be akin to "the godfather of taste making."[25] Hessler also notes that Mubi in the United States has always featured films of multiple languages and different national categories, underscoring each film's "country of origin" as one method of searching through their catalogue.[26] The platform's founding partnership with Celluloid Dreams, a European distribution and sales company, helped localize subtitles, which led to a smoother and more prolific acquisition of foreign films as a basis for their founding logic.[27] Smits and Nikdel observe that "MUBI's binary role as tastemaker and educator is further strengthened by their digital publication, *Notebook*." Displaying "[an] intellectual tone and analytical... writing," *Notebook* attests to "a certain level of cultural capital needed to command the role of tastemaker and cultural gatekeeper, [serving] to strengthen the trust and sense of loyalty that MUBI members invest in the platform."[28] Thus, Mubi has engineered a mode of "credibility as an (art film) industry actor" from its very inception, and its curation has "fill[ed] a market gap" by appealing to those who have long wanted to watch critically acclaimed foreign films online.[29]

23 "MUBI CEO Efe Cakarel on the Streaming Landscape," Bloomberg TV, YouTube, accessed January 15, 2023, https://www.youtube.com/watch?v=2hqwcfMdVZM&ab_channel=MUBI.

24 Mattias Frey, *MUBI and the Curation Model of Video on Demand* (Cham, Switzerland: Palgrave Macmillan, 2021), 54.

25 Jennifer Hessler, "Quality You Can't Touch: Mubi Social, Platform Politics, and the Online Distribution of Art Cinema," *Velvet Light Trap*, no. 82 (Fall 2018): 4.

26 Ibid., 8.

27 Ibid., 8.

28 Smits and Nikdel, "Beyond Netflix and Amazon."

29 Frey, *MUBI and the Curation Model*, 54–6.

Such practices executed on the platform further reinforce the validity and officiality of its subtitles. But, as Paulina Pérez Reyna observes, auteur directors have also started working with Netflix, as in the case of Alfonso Cuaron's Spanish-language film *Roma* (2018), after Criterion ended its contract with Mubi and began its collaboration with Hulu as well as its own streaming platform in 2016 to drive cinephile audiences away from Netflix.[30] As the debate surrounding Netflix's *Squid Game* has shown, subtitling on platforms like Netflix that show mainstream television series are subject to far more criticism relative to the subtitling on Mubi, not surprisingly due to the latter's branding as a trustworthy source of international arthouse cinema, even as its business strategies to exercise more control over distribution share similarities with other streaming powerhouses such as Netflix and Amazon.[31]

While an emphasis on diversity forms the basis of Mubi's "arthouse credo," inclusive of "culture, time period, style, subject and attitude," there is a relative lack of clarity on the platform and its related sites regarding the formulation and reception of subtitles that linguistically deliver content imported from different cultures.[32] When Mubi Social was operational, universalizing ideals prone to Western hegemony commonly informed the point of view of the anglophone discussants on issues such as eliminating differences among non-Western cultures and the dominance of English as a canonical language for transcultural exchange. Yet conversations abounded with users' interest and expertise on foreign films, with "'foreignness' on the site reinforced… as an essential aspect of cinephilia."[33] Hessler assesses that the "negotiation of the national" for films encouraged "users to adapt, contest, or redefine terms like 'national cinema' and 'world cinema,'" creating imagined communities that transcend geographical and cultural boundaries while also inadvertently underscoring a sense of essential national differences among the films.[34] As the digital space for active discussion and reception continues to shift, with many users moving to other

30 Paulina Pérez Reyna, "Auteurs on Netflix: A Reception Study of *Roma* (Alfonso Cuarón, 2018)" (MA diss., University of Oslo, 2019), 25.

31 On the status of Netflix as more akin to "television" and not "cinema," see Diane Burgess and Kirsten Stevens, "Taking Netflix to the Cinema: National Cinema Value Chain Disruptions in the Age of Streaming," *Media Industries* 8, no. 1 (2021): 67–87, which explores the ways the cinema industry can adapt existing value chains to diversify its curations for film festivals and sponsorships beyond its "cultural gatekeeping."

32 Frey, *MUBI and the Curation Model*, 80.

33 Hessler, "Quality You Can't Touch," 10.

34 Ibid., 10. I use the terms "imagined communities" as conceptualized in Benedict Anderson, *Imagined Communities: Reflections on the Origin and Spread of Nationalism* (London: Verso, 1983).

platforms such as Letterbox upon Mubi Social's disintegration in 2015, the primary language of online communication continues to be English, leaving out many non-English-speaking viewers from these discussions that could potentially address the nature of subtitling as well as the problems with notions of "foreignness" and absolute cultural difference between the anglophone world and the rest.[35]

Mass popularity among international audiences and success in multiple film festivals have subdued any debate or criticism over its subtitling, yet *Parasite*'s English subtitles on a global educational platform like Kanopy that makes films, including those from Mubi and Criterion, available for public and academic libraries also harbor room for cultural debate. For example, in *Parasite*, when Yeon-kyo, the mother of a wealthy family, describes her youngest son Da-song's obsession with Native American costumes and features, the English subtitle translates her line, "From last year he's been an Indian otaku" to "From last year he's been an Indian fanatic." While the word "Indian" has not been changed to directly show the wealthy mother's limited cultural understandings of the indigenous peoples of the Americas, the subtitles replace the Japanese word "*otaku*" with "fanatic." According to the *Oxford English Dictionary*, an otaku is "a person extremely knowledgeable about the minute details of a particular hobby (especially a solitary or minority hobby)," specifically "one who is skilled in the use of computer technology and is considered by some to be poor at interacting with others."[36] The use of a word with Japanese roots to describe Da-song's obsession keenly reflects the complexities behind a cultural interest in othered forms within uneven power dynamics, as well as the present-day Korean population's abundant use of Japanese or Japanese-inspired terms as a postcolonial subject. While the substitute "fanatic" crudely reflects the mother's limited cultural understanding and the degree of Da-song's superficial obsession with Native American costumes and features, the English term also obfuscates complex layers of coloniality of different national contexts by eliding the term's Japanese reference used alongside the mother's culturally inappropriate use of the term "Indians" to refer to Native Americans. These subtitles, which are available on the worldwide platform Kanopy, have not been critically debated, despite its widespread availability and global channeling through public and academic spaces for educational and archival purposes.

35 Hessler, "Quality You Can't Touch," 13.

36 "Otaku," *Oxford English Dictionary*, https://www.oed.com/dictionary/otaku_n?tab=meaning_and_use#12271221.

What is also often overlooked in discussions about "foreign" films is how their various translations might diverge. The subtitles of *Decision to Leave* that appear on Mubi differ from those of the theatrical versions of the film that have screened in various countries and languages. Columnist Park also reports that the French subtitle for the phrase "Bring me the head of that kind detective," for instance, is "Bring me the body and heart of that kind detective," which attempts to maintain the "chilling, lewd" tone present in the original Korean expression.[37] However, such subtitles do not appear in the French version on Mubi, suggesting that the translation shared by the columnist was most likely from the theatrical release. But in addition to the gap between translations in theaters and on streaming platforms, subtitles might also vary depending on the platform and the sociopolitical contexts of the agents involved in the subtitling process. This is especially the case with Korean dramas that are viewable only in underground communities and subtitled by fan as opposed to professional subtitlers.[38]

It is important to distinguish fan subtitlers from professional subtitlers assigned by streaming platforms, as the former do not own any legal rights, nor do they have any limitations imposed on them by a multiparty editing process, whereas the latter need to consider the opinions of the platform, distributor, and even director. Fan subtitling in the United States started in the 1980s "in response to a need felt by Japanese anime fans to gain access to… more accurate [translations] of the cultural elements they contained."[39] In Thomas C. Messerli and Miriam A. Locher's study comparing the fan subtitles on Viki and professional subtitles on Netflix for a Korean television series, results revealed that full authorship details for the collaborative subtitling team members was more prominently visible on Viki than on Netflix. Timed comments by viewers

[37] Park, "'Bring Me the Heart.'" Translations by the author of this chapter.

[38] Yuanyuan Chen and Rebecca Crawford, "Danmaku Commenting: A New Interaction Experience on China's Streaming Platforms," in *World Cinema on Demand: Global Film Cultures in the Era of Online Distribution*, ed. Stefano Baschiera and Alexander Fisher (New York: Bloomsbury Academic, 2022), 89–107, also explores the communal commenting experience on streamed online videos in which real-time updates on subtitles appear as user-based comments or input on the content for a "co-viewing experience." Authors state that experts on internet slang in these practices transmit popular subculture entities hinged on another layer of a viewing experience that superimposes itself onto the original media content.

[39] Perego and Pacinotti, "Audiovisual Translation through the Ages"; For a detailed assessment of algorithmically regulated platforms redefining the modes and conditions of public life and the "mass" in mass media, see Aras Özgün and Andreas Treske, "On Streaming-Media Platforms, Their Audiences, and Public Life," *Rethinking Marxism* 33, no. 2 (2021): 304–23. For a detailed analysis and history of fansubbing, see Serenella Massidda, "Fansubbing: Latest Trends and Future Prospects," in *The Palgrave Handbook of Audiovisual Translation and Media Accessibility*, edited by Łukasz Bogucki and Mikołaj Deckert (Cham, Switzerland: Palgrave Macmillan, 2020), 189–208.

accompanying the videos showed "a great [critical] awareness of the subtitling process," and viewers actively expressed "encouragements for and thanks to the subtitling team."[40] This study also revealed that "Netflix translators used more generic European references"—potentially based on guidelines such as those mentioned by Sinha, of adhering to a purported "average reading speed of the viewers"—than the fan subtitlers on Viki. This aspect revealed the tendency among professional translators toward impersonal forms of cultural reorientation in which non-Korean and largely European equivalents are used to cater to anglophone audiences.[41] In his article "Korean TV Serials in the English-Language Diaspora: Translating Difference Online and Making It Racial," Brian Hu additionally discovers the capacity for fan subtitles to call critical attention to issues such as racial discrimination, as in the case of English subtitles for the 2004 Korean drama *A Love Story in Harvard*, by underlining and substituting English words spoken by Asian characters with more emphasis on social justice. Hu asserts that appearing in the beginning of the "Korean wave," *A Love Story in Harvard*, which garnered pan-Asian success, did not find distribution in the United States due to the visually suggested elements of the Caucasian American characters' racism in the drama, which the fan subtitlers later proved as being literally substantial and blatantly apparent through their translations of the Asian characters' spoken words revealing their experiences with racism in the narrative.[42]

Given the difference between fan and professional subtitlers, and given also how subtitles in theaters can vary from those on streaming platforms, one might ponder the extent to which the latter medium offers new options to viewers to pause, reflect, and even search online for terms phonetically inscribed in the target language. In *Decision to Leave*, the motivations and processes for second-language acquisition of both Seo-rae and Hae-jun as they learn each other's respective languages involve curiosity and mimesis that stem from the unknown or mistranslations rather than accurate deliveries via machine-generated tools. These mistranslations and imperfections echo the inevitable gaps in translational interactions.

[40] Thomas C. Messerli and Miriam A. Locher, "Contrastive Analysis of English Fan and Professional Subtitles of Korean TV drama," in *Pragmatics and Translation*, ed. Miriam A. Locher, Daria Dayter, and Thomas C. Messerli (Amsterdam: John Benjamins, 2023), 236.

[41] Ibid., 244.

[42] Brian Hu, "Korean TV Serials in the English-Language Diaspora: Translating Difference Online and Making It Racial," *Velvet Light Trap*, no. 66 (Fall 2010): 36–49.

Furthermore, they hint to international audiences that while what they are reading onscreen may not be literal translations of the source languages in the film, the subtitles are metatextual entities that leave room for the untranslated. As viewers confront italicized terms left phonetically scribed and Romanized in the target text, they, like Seo-rae and Hae-jun, are forced to actively engage with ambiguity and with the unknown.

Conclusion

Decision to Leave delineates how languages and technologies influence and shape one another, and given this thematic focus the film is especially revealing of the ways subtitled cinema reflects both challenges and possibilities in an age of global digital streaming. As much as languages undergo transformation, the circulation of subtitled cinema across streaming services, social media, and video-sharing websites needs further interdisciplinary examination. As this chapter has shown, subtitles are interpretable texts, both signifiers and signified, and works that reflect interlingual and intralingual differences. The continuing growth of global mediascapes and the shifting strategies of global streaming platforms call for critical studies of subtitles, as part of broader discussions of film circulation and distribution patterns.[43] More robust conversation is needed among scholars in cultural studies, translation studies, and media studies in order to unpack the full implications and potential of subtitling practices in the present.

Bibliography

Anderson, Benedict. *Imagined Communities: Reflections on the Origin and Spread of Nationalism*. London: Verso, 1983.

Andrianaki, Evanthia. "An Analysis of the Translation of Culture-Specific Concepts in the Subtitles of *Parasite* and *Squid Game*." PhD diss., Ewha Womans University, 2021.

Baschiera, Stefano, and Valentina Re. "Netflix in Europe: The National in Global Subscription Video on Demand (SVOD) Services." In *World Cinema on*

[43] Roderik Smits, "Circulation Patterns, Abundance, and Scarcity: Film Availability in the Online Era," *Media Industries* 9, no. 2 (2022): 54.

Demand: Global Film Cultures in the Era of Online Distribution, edited by Stefano Baschiera and Alexander Fisher, 69–87. New York: Bloomsbury Academic, 2022.

Bisset, Jennifer. "Still Watching Squid Game on Netflix? Change This Subtitle Setting Immediately." *CNET*, October 26, 2021. https://www.cnet.com/culture/entertainment/watching-squid-game-on-netflix-change-this-subtitle-setting-asap/.

Burgess, Diane, and Kirsten Stevens. "Taking Netflix to the Cinema: National Cinema Value Chain Disruptions in the Age of Streaming." *Media Industries* 8, no. 1 (2021): 67–87.

Chang, Justin."Commentary: 'The 1-Inch-Tall Barrier of Subtitles': Bong Joon Ho Rightly Calls out Hollywood Myopia." *Chicago Tribute*, January 7, 2020. https://www.chicagotribune.com/2020/01/07/commentary-the-1-inch-tall-barrier-of-subtitles-bong-joon-ho-rightly-calls-out-hollywood-myopia/.

Chen, Yuanyuan, and Rebecca Crawford. "Danmaku Commenting: A New Interaction Experience on China's Streaming Platforms." In *World Cinema on Demand: Global Film Cultures in the Era of Online Distribution*, edited by Stefano Baschiera and Alexander Fisher, 89–107. New York: Bloomsbury Academic, 2022.

Cho, Jihyun. "Squid Game and the 'Untranslatable': The Debate around Subtitles Explained." *Conversation*, October 14, 2021. https://theconversation.com/squid-game-and-the-untranslatable-the-debate-around-subtitles-explained-169931.

Díaz-Cintas, Jorge. "The Name and Nature of Subtitling." In *The Palgrave Handbook of Audiovisual Translation and Media Accessibility*, edited by Łukasz Bogucki and Mikołaj Deckert, 149–71. Cham, Switzerland: Palgrave Macmillan, 2020.

Flynn, Niall. "An Intimate Encounter: Negotiating Subtitled Cinema." *Open Library of Humanities* 2, no. 1 (2016): 1–25.

Frey, Mattias. *MUBI and the Curation Model of Video on Demand*. Cham, Switzerland: Palgrave Macmillan, 2021.

Fujii, Moeko. "Refusing to Leave: Misread Language in *Decision to Leave* and the Noriko Trilogy." *Orion Magazine*, April 11, 2023. https://orionmagazine.org/article/late-spring-movie-yasujiro-ozu-reivew/.

Greco, Gian Maria, and Anna Jankowska. "Media Accessibility within and beyond Audiovisual Translation." In *The Palgrave Handbook of Audiovisual Translation and Media Accessibility*, edited by Łukasz Bogucki and Mikołaj Deckert, 57–81. Cham, Switzerland: Palgrave Macmillan, 2020.

Hessler, Jennifer. "Quality You Can't Touch: Mubi Social, Platform Politics, and the Online Distribution of Art Cinema." *Velvet Light Trap* 82 (Fall 2018): 3–17.

Hu, Brian. "Korean TV Serials in the English-Language Diaspora: Translating Difference Online and Making It Racial." *Velvet Light Trap* 66 (Fall 2010): 36–49.

Kiaer, Jieun, and Loli Kim. "One-Inch-Tall Barrier of Subtitles: Translating Invisibility in *Parasite*." In *The Soft Power of the Korean Wave: Parasite, BTS and Drama*, edited by Youna Kim, 90–103. London: Routledge, 2022.

Massidda, Serenella. "Fansubbing: Latest Trends and Future Prospects." In *The Palgrave Handbook of Audiovisual Translation and Media Accessibility*, edited by Łukasz Bogucki and Mikołaj Deckert, 189–208. Cham, Switzerland: Palgrave Macmillan, 2020.

Meili, Angela. "Cinephile File-Sharing: Informal Distribution, Plural Canons and Digital Film Archives." In *World Cinema on Demand: Global Film Cultures in the Era of Online Distribution*, edited by Stefano Baschiera and Alexander Fisher, 31–45. New York: Bloomsbury Academic, 2022.

Lau, Dorothy Wai Sim. "Tang Wei's Stardom in and beyond Sinophone Media Culture: Linguistic Versatility, Public Reception, Chinese-Korean Cultural Imagination." In *Reorienting Chinese Stars in Global Polyphonic Networks: Voice, Ethnicity, Power*, 89–109. Singapore: Palgrave Macmillan, 2021.

Messerli, Thomas C., and Miriam A. Locher. "Contrastive Analysis of English Fan and Professional Subtitles of Korean TV Drama." In *Pragmatics and Translation*, edited by Miriam A. Locher, Daria Dayter, and Thomas C. Messerli, 221–48. Amsterdam: John Benjamins, 2023.

"MUBI CEO Efe Cakarel on the Streaming Landscape." Bloomberg TV, YouTube. Accessed January 15, 2023. https://www.youtube.com/watch?v=2hqwcfMdVZM&ab_channel=MUBI.

"Murder is Lost in Translation in 'Decision to Leave.'" NPR, October 24, 2022. https://www.npr.org/transcripts/1127208739.

"Otaku." *Oxford English Dictionary*. https://www.oed.com/dictionary/otaku_n?tab=meaning_and_use#12271221.

Özgün, Aras, and Andreas Treske. "On Streaming-Media Platforms, Their Audiences, and Public Life." *Rethinking Marxism* 33, no. 2 (2021): 304–23.

Paquet, Darcy. "What I Learned Translating 'Decision to Leave.'" *Notebook*, December 19, 2022. https://mubi.com/en/notebook/posts/what-i-learned-translating-decision-to-leave.

Park Ki-chan. "'Bring Me the Heart,' the Reason Why North Korean Expressions Come Through." *Columnist*, July 30, 2022. https://www.thecolumnist.kr/news/articleView.html?idxno=1192.

Perego, Elisa, and Ralph Pacinotti. "Audiovisual Translation through the Ages," In *The Palgrave Handbook of Audiovisual Translation and Media Accessibility*, edited by Łukasz Bogucki and Mikołaj Deckert, 33–56. Cham, Switzerland: Palgrave Macmillan, 2020.

Pérez Reyna, Paulina. "Auteurs on Netflix: A Reception Study of *Roma* (Alfonso Cuarón, 2018)." MA diss., University of Oslo, 2019.

Shepherd, Jack. "You Could Be Watching Squid Game Wrong—All Because of a Netflix Error." GamesRadar+, October 4, 2021. https://www.gamesradar.com/squid-game-translation-english-korean-closed-caption/.

Shin, Geon-Cheol, and Mark D. Whitaker. "Squid Game between Global OTT Services." In *The Korean Wave in a Post-Pandemic World: BTS, Cosmax and Squid Game*, 527–59. Singapore: Springer, 2023.

Sinha, Ritika. "Lost in Translation: Culture-Bound Lexical Items in English Subtitles of the Rap Songs by Indian Rapper 'Badshah' in Bollywood Movies from 2016–2021." *Rupkatha Journal on Interdisciplinary Studies in Humanities* 13, no. 4 (2021): 1–12.

Smits, Roderik. "Circulation Patterns, Abundance, and Scarcity: Film Availability in the Online Era." *Media Industries* 9, no. 2 (2022): 41–58.

Smits, Roderik, and E. W. Nikdel. "Beyond Netflix and Amazon: MUBI and the Curation of On-Demand Film." *Studies in European Cinema* 16, no. 1 (2019): 22–37.

Storm, Catalleya. "Squid Game Exposes Issue with English vs. English CC." *Hearing Like Me*, December 22, 2021. https://www.hearinglikeme.com/squid-games-exposes-issue-with-english-vs-english-cc/.

Thorton, Niamh. "Bridges, Streams and Dams: The Multiple Negotiated Strategies of Distribution and Access in Latin American Cinema." In *World Cinema on Demand: Global Film Cultures in the Era of Online Distribution*, edited by Stefano Baschiera and Alexander Fisher, 47–66. New York: Bloomsbury Academic, 2022.

Výmolová, Magdalena. "Vocabulary Enhancement through Watching Korean TV Series with English Subtitles." BA diss., Masaryk University Faculty of Arts, 2018.

10

Streaming the Transmedial and Transcultural: Contemporary African Cinemas and the Netflix Viewing Environment

Alexander Fisher

The past ten years have proven to be something of a golden age for African film, peaking with the coming of age of streaming distribution and the consolidation of the now-familiar global giants of Netflix, Amazon Prime Video, and AppleTV+, alongside regional SVOD services such as Citizen TV+ and Showmax. During this period, the emergence of SVOD has had a radical effect on the international distribution of the continent's cinema, vastly increasing the global availability of a heavily marginalized area of film culture. This phenomenon has closely followed the digital transformation of film production in Africa, which has been the fundamental driver in the emergence of nationally based, domestically oriented film industries where the likes of Kenya's Riverwood and Uganda's Ugawood have followed Nollywood's model to establish financially self-sustaining national film cultures.[1]

However, while the SVOD phenomenon has certainly improved international availability of African film, questions remain regarding the curation decisions of the leading global platforms, which ultimately determine how their users encounter the continent's output. On the one hand, the SVOD giants have provided international, highly accessible platforms for releases from relatively undeveloped national film industries, releases that might otherwise have been confined

[1] For accounts of these domestic industries, see M. Adejunmobi, "Streaming Quality, Streaming Cinema," in *A Companion to African Cinema*, ed. K. Harrow and C. Garritano (Hoboken, NJ: Wiley-Blackwell, 2019), 219–43; J. Haynes, "Anglophone West Africa: Commercial Video," in *African Filmmaking: Five Formations*, ed. K. W. Harrow (East Lansing: Michigan State University Press, 2017), 81–116; and O. Okome, "Women, Religion and the Video Film in Nigeria," *Film International* 2, no. 1 (2004): 4–13. For an analysis of Nollywood distribution, see R. Lobato "Creative Industries and Informal Economies: Lessons from Nollywood," *International Journal of Cultural Studies* 13, no. 4 (2010): 337–54.

within the boundaries of "domestic cinema." On the other, the international market demands of SVODs, which vary their inventories by region and rely heavily on preference algorithms to organize their content, have in some ways led to a homogenization of African output, with the search terms "Nollywood" and "African Movies" typically acting as catchall phrases for a regionally diverse range of releases. Thus, paradoxically, the emergence of regionally specific film industries in Africa has in many ways been obscured, as increasingly diverse content reaches SVOD users under the homogenous label of "African Movies."

This tendency is clearly evident in the nondomestic inventories of Netflix, the leading multinational platform for African film distribution, which is not only accessible in all African countries but has also made substantial investment in African film and television production, resulting in a number of "Netflix Originals" being produced on the continent (such as *Queen Sono* [dir. Kagiso Lediga, South Africa, 2020]).[2] For this reason, Netflix is emblematic of the international online distribution of African film, and for that reason will form the focus of this study. Moreover, the analysis here will build on a previous study I undertook during the period 2020–2022, in which I examined the platform's role in raising the global visibility of African cinema (with a focus on its UK site) and the ways it has generated an increasingly expansive definition of "African Movies" that encompasses not only domestically produced releases but also foreign-produced work that lays claim to the "African Movies" search tag by virtue of its African setting or subject matter (for example, *Beasts of No Nation* [dir. Cary Joji Fukunaga, 2015, United States] and *The Ivory Game* [dir. Richard Ladkani and Kief Davidson, United States, 2016]). These movies were also promoted as part of Netflix's curated inventory of African titles under the heading "Made in Africa."[3] Two years later, a search for "African Movies" on Netflix's UK site reveals a similar scenario but with a wider range of both domestically and foreign-produced material rubbing shoulders with "Netflix Originals."[4] Thus, while the circulation and availability of African films has

[2] In contrast, Showmax is geo-blocked so that it is only accessible to its African markets.

[3] A. Fisher, "Netflix and Africa: Streaming, Branding and Tastemaking in Non-domestic African Film Markets," in *World Cinema On Demand: Global Film Cultures in the Era of Online Distribution*, ed. S. Baschiera and A. Fisher (London: Bloomsbury Academic, 2022), 189.

[4] While it remains to be seen how sustained this growth will be, it should be noted that a number of the big players have recently shelved major projects, including the Zimbabwean animation *Tunga* (M. Basilan, "Netflix to Cut Animation Unit, Streaming Giant Looks to Outsource Producers," *International Business Times*, December 10, 2023, https://www.ibtimes.com/netflix-cut-animation-unit-streaming-giant-looks-outsource-producers-report-3715005). It is likely that this is due to shifting priorities in terms of Netflix's brand positioning, a possibility analyzed in relation to African output in Fisher "Netflix and Africa."

continued to develop, many of the issues identified in my previous study remain, particularly with regard to the loss of regional specificity.

In both of these analyses of Netflix's UK site, it is quite evident that the platform's categorization and preference systems simultaneously generate and obscure regional specificity in African cinema. As a result, from a nondomestic perspective at least, African cinema has become ever more expansive as a term, at the same time as the plurality of the continent's production is becoming increasingly recognized. Although there have been further developments in this regard over the past two years, which could form the basis of a further study, this chapter will make a departure from the analysis of distribution and instead explore the relationship between Netflix's international exports of African film and their cultural specificity at a more finely grained, textual level, examining three films from different anglophone African countries, all of which were available for streaming during 2020–4. These are *Nairobi Half Life* (dir. David "Tosh" Gitonga, Kenya, 2012), *The Girl in the Yellow Jumper* (dir. Loukman Ali, Uganda, 2020), and *Gold Coast Lounge* (dir. Pascal Aka, Ghana, 2020).[5] While the films are aesthetically diverse, they each fall broadly into the genre of crime film, and it is this global generic legibility, and the ways it operates in relation to cultural specificity, that in part propels the selection of these titles. They are also movies that demonstrate a characteristically postmodern self-awareness of their positions as African films within the global film ecosystem, each featuring an abundance of references to Hollywood film at the same time as they articulate aspects of cultural specificity. Thus, what singles out each of these movies is an overt self-awareness of the dichotomous relationships between the global perspective of their genres, which are aesthetically indebted to Hollywood, and the cultural specificity that marks their narratives, characters, and locations. It is the contention of this chapter that Netflix offers a particularly hospitable environment for this duality, an environment that is shaped by the dynamics of its business model that simultaneously provides a platform for niche and marginal cinemas and in turn offers a cultural heterogeneity of film experience, at the same time as its system of preference algorithms generates "new media" characteristics of interconnectedness and intertextuality that concurrently induce culturally homogenizing effects.

[5] Nigerian examples were specifically excluded from this study since the region's films have been subject to significant analysis elsewhere, and also because none of the selected regions' film output is of a comparable scale to Nigeria's.

Before proceeding further, it should be noted that the following discussion relates to Netflix's UK site, although many of the findings may be extrapolated in relation to its nondomestic sites elsewhere, particularly in the anglophone world (although numerous small-scale studies at various regional levels are needed to build an assessment of Netflix's global impact). Indeed, Amanda D. Lotz explains that despite its global reach Netflix cannot be understood as a single site but instead as a "federation of national services" consisting of "nearly 200 national or regional libraries of content."[6] As Ramon Lobato and Lotz previously noted, "Netflix's catalog, cultural status, brand recognition, and market power also vary enormously from country to country."[7] For Lobato and Lotz, this presents media scholars with a major challenge when attempting to understand Netflix since it offers significantly different inventories to different regions.[8] As they observe,

> Critically locating Netflix in a global context requires holding two contradictory realities in balance: Netflix is a single company that has direct-to-consumer subscription relationships with 150 million customers worldwide. This makes it arguably more global than any previous screen producer and distributor. But to make any claim about Netflix requires locating it in a particular place—in a country-specific catalog; in a nation-state with particular technological infrastructure, competing and complementary services, and regulatory regimes; and in markets characterized by different audience expectations, preferences, and cultural norms.[9]

Moreover, these national inventories are linked not only by a single company but also by a "common platform architecture"[10] that is in itself a globalizing factor; thus, the operation of the platform is globally uniform, creating a branded user experience that has a homogenizing influence in the way it shapes consumer viewing tastes and patterns, offering titles via an ostensibly precise preference algorithm that becomes more finely tuned with use. In this way, the targeting of regional markets works in tandem with the targeting of taste patterns, marking a radical shift in the way movies circulate. As Graeme Turner argues, "The customization of consumption that marks the new environment changes

6 A. D. Lotz, "In between the Global and the Local: Mapping the Geographies of Netflix as a Multinational Service," *International Journal of Cultural Studies* 24, no. 2 (2021): 196–7.

7 R. Lobato and A. D. Lotz, "Imagining Global Video: The Challenge of Netflix," *Journal of Cinema and Media Studies* 59, no. 3 (2020): 132.

8 Ibid.

9 Ibid.

10 R. Lobato, "Rethinking International TV Flows Research in the Age of Netflix," *Television & New Media* 19, no. 3 (2018): 245.

everything; while the choices being made are certainly organized around the demands and patterns of everyday life, this is 'not necessarily a shared everyday life.'"[11] As a result, analysis of Netflix requires that both the global dynamics of the platform's functionality and the regional specificities of its inventories are taken into account, which allow it to foster particular patterns of consumption. As Lotz goes on to state,

> *Rather than trying to aggregate a nation-bound audience that can be sold to advertisers who conceive of audiences aggregated in national clumps, Netflix targets subscribers based on tastes and sensibilities that are often not sufficiently popular to be addressed by services aiming for a national "mass" audience.*[12]

Thus, the "contradictory realities" between regional specificity and a "common platform architecture" structure modes of viewing that involve particular types of encounters between the global and the local. It is this platform specificity that shapes the analysis in this chapter, which aims to broach the function of African films as global exports within the Netflix viewing environment (in this case, as found on the UK site during 2022–4). In this respect, the study aims to begin bridging the gap identified by Lotz, who asserts that "approaches are needed for investigating, analyzing, and theorizing beyond the single-nation case. Frameworks that moderate the poles of multinational scale and national specificity of these services must be developed."[13] In this spirit, the analysis here will examine more than the poles of "multinational scale" and "national specificity," expanding its scope to develop a more plural account of Netflix as a "transcultural" viewing space.

The tensions identified by Lobato, Lotz, and Turner may be identified through a cursory inspection of Netflix's UK inventory, which offers a powerful insight into SVODs' wider effects on the international distribution of African movies. In addition to the increasing range of African titles available on the various platforms, there have been a number of developments in the categorization and marketing of these titles. Although Nigeria remains by far the most heavily represented region, there is a notable diversification of search tags that enables more nationally focused searches, in some cases categorizing titles by country of production (although the definition and accuracy of "country of production"

[11] G. Turner, "Approaching the Cultures of Use: Netflix, Disruption and the Audience," *Critical Studies in Television* 14, no. 2 (2019): 227–8.
[12] Lotz, "In Between the Global and the Local," 207, author's emphasis.
[13] Ibid., 197.

remain major variables). For instance, whereas as recently as 2021 the only identifiable category tags on the UK site were "Nollywood" and "African Movies," these have now been joined by "Kenyan Film and TV" and "Ghanaian Movies and TV" (although these are used sporadically and do not always appear in the tag lists for the relevant region's content). In contrast, at the time of writing, there is no tag for "Ugandan Film" or "Ugandan Film and TV," and a search for "Ugandan Film" currently only throws up four titles, including *The Girl in the Yellow Jumper*, *Aki and Pawpaw* (dir. Biodun Stephen, Nigeria, 2021), the "Netflix Original" TV series *Blood Sisters* (dir. Biyi Bandele and Kenneth Gyang, Nigeria, 2022), and *The Boy That Harnessed the Wind* (dir. Chiwetel Ejiofor, UK/Malawi, 2019). However, the first two titles are in fact Nigerian productions, while *The Boy That Harnessed the Wind* is a UK/Malawi coproduction. In other words, a search for "Uganda" raises a much wider range of content, but it often has little or no association with Uganda itself.

Much of this is complicated by shifting trends in viewership for African film, which Netflix categories both follow and condition. Particularly notable is the move from homogenous notions of "African cinema," represented by an auteur-based, international art house cinema, which sought, in part, to forge a pan-African identify and was typically viewed by an educated cosmopolitan audience, to a pluralized pattern characterized by national film industries catering to a wider and more popular viewership. However, this move is certainly not spurred by Netflix; as far back as 1997, Olivier Barlet called for a notion of "African cinemas"[14] as regional modes of production began to usurp pan-African models, and the influence of SVOD emerged at a late stage in this process. Indeed, in expanding the definition of "African Movies," Netflix complicates and potentially unwinds some of the work of African filmmakers and critics in establishing and defining specifically "African" films, depoliticizing the term to the extent that it identifies little more than a location of production, as opposed to culturally specific modes of production.

The example of Netflix's UK site, then, demonstrates how the platform generates both heterogeneity and homogeneity in terms of its inventory and categorization, in which regional tags are becoming increasingly specific at the same time as they may be deployed in a highly generic or nebulous fashion. However, these national markers are used in the context of a platform where preference algorithms are increasingly rendering ideas of regional or national

[14] O. Barlet, *African Cinemas: Decolonizing the Gaze* (London: Zed Books, 1997), viii–ix.

cinema as undefined (or even irrelevant). What this generates is a particular kind of viewing environment that establishes and fosters new forms of encounter between marginal film cultures and nondomestic exhibition contexts, structuring a viewing experience that is quite distinct from those associated with orthodox modes of distribution. In this way, the platform architecture and business model of Netflix produce a scenario that problematizes orthodox distinctions between the national and the international, the local and the global, and the periphery and the center, constructing a transcultural space for exhibition that fosters—and may even generate—a transcultural or even transmedial aesthetic at the level of individual film texts.

An aspect of Netflix's business model that possesses a particular significance in relation to African cinema is its ability to take advantage of what Chris Anderson calls "long tail" markets, where material accrues profitability over the course of an extended distribution window.[15] This in turn helps Netflix fulfill its requirement to populate inventories to the degree required to effectively feed its preference algorithms. These vast catalogs, and the recommendation systems that organize them, are closely intertwined with the phenomenon of (for want of better terms) "binge viewing" or "binge watching," which is a fundamental notion in the conceptualizing and analysis of the "Netflix experience." While this behavior is often framed in pathological terms (the term "binge" evokes strong connotations of substance abuse), it also offers significant scope for audiences to (consciously or otherwise) encounter marginal and niche content, fundamentally distinguishing the platform from orthodox forms of linear distribution in terms of the viewing experience it offers its users. This development is particularly relevant to African film, which has been heavily underrepresented across the spectrum of film exhibition, criticism, and scholarship. With preference algorithms determining access to vast inventories of content, orthodox modes of viewing, in which viewers seek individual titles, are replaced by something more akin to a highly malleable and personalized "viewing strip," where the end of one text segues into the beginning of another, often offered on the basis of the individual subscriber's previous search and viewing history.[16] The fortunate side effect of this is the vastly increased marketability of niche and marginal

[15] See C. Anderson, *The Long Tail: Why the Future of Business Is Selling Less of More* (New York: Hyperion, 2006).

[16] See Z. Snider, "The Cognitive Psychological Effects of Binge-Watching," in *The Netflix Effect: Technology and Entertainment in the 21st Century*, ed. K. McDonald and D. Smith-Rowsey (New York: Bloomsbury Academic, 2016), 117–28, for a detailed analysis of the psychology of "binge watching" behavior.

content at a global level, where the long tail model and recommendation based patterns of consumption allow these titles to accrue financial viability, resulting in a situation that has allowed African filmmaking to acquire an unprecedented level of global accessibility.[17]

Thus, despite the negative connotations of the term, "binge watching" represents a practice that requires a substantial rethinking of the ways we understand film consumption in the internet age, where notions of the "closed text" are increasingly being challenged. As Sheri Chinen Biesen has noted (in an analysis of Netflix's inventory of film noir), the features of the platform's distribution model have created "a convergent new media viewing environment that fosters 'binge-watching' of film noir in a way that reimagines traditional cinematic reception and distribution."[18] For Biesen, noir foreshadows the media convergence of Netflix in the way its visual and narrative codes link texts.[19] This position may be extended to film genres more generally, particularly in view of their key roles in shaping user behaviors such as "binge watching." As such, in the viewing environment of Netflix (and SVODs in general), genre forms a ready-made system of textual interconnectedness whose intersections operate as enabling aspects of "new media" ontologies, in which genres have evolved to feed preference algorithms. These algorithms then produce ever more precisely targeted recommendations, shaping viewing practices (such as, but certainly not exclusively, "binge watching") and in turn the experience of film spectatorship to one that creates openings for niche content (such as that from hitherto marginalized regions).

Biesen's analogy for this pattern is an expansion of Henry Jenkins's notion of "transmedia storytelling," which Jenkins defines as "the art of world making."[20] While the most widely understood aspect of Jenkins's formulation concerns the articulation of a story world across a range of media and platforms, Biesen develops Jenkins's further idea that "transmedia phenomena also provide audiences with new ways to engage with media texts and to create new meanings by virtue of the intersection between different platforms and contexts," which

17 See S. Baschiera and A. Fisher, "Introduction," in *World Cinema on Demand: Global Film Cultures in the Era of Online Distribution*, ed. S. Baschiera and A. Fisher (London: Bloomsbury Academic, 2022), 1–11; Fisher "Netflix and Africa."

18 S. C. Biesen, "Binge-Watching 'Noir' at Home: Reimagining Cinematic Reception and Distribution via Netflix," in *The Netflix Effect: Technology and Entertainment in the 21st Century*, ed. K. McDonald and D. Smith-Rowsey (New York: Bloomsbury Academic, 2016), 129.

19 Ibid., 131.

20 H. Jenkins, *Convergence Culture: Where Old and New Media Collide* (New York: NYU Press, 2006), 21.

extends the idea beyond the traditional ontological boundaries of world making to include broader characteristics across texts, in particular, those of genre. As Biesen continues,

> In this regard, noir can be seen as a precursor to this new media development. Film genres, especially in the case of film noir, entail audiences identifying narrative and stylistic patterns across multiple texts.... In a certain sense, what Jenkins describes as transmedia convergence is built into the way Netflix as a platform connects viewers with media. It assumes that its users want to be able to inhabit a certain type of story world, one in which there are common elements that both reinforce and expand one another across individual texts. It also assumes that what is pleasurable about consuming media is the ability to decipher these types of patterns whether they be narrative, thematic, stylistic, or extra-textual.[21]

From this standpoint, "transmedia" may be understood as a way of characterizing the Netflix environment, where film, television, and "new media" converge in a way that establishes new patterns of taste and viewing habits, generating a particular kind of viewing experience, of which "binge watching" forms a part. None of this is to suggest that "binge watching" is intrinsic to the "Netflix experience," and of course it is eminently possible, and indeed common practice, to search for and view individual titles in the Netflix system. However, "binge watching" is helpful as a concept that represents a well-known viewing pattern that has been shaped by the platform (even if it has been conceptualized in pejorative terms), a pattern that vividly represents the myriad pleasures that SVODs offer, and the ways these challenge traditional ontologies of film experience. Thus, it is these new viewing patterns and their relationship to the notion that "users want to be able to inhabit a certain type of story world" that is key to understanding how users may experience the three African films under scrutiny here. As films that fall into various variants of the "crime film," the algorithm already "knows" they will offer "a certain type of story world," regardless of the regional origins of the films themselves.

Intersecting with these new media characteristics is Netflix's contradictory categorization of African cinema within culturally homogenous terms. Netflix's tendency to group films under the "African Movies" category tag potentially positions a culturally diverse range of content within a stereotyped notion of difference (for example, "Black" versus "White" cinema), sidelining cultural

[21] Biesen, "Binge-Watching 'Noir' at Home," 131.

specificity so that the very real differences between "Ugandan Cinema" or "Kenyan Movies and TV" may fail to reach a viewer encountering these screen cultures via the platform. Yet it is the same category tagging system that may lead a viewer to unintentionally encounter (and subsequently enjoy) an African film, paradoxically fostering a culturally heterogeneous viewing environment. Thus, in the context of African film streaming, the Netflix viewing environment proves to be extremely complex, where the characteristics of new media shape viewing behaviors in a way that results in a particular transcultural experience, built around homogenous notions of "African Movies" yet creating a potentially heterogeneous viewing experience.

This scenario may be compared to the paradoxes of transculturality at large, whose characteristics may be summarized via the words of James Clifford, which seem ever more apt as new media continue their advance:

> In cities on six continents foreign populations have come to stay—mixing in but often in partial, specific fashions. The "exotic" is uncannily close. Conversely, there seem no distant places left on the planet where the presence of "modern" products, media, and power cannot be felt. An older topography and experience of travel is exploded. One no longer leaves home confident of finding something radically new, another time or space. Difference is encountered in the adjoining neighborhood, the familiar turns up at the ends of the earth…. "Cultural" difference is no longer a stable, exotic otherness; self-other relations are matters of power and rhetoric rather than of essence. A whole structure of expectations about authenticity in culture and in art is thrown in doubt.[22]

The similarities between Clifford's analysis and Netflix's viewing environment are not difficult to identify, when the accessibility of a heterogeneity of content means the "'exotic' is uncannily close." With the logic of SVODs' preference algorithms constantly opening avenues for culturally diverse film exhibition, the platforms' viewing environments come to resemble the cultural "melting pots" of the global metropolises of London, Paris, or New York, where "difference is encountered in the adjoining neighborhood," while the platforms' global reach means "the familiar turns up at the ends of the earth." This is particularly apt when users may be viewing in London or Lagos, New York or New Delhi (on occasion, starting a film in one of those cities and finishing it in another), while engaging with a culturally heterogeneous range of films drawn from all over the

[22] J. Clifford, *The Predicament of Culture: Twentieth-Century Ethnography, Literature, and Art* (Cambridge MA: Harvard University Press, 1988), 13–14.

globe. Crucially, African titles do not simply circulate within this transcultural environment but instead form fundamental components, their reputation for scarcity generating "'cultural' difference" within an environment that constantly undermines "a stable, exotic otherness."

It is within this concurrently transcultural and transmedial environment that I will now situate *Nairobi Half Life*, a particularly interesting case in that its original release in 2012 predates Netflix as the streaming giant we are familiar with today, and it is only in 2020 that it received a release on the platform.[23] The film's emphasis on location (set in the eponymous city of Nairobi, Kenya) also positions it as the most overtly "local" of the three examples, but its deployment of crime narrative and intertext means that the film exemplifies many of the characteristics of the Netflix viewing experience. The story of an aspiring young actor by the name of Mwas (Joseph Wairimu), *Nairobi Half Life* follows our protagonist as he pursues a stage career in Nairobi and concurrently becomes ensnared in the city's criminal underworld. In terms of its subject matter and location, the film exhibits clear parallels with previous movies about young street criminals set in slums. The youthful and naive Mwas prompts comparisons with the protagonists of *City of God* (dir. Fernando Meirelles and Kátia Lund, Brazil, 2002), *Pixote: A Lei do Mais Fraco* (dir. Hector Babenco, Brazil, 1981), and *The Young and the Damned* (dir. Luis Buñuel, Mexico, 1950), while his journey from rural Kenya to urban Nairobi recalls the likes of Perry Henzell's *The Harder They Come* (Jamaica, 1972). As with its forerunners, *Nairobi Half Life* depicts the adverse poverty and ruthless criminality of its urban location, where gangsters and corrupt police officers rule the streets with impunity.

The movie is overtly self-aware of film and performance culture in general; Mwas sells Hollywood action films on VCDs for a living, his sales pitch involving dramatic reconstructions of famous set pieces. Indeed, the movie's opening scene involves an extended movie reference, in which Mwas re-creates the "five point palm exploding heart technique" from Quentin Tarantino's *Kill Bill* (United States, 2003). Later, when a small boy points to a marketing image for *300* (dir. Zack Snyder, United States, 2007), Mwas performs a theatrical synopsis of the film, mimicking the voice-over narration of Spartan soldier Dilios (David Wenham) and invoking the young Leonidas's (Tyler Neitzel/

[23] See E. Ngigi, "*Nairobi Half Life* Is Set to Go Global," *Star*, October 18, 2022, https://www.the-star.co.ke/sasa/2022-10-19-nairobi-half-life-is-set-to-go-global; and The Sauce, "Netflix Breathes New Life on *Nairobi Half Life* on Its 10-Year-Anniversary," October 24, 2022. https://www.capitalfm.co.ke/thesauce/netflix-breathes-new-life-on-nairobi-half-life-on-its-10-year-anniversary/.

Eli Snyder) brutal upbringing, presaging Mwas's own rites of passage into the unforgiving Nairobi underworld. Later, Mwas appears visibly moved as a film screening closes at Kenya Cinema Plaza, and as he begins to establish himself as an actor his narrative alternates between scenes of theater rehearsal and ruthless street criminality, eventually culminating in Mwas's impassioned monologue to a transfixed full house.

Mwas's cinephilia and love of dramatic performance also establishes a self-reflexive relationship between the character and the circumstances of the film's production in the sense that *Nairobi Half Life* is the result of a training project as opposed to a commercial enterprise. The film emerged from the activities of One Fine Day Films (OFDF), a project founded by German filmmakers Tom Tykwer and Marie Steinmann and backed by various capacity-building initiatives. *Nairobi Half Life* was subsequently born out of a number of filmmaking workshops run by OFDF.[24] Thus, the film's production was a labor of love for many of its participants, and this is echoed in its protagonist's enthusiasm for Hollywood film, whose self-reflexively brings together the global and the local in a manner that readily translates into the transmedial and transcultural space of Netflix's viewing environment.

A similarly self-conscious intertextuality is found in *The Girl in the Yellow Jumper*, which in 2022 became the first Ugandan film to be distributed on Netflix. A violent thriller that recalls the work of Hollywood auteurs such as Spike Lee, Quentin Tarantino, and the Coen Brothers, *Girl* depicts the brutal twists that follow after Patrick (Maurice Kirya), an off-duty cop, offers a lift to the mysteriously abandoned and injured Jim (Michael Wawuyo Jr.), whom he finds wandering a highway in the middle of nowhere. Meanwhile, a murderer known as the "Cigarette Butt Killer" is on the loose. Within this familiar scenario, the movie wears its film references on its sleeve, most visibly during the opening shot, where the camera tracks a row of DVD spines whose titles include *Reservoir Dogs* (dir. Quentin Tarantino, United States, 1992), *X-Men* (dir. Bryan Singer, United States, 2000), and *Fargo* (dir. Joel and Ethan Coen, United States, 1996), all key movies from the Hollywood "independent" auteurs who clearly influence the film. Alongside this, sly references to new media (notably, the ruthless online "Addicting Games" craze of "Kitten Cannon") flesh out the film's popular culture

[24] R. Steedman, "Nairobi-Based Middle-Class Filmmakers and the Production and Circulation of Transnational Cinema," *Poetics* 75 (2019): 7.

allusions. The overt nods to contemporary Hollywood presage numerous plot twists, triggered largely through the unreliable accounts of Jim—much of which are told in flashback—until Jim's dishonesty is unmasked in a twist that recalls *The Usual Suspects* (dir. Bryan Singer, United States, 1995) (itself followed by a further twist that recalls *Saw* [dir. James Wan, United States, 2004]). The drama is interrupted twice by Spike Lee–esque vox pops where an anonymous "Sketch Artist" (Kavumba Mustafa Ali), speaking in a pronounced US American accent, proffers animated illustrations of topics raised in Patrick and Jim's conversation: first, the rules of the traditional and largely extinct children's game known as Bladder (which presage the narrative's structural development), and later, a graphic demonstration of why being shot in the kneecap is so excruciatingly painful (emphasizing a discussion that will lead Jim to sustain the same injury during the film's final act).

This referentiality is sustained as the story unfolds; in one scene, Jim cites the actor Jim Carrey as the inspiration for his name, a reference that is lost on Patrick but which not only hints at Jim's overactive imagination but also his reliance on Hollywood to imagine and articulate his experiences. This is developed later in the scene; finally persuaded to explain his predicament and (more urgently) his severe hand injury, Jim describes how he "woke up tied to a chair, just like in the movies." Through these ongoing references to Hollywood cinema, *Girl* not only lays the foundations for our unreliable narrator-protagonist but also generates a postmodern irony in which the norms of Hollywood storytelling are self-consciously referenced to form a critique of its imperialist stranglehold over global film culture. From this perspective, the film's "quotations" of Hollywood alternate between homage and satire, an idea that may be extended to its narrative structure more generally, which ironically adopts the generic norms of its target in what may be construed as the articulation of an anti-imperialist cultural critique. This reading is lent credence by the film's regionally specific details within its "global" film aesthetic. Notably, Patrick lectures Jim about the imposition of Christianity on Africa through colonialism, proffered as an explanation as to why Jim should not have been surprised to discover the existence of Christian fundamentalists. Elsewhere, as the pair eventually reach Patrick's home, Jim notes the local custom of transporting raw fish by hanging it over the car's radiator grill, a practice that, according to Patrick, prevents the fish overheating.

However, it is during the first vox pop scene, in which the Sketch Artist explains the rules of Bladder, that we witness an overt collision of local culture and the global "new media" environment, as the Sketch Artist closes the segment with the dismissive line, "Unfortunately today, you won't find anyone playing this game anymore as it's not on a downloadable app." This overt reference to new media, emphasizing its diluting effects on indigenous culture, is apt in the context of this film, in which a postmodern aesthetic largely adopted from Hollywood overwhelms its locally specific dimensions. Just as the pastime of Bladder is succeeded by the downloadable diversion of Kitten Cannon, Africa's film aesthetics move ever closer to Hollywood's, with both phenomena ostensibly asserting the same imperialist stranglehold. Thus, while *Girl* appears to pay homage to those Hollywood auteurs who have broken both filmic and social convention (Tarantino, Singer, Lee, etc.), it also exhibits an acute awareness of its own position within the global film ecosystem, where SVOD has arguably shackled regional cinemas even more tightly to Hollywood's formations of genre. In this sense, *Girl* might be understood as a text that negotiates the tensions of the transmedial and transcultural environment structured by Netflix, self-reflexively engaging with the Hollywood genres that form a backbone of Netflix's preference system and thus revealing how an imperialist balance of power is promulgated within the platform itself.

Of the three films under analysis here, *Gold Coast Lounge* is the most overtly postmodern, deploying a heightened, self-conscious aesthetic characterized by nostalgia and pastiche. Its period setting (colonial Ghana at the cusp of independence), high-contrast black-and-white cinematography (in a cinemascope aspect ratio), period costume, dramatic music, and low-key lighting all position the film as neo-noir, at the same time as there are overt references to specific movies and franchises. Narratively, *The Godfather* is an obvious touchpoint, but other references are numerous; for instance, the opening pretitle sequence culminates in John Barry–esque orchestral chords, which give way to an elaborate title sequence recalling, or even directly referencing, the Bond movies, with a female "diva" vocal accompanying extravagant, highly stylized animation. Elsewhere, digitally "aged" images (adding celluloid scratches, dust, and missing frames) mark dramatic turning points, while the hyperbolic incidental score often recalls the spaghetti westerns of Sergio Leone.

This stylized aesthetic forms a suitably self-conscious conduit for a familiar epic narrative depicting the rise and fall of a crime syndicate, headed by the ruthless John Donker (Adjetey Anang) and his "adopted" sons Wisdom (Pascal Aka) and Daniel (Alphonse Menyo), within whom he installed a fierce rivalry. Between them they run the Gold Coast Lounge, a private members' club frequented by much of Ghana's criminal contingent. The rivalry between Wisdom and Daniel develops into insurrection and the downfall of the syndicate, although the Gold Coast Lounge remains open, apparently bought out by its American beneficiary.

The crime/family drama is intrinsically tied to the political context surrounding the closing days of colonial Ghana, with the fate of Donker's syndicate and the Gold Coast Lounge reflecting broader issues of government, trade, and commerce. We learn that the lounge—along with its criminal clientele—has thrived under colonial rule but is now under the threat of closure by the new independent government, who are deemed to be intolerant of corruption.

The narrative articulation of Ghana's shift to independence is also articulated visually, particularly in the uses of space, which establishes an overt dichotomy between rural and urban environments. This dichotomy is established early in the film, where the neo-noir aesthetic gives way to the iconography of rural Africa during a flashback sequence in which Daniel relates his recruitment by Donker, transporting us to his childhood village and his brutal path to mob criminality in urban Accra. Elsewhere in the film, traditional customs are distorted and put to murderous use by Wisdom, whose attempts to kill Daniel involve the acquisition of poison from a juju man and the attempted use of a traditional mortar and pestle as a method of execution.

The broader political landscape depicted by *Gold Coast Lounge* is one of a shift from colonialism to neocolonialism, both of which are accompanied by cultural imperialism, and the film's hyperreal style and intertextual pastiche represent a distinctly self-reflective articulation of the movie's own place within this particular formation of transculturality, an articulation comparable to that found in *The Girl in the Yellow Jumper*. However, where *Girl* articulates its irony through overt references, *Gold Coast Lounge* articulates its irony through stylistic excess.

However, this postmodern use of genre, in which genre coding becomes a thematic motif in itself, means that *Gold Coast Lounge* is the most overt of the three case studies in terms of how it embodies the dynamics of the Netflix viewing environment, where users are invited "to inhabit a certain type of story world." Thus, *Gold Coast Lounge* not only demonstrates a postmodern engagement with Hollywood's imperialist grip on film culture but also represents an embodiment of the transmedial and transcultural characteristics of the Netflix viewing environment.

In conclusion, while *Nairobi Half Life*, *The Girl in the Yellow Jumper*, and *Gold Coast Lounge* are all quite different iterations of crime narrative, varying radically in terms of their characters, settings, and subject matter, they are all movies that may be characterized as aesthetically postmodern in that they adopt, to differing degrees, globally legible, referential, and intertextual styles that render them as almost emblematic of film tastes in the age of SVOD, in which preference algorithms structure an interconnected viewing experience involving increasingly diverse content that can cross cultural borders with unprecedented ease, shaped in part by the pragmatic necessity of SVOD platforms to populate their inventories in order to feed their preference algorithms. In this sense, these three films embody the spirit of the streaming age, simultaneously emulating the transmedial and transcultural dynamics of SVOD viewing environments in which a "whole structure of expectations about authenticity in culture and in art is thrown in doubt."[25] As a result, the ways these movies are encountered on Netflix means that they offer a quite particular experience of "African cinema" that explodes both expectations of cultural difference and the ways we encounter it, challenging both ontological and cultural boundaries between film and video on one hand, and African and non-African contexts on the other.

Moreover, each of these films demonstrates an overt self-reflexivity that suggests a self-awareness of their own positions as African films within the global film ecosystem, and in the cases of *The Girl in the Yellow Jumper* and *Gold Coast Lounge* they exhibit not only a suspicion of Hollywood's ongoing imperialist dominance but even a suspicion of the particular dynamics of power at play within new media viewing environments, recognizing that the "transmedial" interconnectedness of SVOD in fact fosters a "transculturality" that is far from equitable, even while it represents a vast improvement on the *status quo ante*. Thus, the relationship between Netflix and African cinema generates a multifaceted set of paradoxical tensions within which African films are finally able to achieve global exhibition, but within terms that sustain, and perhaps even intensify, the imperialist power relationships that characterize global film culture.

[25] Clifford, *Predicament of Culture*, 14.

Bibliography

Adejunmobi, M. "Streaming Quality, Streaming Cinema." In *A Companion to African Cinema*, edited by K. Harrow and C. Garritano, 219–43. Hoboken, NJ: Wiley-Blackwell, 2019.

Anderson, C. *The Long Tail: Why the Future of Business Is Selling Less of More*. New York: Hyperion, 2006.

Baschiera, S., and A. Fisher. "Introduction." In *World Cinema on Demand: Global Film Cultures in the Era of Online Distribution*, edited by S. Baschiera and A. Fisher, 1–11. London: Bloomsbury Academic, 2022.

Barlet, O. *African Cinemas: Decolonizing the Gaze*. London: Zed Books, 1997.

Basilan, M. "Netflix to Cut Animation Unit, Streaming Giant Looks to Outsource Producers." *International Business Times*, December 10, 2023. https://www.ibtimes.com/netflix-cut-animation-unit-streaming-giant-looks-outsource-producers-report-3715005.

Biesen, S. C. "Binge-Watching 'Noir' at Home: Reimagining Cinematic Reception and Distribution via Netflix." In *The Netflix Effect: Technology and Entertainment in the 21st Century*, edited by K. McDonald and D. Smith-Rowsey, 129–42. New York: Bloomsbury Academic, 2016.

Clifford, J. *The Predicament of Culture: Twentieth-Century Ethnography, Literature, and Art*. Cambridge MA: Harvard University Press, 1988.

Fisher, A. "Netflix and Africa: Streaming, Branding and Tastemaking in Non-domestic African Film Markets." In *World Cinema on Demand: Global Film Cultures in the Era of Online Distribution*, edited by S. Baschiera and A. Fisher, 189–204. London: Bloomsbury Academic, 2022.

Haynes, J. "Anglophone West Africa: Commercial Video." In *African Filmmaking: Five Formations*, edited by K. W. Harrow, 81–116. East Lansing: Michigan State University Press, 2017.

Jenkins, H. *Convergence Culture: Where Old and New Media Collide*, New York: NYNYU Press, 2006.

Lobato, R. "Creative Industries and Informal Economies: Lessons from Nollywood." *International Journal of Cultural Studies* 13, no. 4 (2010): 337–54.

Lobato, R. "Rethinking International TV Flows Research in the Age of Netflix." *Television & New Media* 19, no. 3 (2018): 241–56.

Lobato, R., and A. D. Lotz. "Imagining Global Video: The Challenge of Netflix." *Journal of Cinema and Media Studies* 59, no. 3 (2020): 132–6.

Lotz, A. D. "In between the Global and the Local: Mapping the Geographies of Netflix as a Multinational Service." *International Journal of Cultural Studies* 24, no. 2 (2021): 195–215.

Ngigi, E. "*Nairobi Half Life* Is Set to Go Global." *Star*, October 18, 2022. https://www.the-star.co.ke/sasa/2022-10-19-nairobi-half-life-is-set-to-go-global.

Okome, O. “Women, Religion and the Video Film in Nigeria.” *Film International* 2, no. 1 (2004): 4–13.

The Sauce. “Netflix Breathes New Life on *Nairobi Half Life* on Its 10-Year-Anniversary.” October 24, 2022. https://www.capitalfm.co.ke/thesauce/netflix-breathes-new-life-on-nairobi-half-life-on-its-10-year-anniversary/.

Snider, Z. “The Cognitive Psychological Effects of Binge-Watching.” In *The Netflix Effect: Technology and Entertainment in the 21st Century*, edited by K. McDonald and D. Smith-Rowsey, 117–28. New York: Bloomsbury Academic, 2016.

Steedman, R. “Nairobi-Based Middle-Class Filmmakers and the Production and Circulation of Transnational Cinema.” *Poetics* 75 (2019): 1–10.

Turner, G. “Approaching the Cultures of Use: Netflix, Disruption and the Audience.” *Critical Studies in Television* 14, no. 2 (2019): 222–32.

11

Invoking the Shadow Archive: Three Documentaries That "Look Back" from 1970s Beirut

Samirah Alkassim

This essay dwells in the realm of the "unseen" or shadow side of discourses on streaming and cinema. My intervention follows a theoretical and textual exploration of the role of archival material in three documentaries that are hard to access but selectively available through streaming media. Made in 1970s Beirut, these are *Les Femmes Palestiniennes* (The Palestinian Women) by Jocelyn Saab, 1974; *Atfal Walakin* (Children Nevertheless) by Khadijeh Habashneh, 1979; and *Khotwa Khotwa* (Step by Step) by Randa Chahal, 1979. In these films the archival fragment and its indexical value (the assumption that archival documentary material represents reality) bear heavily on processes of signification and speak uncannily to contemporary political crises in the Arab Middle East. By focusing on these "old" documentaries that engage mostly "vanished" or plundered archives, and that can only be found on the margins of the contemporary streaming landscape, I am highlighting the dissonance between what we expect filmic archives, and streaming media curation, to allow and what they tend to exclude. This dissonance compels a serious look and, on closer examination, reveals that streaming media is sometimes capable of delivering, in surprising though inconsistent ways, the type of films and media that it more often excludes. Crucial to this typically excluded media is the idea of traveling between past and present, through rhetorical strategies in archive-based films that use documents from a previous era. When these documentaries are streamed, time travel is possible not only because of the processes of signification unleashed in the films but also because of the service that streaming delivery provides, as a "sharing" medium and, in some cases, platform. Although we are always time traveling when we watch archive-based films that present material taken from another time and context of use, here the streaming service as a sharing medium becomes a *vehicle* that allows for this travel. This gains renewed significance in the current context of imperial wars and the accompanying destruction of archives in specific countries (Gaza, Lebanon, Syria, and Iraq) of the Middle East.

But, first, it is important to acknowledge the role of archives in cinematic research, curation, and filmmaking. In an ideal world, the relation between films and archives is reciprocal and transformative. We appreciate archival depositories because of the films they house, and we attribute cultural power to films because of the archives to which they belong. But archives are always incomplete, however potentially unending, and in the Arab world they are "uneven" at best—ranging from closely guarded personal collections to insufficiently maintained state institutions and precariously funded nongovernmental cultural centers. In the resulting gaps, archives are remembered, imagined, and sometimes imaginary projects. The internet adds further dimension to the filmic archive, and this allows us to consider the functionality and role of web-based streaming platforms in archival work. What do we make of the unevenness of access, when a film once viewable via password on Vimeo disappears and reappears as a link on a website, and when we must await the release of recently restored films or draw on our social networks to request password-protected access to controversial, politically engaged, or disappeared films that are safeguarded by those who retain rights over them? In the absence of access and in the necessity of waiting, we can theorize about the epistemologies of absence/restriction, but we must work hard to turn a negative into a positive. This in turn reasserts the power of the scholar-curator, who, by sharing online access to a forgotten or lost text, can become a gatekeeper and an archivist in the age of streaming.

We may also consider platforms like Mubi, the Beirut-based Aflamuna, and even Amazon and Netflix as cultural gatekeepers, given that such platforms afford new possibilities to curate international films.[1] Exemplary of such curation was the 2021 Netflix program "Palestinian Stories," which made available for the first time several films by award-winning Palestinian directors to a regional as well as global fee-paying membership.[2] Before its disappearance in 2024, explained by Netflix as a simple matter of expired licensing agreements with

1 See the "About" page on the Mubi website, https://mubi.com/en/us/about, and the Aflamuna Online web page, https://aflamuna.org/aflamuna-online/.

2 The "Palestinian Stories" program on Netflix was curated by Front Row Filmed Entertainment, a film distribution company based in Dubai, the website of which boasts a library of 1,800 films and claims to "acquire" between 90 and 120 titles per year.

distributors,[3] a quick scan through the list of titles originally featured as part of this collection revealed a formidable selection of recent award-winning auteur films, both fiction and nonfiction, such as Mai Masri's *3000 Nights* (2015), Najwa Najjar's *Eyes of a Thief* (2014), Hany Abu Assad's *Omar* (2013), Mahdi Fleifel's *A World Not Ours* (2012), and Elia Suleiman's *Divine Intervention* (2002), among many others. But the artistic and critical value of these films notwithstanding, their selection indicates also the exclusion of other sorts of films, such as the independently produced experimental documentary *Palestinian Stories* by French filmmaker Dominique Dubosc made in 2008. In spite of the title's resemblance to that of the Netflix program, we will not find Dubosc's film on Netflix. You might, however, come across this film through the Palestine Film Index, a database containing links to six hundred Palestinian and Palestine-solidarity films, and several publications about them, created by a group of international volunteer artists, curators, and filmmakers who have been working on it since 2021.[4] The existence of this index and the effort it represents expose some of what is excluded by Netflix, which raises questions about its curation criteria, even if we understand how the chain of opportunity for Palestinian and Palestine-solidarity filmmakers may well advance from film festivals to distribution via streaming platforms with both regional and global reach.

The story of Netflix's "Palestinian Stories" is unique in that it belatedly acknowledges the arrival of Palestinian cinema on the world cinema map (as deemed by film festivals), but it also suggests the ways the streaming industry is changing media consumption landscapes in the Middle East. An interactive study of media consumption patterns in the Arab world conducted by Northwestern University–Qatar in 2019 informs us about increasing consumption of media through the internet, with the greatest levels found in the richest countries of the region (Saudi Arabia, Qatar, and United Arab Emirates).[5] Included in this consumption of what presumably consists largely of information and entertainment is film programming, typically featuring the latest offerings from the commercial Arab film industry and international box office successes, with the occasional classic Egyptian film from the golden age (1940s–60s).

[3] See "Netflix Wiped Most of Its Palestinian Stories Collection—and Erased the Whole Thing in Israel," *Intercept*, October 25, 2024, https://theintercept.com/2024/10/25/netflix-palestinian-stories-israel-movies/?utm_source=twitter&utm_campaign=theintercept&utm_medium=social.

[4] The Palestine Film Index is a volunteer-created "list of films from and about Palestine and the Palestinian struggle for liberation, made by Palestinians and those in solidarity with them."

[5] See "Media Use in the Middle East, 2019," https://mideastmedia.org/survey/2019/interactive/media-use-by-platform/who-use-the-following-streaming-services.html.

A more recent media analysis addresses shifting production patterns. According to a 2022 Egyptian market trend report, media production in the Middle East has moved to the age of "narrowcast," where content is produced for smaller, specialized audiences through subscription-based channels or platforms.[6] This narrowcasting is enhanced by the multitude of devices streaming such media and new regional business ventures and platforms like Shahid VP and Netflix producing specialized content for local audiences. We are promised that this trend will only increase access to both old and new programming, of both the commercial and arthouse varieties.

And yet, even as streaming technologies offer greater possibilities for curating and circulating noncommercial content, including arthouse cinema and films from earlier time periods, there seems to be no space for the explicitly politically engaged, anti-imperialist films of the past, such as the films examined in this essay. Despite the "Palestinian Stories" example that makes some space for documentary films, the more overtly radical or oppositional films of the Third Worldist cinema movement, including documentaries, remain unavailable on any commercial TV channel and have very limited availability on streaming platforms as well.[7] In the "now" of neoliberal policies and multinational corporations that govern world politics, the arms and fossil-fuel industries, military industrial complexes, and resulting lopsided wealth inequalities, where do we find the "then" of older leftist films that have not been preserved by publicly accessible archives?

Laura Marks reminds us that to research and teach the (subtitled) history of Arab cinema requires an encounter with "shadow archives," which survive through memory and imagination, in response to an archival fragment or the lack thereof.[8] Here she is gesturing to Akira Mizuta Lippit's exploration of the link between cinema, twentieth-century postwar Japanese imaginaries, and archives in the absence of what could be "seen" in the atomic age.[9] Marks's and Lippit's ideas are helpful in theorizing lost or forgotten Third Worldist cinema

6 See Enterprise, "The Age of Narrowcast Is upon Us: How Streaming Is Changing the Middle East's Media Landscape," February 20, 2022, https://enterprise.press/stories/2022/02/10/the-age-of-narrowcast-is-upon-us-how-streaming-is-changing-the-middle-easts-media-landscape-64729/.

7 Both Mubi and Aflamuna have programmed limited-time screenings of Palestine-solidarity films and militant films from the Palestine Film Unit and the Palestine Cinema Institute, particularly during the Covid pandemic and at crucial moments since 2020 to express solidarity with the Palestinian people during Israeli Occupation Forces aggressions.

8 Laura U. Marks, *Hanan al-Cinema: Affections for the Moving Image* (Cambridge, MA: MIT Press, 2015), 193.

9 Akira Mizuta Lippits, *Atomic Light (Shadow Optics)* (Minneapolis: University of Minnesota Press, 2005).

from the Middle East. These ideas also assume fresh significance in the aftermath of the 2023–2024 Israeli bombings in Gaza and recent bombings in Lebanon, which have not only surpassed the nuclear payloads of both Hiroshima and Nagaski but also resulted in genocide, scholasticide, and massive decimation of historic, governmental, and cultural archives.[10] In the aftermath of this war, will we find the photography studio of Haj Salameh documented in Abdessalam Shehadeh's film *To My Father* (2008), which was dedicated to Gaza's pioneering photographers? Shall we assume that most of the faces encountered in the film are deceased, or will have been killed in one of the twenty-first-century wars in Gaza? Where will we find Shehadeh's film, or other films we have heard of or read about but not seen? To start, we can search online for a trace or fragment of a digital version that someone might have uploaded or that might be housed in an online collection. We might even return recursively to YouTube, which, scholars have debated, might be seen as an archive given the vast quantity of films and television series, often unsubtitled, that it hosts, often free of charge.[11]

To research and teach Arab cinema, then, depends on the capacity of material and virtual archives to survive the ravages of war and looting (as in the case of the Palestinian Cinema Institute), self-censorship (in the face of repressive states), and the capitalist dictates of streaming media corporations (including not just the abovementioned sites but also video hosting and file-sharing platforms like Vimeo, Dropbox, and WeTransfer). Not only do these pressures produce gaps in our efforts to conduct research and teach films that appear and disappear, but they also erase and elide great amounts of labor that go into building film histories and critical documents about a history in process. To correct this, we must conduct archaeologies of archives, in addition to finding the means to support the building of new ones. This must be a collective effort.

In the recent post-Arab "spring" archive fever, inspired by the collective work of people demonstrating their right to have rights, and in part motivated by the lack or moribund state of actual archives in the SWANA region, people have been working in "hybrid, fluid networks" to organize and process "re-readings of the past," creating "an archeology of knowledge" as a way forward.[12] This activity

10 See "UN Experts Deeply Concerned over 'Scholasticide' in Gaza," on the website of the UN Human Rights Office of the High Commissioner: https://www.ohchr.org/en/press-releases/2024/04/un-experts-deeply-concerned-over-scholasticide-gaza.

11 Rick Prelinger, "The Appearance of Archives," in *The Youtube Reader*, ed. Pelle Snickars and Patrick Vonderau (Stockholm: National Library of Sweden, 2010), 268–74.

12 Sune Haugbolle, "Archival Activists and Hybrid Archives of the Arab Left," in *The Arab Archive: Mediated Memories and Digital Flows*, ed. By Donatella Della Ratta, Kay Dickinson, and Sune Haugbolle (Amsterdam: Institute of New Cultural Works, 2020), 7–19.

reflects a strong desire to connect the past of the Left in the Arab world, with revolutionary groups across the region investigating their own practices and that of their predecessors. An extension of this has been a renewed interest in the militant filmmaking of the past, particularly aligned with the Palestinian resistance movement of the 1970s. The few, mostly documentary films that have been made available, some of which have been restored privately by scholars, activists, filmmakers, and friends of the deceased, have been curated in segments of Arab film festivals, both online and in person, and featured on online platforms like Mubi and Aflamuna that focus on art-house cinema, as well as Palestine-solidarity screening events.

Emerging from this context of interest, this chapter focuses on three such films made by women filmmakers, recognized belatedly as pioneers of their generation, each of whom had close connections to and support from the Palestine Cinema Institute in Beirut. Produced in the contexts of Palestinian liberation and the early stages (1974–8) of the Lebanese civil war (which lasted until 1991), and significantly reliant on archival footage, these are films you will not see on state or satellite television because they present a political history, temporality, and point of view of resistance fighters that neither state nor commercial media channels, including Netflix, will show.[13] Thus, this chapter aims to transmit from the shadow archive (which the reader might be unable to see because of material limitations) a textual rereading of these films, reinscribing both filmmakers and films, historically and epistemologically, into the histories of 1970s Arab cinema. Doing so participates in rescuing them from being lost and disconnected texts, while simultaneously calling attention to the continued value of content that remains relatively inaccessible, despite the abundance and connectivity created by streaming media. Because these films are still largely unavailable to the public without the intervention of curators, it is incumbent on researchers, educational and cultural institutions, and even possibly the gatekeepers of narrowcast media landscapes to find ways to make them public not only for the knowledge they afford and its relevance in the present but also for their contributions to vital, internationalist, pan-Arab, feminist film histories.

[13] Increasingly, cultural organizations like Aflamuna and Maskoon (both based in Beirut), film websites like Mubi, and international film festivals will screen older Arab films, including the militant films made by the different factions of the PLO. As of the time of this writing, it is uncertain to what extent Israel's 2024 bombings in Beirut will disrupt the programming of Lebanese-based organizations.

From the Remains of a Ruined Archive

Crucial to the significance of these films is their mode of activating audiences through felt engagement with the shadow archive of the pan-Arab Palestinian resistance that was shattered by the early 1980s. Accompanying this shattering was the disappearance of the archive of the Palestine Cinema Institute (PCI) in the wake of the Israeli siege of Beirut in 1982 and the events that followed the attempted dissolution of the PLO. Indeed, emerging in 1974 from its evolution from the Palestine Film Unit (PFU), after an intermediary experimental effort of the Palestine Cinema Group in 1973, the PCI archive was created in response to the absence of representations or evidentiary documents about Palestinians made by Palestinians. By the time of its disruption in 1982, the institute had built an internationally connected Third Worldist center of films and filmmaking that documented and narrated the Palestinian resistance movement.[14] Drawing on the PCI archive, these films are documents of a specific, not completely forgotten, historical time and place; they convey informative and "timely" political arguments; and they rely on archival material to "awaken" the viewer into a critical mode of reception that enables reflection on how the past lives on in the present.

The awakening in the viewer calls to mind what Walter Benjamin theorized about the dialectical image, as a recognition of a historical truth that emerges quickly or "as a flash" arising from an image or image sequence. This awakening is not only political but also affective, and it is triggered by the encounter with archival fragments within a film's overall diegesis or narrative.[15] The dialectical image may or may not be the material image that one sees, but it is imprinted on, and is produced through, the critical engagement of the viewer. Put another way, because the archival fragments were skillfully selected and placed in the films under discussion, we are able to "see" or experience and intellectually apprehend dialectical images.

These films do more than demonstrate the synergistic relationship between documentary filmmaking and Palestinian activism in Lebanon in the 1970s, underscoring the importance of the Palestinian struggle in connecting the politics of Lebanon to the larger Arab world in that moment. They also respectively summon our recognition of certain historical truths (apprehension

[14] Khadijeh Habashneh, *Knights of Cinema: The Story of the Palestine Film Unit* (Cham, Switzerland: Palgrave Macmillan, 2023).

[15] Eli Friedlander, *Walter Benjamin* (Cambridge, MA: Harvard University Press, 2012).

of the dialectical image) through the processes of their filmic statements, which include use of archival moving and still images: the agency of Palestinian women as partners in the Palestinian cause (in Saab's *Les Femmes Palestiniennes*), the rights of Palestinian children as enshrined in the UN Declaration of the Rights of the Child that was drafted in 1959 (in Habashneh's *Atfal Walakin*), and the carefully planned US-led orchestration of the dismantling and evisceration of unity and progressive alliances between Arab countries and peoples that began with the Sadat-Kissinger negotiations of 1974 and 1975 (in Chahal's *Khotwa Khotwa*).[16] It is not surprising, then, that these films would be difficult to see on conventional satellite, cable, or terrestrial television, considering the leftist positionalities they reveal, which are out of sync with the market-driven mandates of broadcasting, even if they may speak to the current global resurgence of support for Palestinian sovereignty and rights and condemnation of the ways the United States and world powers and institutions enable Israel's violent actions and racist settler-colonial ideology.[17]

As I will endeavor to explain, archival imagery in each film works to reanimate a moment in historical time, thereby inviting us to engage with and reflect on processes of reading a past that haunts the present of unresolved conflicts and imperial violence. In the process of interpreting and reading these texts we become—like the viewer that Catherine Russell, building on Benjamin, imagines as the ideal recipient of found footage films—"not simply a historian but a participant in historical construction."[18] As a result, each film demonstrates the importance of archives not only as *lieux de memoire* or containers of history and inscriptions of memory but also as entities that belong to, feed, and summon cultural memory, processes that are uniquely suited to cinema.[19] The archival material in these films thus delivers the mise-en-scène of a participatory cultural memory, which then allows us to "feel the archive."[20]

16 Of the numerous accounts of these negotiations, see Edward R. F. Sheehan, "How Kissinger Did It: Step by Step in the Middle East," *Foreign Policy* 22 (1976): 3–70, https://doi.org/10.2307/1148069.

17 This is not specific to Western cultural or media hegemony. Most Arab governments do not care about Palestinian sovereignty; the government-owned terrestrial media companies in Arab countries only pay lip service to the Palestinian issue when it is a humanitarian issue. Even the Roya satellite channel, which is owned by a Palestinian Jordanian and which programs Palestine-themed content, would probably not air the films discussed in this chapter because of the leftist politics and filmmaking aesthetics these films deploy, which might be deemed by programming gatekeepers as too much information and not enough entertainment.

18 Catherine Russel, *Archiveology: Walter Benjamin and Archival Film Practices* (Durham, NC: Duke University Press, 2018), 223.

19 Inez Hedges, *World Cinema and Cultural Memory* (New York: Palgrave, 2015).

20 Pad.ma, "10 Theses on the Archive," in *Dissonant Archives: Contemporary Visual Culture and Contested Narratives in the Middle East*, ed. Anthony Downey (London: I. B. Taurus, 2015), 352–63.

Feeling the Archive

Before launching into case studies of each film, it is noteworthy to consider that part of the indexical power of these films (based on the indexical trace of reality afforded by documentary footage) produces the sense that we are not just "looking back" at a moment in time but also that these films "look back" at us. An arresting example of this, especially for how it engenders an active viewer and "felt engagement" with historical archives, is a sequence at the beginning of Habashneh's *Atfal Walakin*. Just under four minutes into the film the camera zooms into a close-up of a girl's face sitting in the garden of Beit Atfal Al- Sumoud, the home and school created for orphans of the 1976 massacre of the Tal Al-Zaatar refugee camp by Christian phalangist forces in Beirut. The girl in the shot, surrounded by a group of younger children, is looking back at the camera, through children's arms and other movement that come and go from the frame (Figure 11.1). Her return gaze inaugurates a flow of archival images, sounds, and information, which becomes a pattern of layered documentation of past events.

What unfolds is precisely remnants of the preceding massacre's impact on children that have led to the moment of this look: a still image of two children staring back, through the railings of a truck; footage of girl carrying a bag on her

Figure 11.1 The young girl looks at the camera. *Atfal Walakin* (Children Nevertheless), directed by Khadijeh Habashneh Abu Ali © Palestinian Cinema Institute and the General Union of Palestinian Women 1979.

shoulder who looks up at the camera from a scene of survivors of the massacre. This mise-en-scène of a past moment is delivered not only by images but also by sounds that form a provisional representation of what the girl with the look might remember: a woman's voice-over informs us that hundreds of children were killed or perished from hunger and thirst resulting from the siege and massacre; she tells us that the camp was totally destroyed, and fifteen thousand people were forced to migrate, half of whom were children, among whom three hundred had lost all family members. A general wall of sound envelopes the scene of chaos, from people trying to communicate and the wailing of children. This is followed soon after by an intertitle in Arabic against the static of a silent soundtrack, declaring Article 6 of the Declaration of the Rights of the Child, which the subtitles translate as, "The child must grow up under the care and supervision of his parents and should benefit from financial and moral security." The sonic chaos resumes over footage of children asleep or dead on the ground, followed by a still image of a woman crying while clutching tightly onto two crying children, and then footage of a group of adults walking around in the aftermath of the massacre, with another woman crying into a handkerchief.

This flow of images and sounds relies on highlighting temporal disparities between the present and the past, or "then" and now, which are some of the most common and easily apprehended usages/meanings of archival images. This is what Jamie Baron calls the "archive effect," a classic example of which is when a person or place is shown at different ages within the context and duration of one film.[21] This flow also breaks into our own time of viewing and delivers the aesthetic and political argument of the film. Through these ruptures, we are forced to watch, listen, and read the traces of what is left after an unrepresentable event. However ambiguous the inaugurating look might seem, its dual meaning is clarified by the sequence it unleashes: the girl is not just looking back at the camera; we are also positioned to look back at events that have led to her reality. One meaning produces the other.

The dual "looking back" that relies on archival images engages a process that has been theorized elsewhere, by drawing on Walter Benjamin's oeuvre and heavily reliant on his theory of the dialectical image. Russell explains this process and practice as one of archiveology, or a "language of images."[22] She

[21] Jaimie Baron, *The Archive Effect: Found Footage and the Audiovisual Experience of History* (New York: Routledge, 2014).

[22] Russell, *Archiveology*, 219.

sees Benjamin's theories as uniquely suited to archive-based filmmaking due to, among other things, the ways such films allow ideas to speak for themselves in a material presentation of historical revision that calls into question their historical construction. In Russell's formulation, the dialectical image is not restricted only to usages and effects of archival footage, nor only to words, images, or representation. Rather, as she articulates Benjamin's view, we encounter the image across the presentation of representation. By this she means the fragments of films that have been taken out of their original contexts of representation, which are representations of a kind of reality (whether fiction or documentary), are then presented in new contexts in cameraless, archive-based, possibly found-footage films. The act and meaning of representation itself are foregrounded in such films. It is from this presentation of representation that the dialectical image emerges, through a recognition in the viewer that is distinct from seeing or visibility. Recognition comes about through a process of movement and freezing of thought, which results in a presentation of an idea, what Benjamin called a "constellation." While dialectical images are not restricted to avant-garde or documentary films, nor even necessarily to the moving image itself, Russell's explanation of how they erupt from archive-based films has unanticipated resonance in *Les Femmes Palestiniennes*, *Atfal Walakin*, and *Khotwa Khotwa*: they too use historical fragments (archival images) to interrogate textuality in a deconstructive practice, disrupting historical narratives that made little space for or brought little visibility to the perspective of the Palestinian refugee.

It is, however, important to note a distinction between films based on archival footage (films that use moving imagery derived from the holdings of an archive) and the films Russell discusses, which are avant-garde found-footage films that appropriate scenes from mostly classical European and American cinema and repurpose them in completely different contexts. All the archival fragments of the films discussed in this chapter belonged at one time to an archive (the PCI, or the Iraqi National Archive), whose purpose was to gather information relevant to the histories the archives were preserving. But, as I shall argue, these films *do* participate in cinematic processes of archiveology, creating new meanings through a language of archival images. We can see this by the way the selection and arrangement of historical fragments (or archival images) participate in a deconstructive practice that emphasizes historical disjunctions for Palestinians, at a moment in the 1970s when there was great interest and investment among leftist political groups in the transformative potential of film to communicate and transmit their cause, in solidarity with other national liberation movements and to the world.

All three films visualize ruptures between successive layers of a past that are irretrievable. This introduces a different kind of archival affect (theorized by Baron as emerging from our interaction with archive effects),[23] unintended by the filmmakers but inherent to the historicizing of their work: that of feeling the archive beyond interpreting it, which may include a recognition of what has been lost in the wake of the "failed" Palestinian revolution and the events that have led to the present moment. By paying attention to the process launched by archival images, not only might we understand how these films perceived their own past and initiated new histories, but we are also reminded of the centrality of the Palestinian struggle and resistance movement in the politics of the past—and indeed of the dystopian present in the Middle East.

The films discussed in this chapter belong to different histories and contexts than those attended to by Russell and Baron, but archive effects and archive affects occur in their instances of appropriation to reinscribe history and summon memory. Where Russell's study becomes additionally relevant is in her discussion of how allegory and indexicality work together in archive-based films. Her attention to the indexical and allegorical power of archival footage draws from linguistic studies that have been established in film theory, where documentaries have been historically (mis)recognized for the indexical power of the image (it is what it shows), whereas fiction films have been historically discussed in relation to their allegorical functions.

Bruzzi and others have debunked such limiting distinctions, arguing that documentaries, through their assumed links to the real, historical world, can be read allegorically.[24] This is possible because each fragment of material, whether original footage shot for the film or archival material, has context-bound indexical meaning on its own, which, when composed in combination with other shots and formed into edited sequences of material, produces meanings that advance the allegorical (message-bound) content of the films.[25] This occurs in these films by Saab, Habashneh, and Chahal, where archival footage emphatically references the defining historical ellipsis of the Palestinian *nakba*, its later iteration in 1967, and the revolutionary period of the *longue durée* of the 1970s. It does so on the levels of index and allegory. We can better understand how archive effects work and how they animate archive affects by closely reading select sequences from the films.

23 Baron, *Archive Effect*.
24 Stella Bruzzi, *New Documentary*, 2nd ed. (New York: Routledge, 2006).
25 Russell, *Archiveology*, 98.

The Power of Ellipses from Archival Fragments

In each of these films, the off-screen presence of the filmmaker is expressed as voice-over and at times as interlocutor within the diegesis of the film. But the filmmakers do not center their subjective voices within their films, even though their films have been shaped by subjective choices that we may feel. This feeling for the filmmaker who is no longer there, or whose younger voice registers their material presence in a past temporality, may be another way the archive affect touches us. If the archive affect is our emotional investment, as Baron contests, in the feeling of loss when viewing the archival trace, this adds another layer to our reading of these films that respectively illuminate women's involvement in a once living political project, the consequences of war on the lives of displaced and traumatized Palestinian children, and the outcome of an externally brokered agreement to dissolve Arab unity. Our affective responses might then be mobilized epistemologically or toward positive action, whether we are learning about these subjects for the first time or revisiting them.

In its ten-minute duration *Les Femmes Palestiniennes* adheres to a straightforward style of exposition, with Saab's voice interpolating the footage about women who participated in the Palestinian revolutionary movement and who were ordinary people, constitutive of the fabric of Palestinian society in the pre-civil war Lebanon of the early 1970s. Her voice invites us to see these women as engaged in more than one struggle within the Palestinian resistance, fighting both for Palestinian rights and for women's emancipation. We are shown images of *fedaeeya* or women freedom fighters in combat training, climbing walls on ropes, loading guns (Figure 11.2). We hear from women university students who speak candidly to the off-screen filmmaker about the challenges they face and their commitment to the Palestinian resistance.

From today's perspective this is an extraordinary historical text, even though its narrational strategies are somewhat conventional. But what seems like a familiar usage of archival footage to bridge temporal and geographic gaps also sutures and simultaneously reveals an unexplained disparity. A sequence of black-and-white footage shot by an unknown source, possibly a British or French news source, or even by PFU cofounder Hani Jawharieh,[26] shows the crossing of Palestinian refugees on the broken Allenby Bridge in 1967, with the

[26] The source of this footage of refugees crossing the Allenby Bridge is still unverified. It is possible that it was shot by Hani Jawharieh, as he was involved with the PCI, which supplied this footage to Saab.

Figure 11.2 *Fedaeeya* in combat training. *Les Femmes Palestiniennes* (Palestinian Women), directed by Jocelyne Saab © Antenne 2 (France 2) Public National Television 1974.

camera lingering on the exasperated facial expression worn by a teenage girl who becomes a protagonist in the film soon after. But before this happens, there are several shots of refugees, men, women, and children, carrying belongings and helping each other reach land on the Jordan side, after which the film suddenly shifts to color stock and we are in Lebanon. This suggests that once we leave the archival footage, we have left Jordan and entered Lebanon, where the Palestinian resistance had relocated after the events of Black September.[27] There is no explanation for the ellipsis, but we can infer that the skipped interlude, while not necessarily inconsequential, is not as important as the connection between the trauma of 1967 and the subsequent period of activism in Lebanon that the bridge sequence signifies. The girl whose facial expression this sequence prioritizes to represent the mass fleeing of people has become a woman politicized through trauma and rupture and has wed a branch of the Palestinian cause in her work and life in the refugee camp in Lebanon. This sequence also demonstrates the archive effect because it signals a temporal break within one singular text (i.e., the young girl ages into a young woman through the ellipsis).

[27] Black September, commonly referred to as Jordan's civil war, was an armed conflict that erupted between the PLO and the Jordanian army in September 1970 and lasted a little under a year. It resulted in major destruction of refugee camps, thousands of deaths of Palestinians, and the expulsion of the PLO from Jordan.

The archive effect works through the bridge-crossing sequence in relation to that which follows, precisely because we recognize the disparities, temporal and intentional, that are evident in the text.[28] As previously explained, it produces an awareness in the viewer of a disparity between times and, in this instance, spaces that makes the viewer more actively engaged in understanding how subjects are produced and politicized through settler colonialism and war. After this we follow the woman protagonist in the camp, who unlike most of the women we see wears a pantsuit without a headscarf, as she walks, visiting people, clutching a notebook while she describes the challenges of living in the camp—the weak education system in the UN schools, poor sanitation, unclear water, no light—and, having made it to university, realizes the determining class difference between herself and most of the other students. The film's ending happens somewhat abruptly after a focus on what Saab calls the "impulsive, passionate" ones: a villager addressed by the off-screen Saab as "Um Anta" who recounts in a field of rubble that a free Palestine can only be achieved through armed struggle; a university student who summarizes the social, economic, and political challenges Palestinian women face, and who similarly concludes that the liberation of Palestinian women will only be achieved through armed struggle; and a young militant who recounts several secret operations she conducted in Palestine and the unspeakably harsh treatment she received in Israeli prisons. Her testimony, given in the same room as the first woman we encountered in the Lebanon sequence, framed in shallow space with a wall behind her displaying a poster of a *fedaee*, ends with a series of charcoal-and-ink drawings of refugees. This figuratively raises the image of the impulsive and passionate *fedaeeya* to the level of allegory and gives her the last word, suggesting, without saying, "to be continued." The archive affect may be enhanced by our realization, in the context of unending occupation and oppression of Palestinians, that these courageous women, if still alive, are living very different realities from the ones they imagined and fought for decades ago.

Habashneh's *Atfal Walakin* is more poetic than *Les Femmes Palestiniennes* in its use of archival fragments. It is composed of moving images from the early days of UNRWA in the 1950s (probably shot by a European cameraman), photographs, newspaper clippings, and footage shot in the occupied Palestinian territories that are combined with original footage shot for the film. The film is full of sequences in which allegorical, poetic, and indexical meanings arise from

[28] Baron, *Archive Effect*, 22.

combinations of images and between images and sounds. We already discussed the sequence "representing" or standing in for the Tal Al-Zaatar massacre, which is soon after followed by the footage of post-1948 refugee tents and camps in Jordan. Similar to *Les Femmes Palestiniennes*, there is no explanation for the ellipsis between Jordan and Lebanon, or Lebanon and the occupied territories, just a visual cue, which is always, in this film, a child's drawing or notebook that serves as a bridge across temporal and geographic disparities.[29] At seven minutes and twenty-five seconds into the film, a succession of three children's drawings (of buildings in a camp, an airplane bomber, and a *fedaee*) connects a preceding poetic sequence of archival footage from the past (of tents in a refugee camp ostensibly in Jordan from the 1950s) to the profilmic present of children drawing in a classroom at Beit Atfal Al-Summoud in Beirut, before returning once more to the archival footage. The interplay between then and now, despite or because of spatial ellipsis, suggests that the past inhabits the present, one producing and coexisting with the other. But further, whoever shot the archival footage was a skilled cinematographer or had the good fortune of good exposure on a day of spectacular light, as we see in a close-up of layered tent fabric, pulled here and there by the wind. It's a visually arresting shot for its graphic details, revealing gradations of tone between black and white. The accompanying lines from an uncredited poem by Ahmed Dahbour, who is later credited in the end for being a poet, speak to the degradation the image indexically presents despite the incommensurable visual aesthetics.

What is additionally unique about this film is its inclusion of testimonies of child survivors of routine violence from Israeli occupation forces that persists until now, but which had not been previously captured in a documentary. This is another way the film contributes to cultural memory. A scene that drives this home takes place soon after fifteen minutes, from a sequence of black-and-white footage showing the juxtaposition of daily life activities with civilian resistance: boys and men throwing stones, fire and smoke around them, people moving about their business, schoolgirls walking to or from home, Israeli soldiers jumping down from a truck with batons swinging, followed by a reverse shot of a crowd of boys and kids throwing stones, and Israeli soldiers in an archway chasing the demonstrators. There is a black frame that inaugurates the emphasized

[29] The use of children's drawings to express what Palestinian children experienced began in a different context, most notably in an exhibition of children's drawings organized by Jordanian artists Hani Hourani and Mona Saudi and gathered from refugee children who experienced the 1967 war, displaced in the Baqaa camp in Jordan.

scene of three or four soldiers knocking down a teenage boy they were chasing and beating him. We hear his voice, "My name is Mohamed Khalid Abdullah." But soon after this, the scene is repeated, slowed down and slightly brighter in exposure. As we see him being caught by the first soldier we hear, "I'm thirteen years old. When they imprisoned me in Ramallah, they left me standing all night long." The scene repeats, of soldiers circling his body on the ground, punching, beating, kicking him. He tells us, "When I tried to sit down the interrogator beat me. After that he started beating me and hit my genitals. He threatened me, if I don't confess, he would rape my sister and mother in front of me. They forced me to drink dirty water." His last line, "Because of this I confessed to throwing stones at the jeep," overlaps a color shot of kids in a class under a white tent, learning how to defend themselves, which provides a rebuttal to the accusation often made by Israeli propagandists and their supporters that Palestinians turn children into armed fighters, by showing this as the natural consequence of a vicious and militarized occupation that uses the most sophisticated military technology to police, terrorize, and kill Palestinians.

Reading the Dialectical Image

In truth, archive affects and dialectical images are subjectively and contextually determined, like the *punctum* in photography, and while Russell limits archiveology to a language of images, it can equally pertain to a language of sounds and combinations of sound and image. What is important is the process through which meaning is made using archival fragments, which is incomplete without the active and critical reading and interpretation of the viewer. As such, and consistent with the intervention of this essay, I will offer my own reading of a few filmic sequences as an exemplification of archive affects and the dialectical image produced in the context of Israel's war against Palestine and Lebanon.

The closing section of *Atfal Walakin* begins with an image and quote from former Israeli prime minister Golda Meir, after we have seen a *dubka* dance of the children performed for an audience, in the front row of which sits Yasser Arafat and his second-in-command, Salah Khalaf (Abu Iyad). We hear children singing, on the extradiegetic level, "After I had the rose in my hand, now it's time for someone else, and I am left deprived." The image of Meir appears with the quote in Arabic, "I don't want to wake up every morning and endure the panic of the birth of a new Arab child. I want a nation with massive Jewish majority—Golda Meir, *Maareef Newspaper*, October 25, 1973." This is followed by a photo of Meir speaking with her mouth open into a microphone and her hand raised as if to emphasize the words. Her "look" initiates a sequence of images of dead Palestinian children. The first is a close-up of a dead child's face and after zooming out pans across a row of six dead children. The following sequence places emphasis on hands: it begins with a shot of women crying and publicly mourning this atrocity, followed by more close-ups of dead children's bodies and hands, suggesting that the hand that once held the rose in the preceding song has lost its agency. While watching this sequence, we cannot avoid thinking of the present, of every image of violence we have ingested of dead children in Gaza and the West Bank, since the early 2000s and more specifically since October 2024. We are painfully reminded of the necessity of material, economic, and political changes that are crucial to ending the disabling necropolitics of the Israeli project of settler colonialism. Reading the sequence in this way demonstrates how the dialectical image arises as a flash of historical truth.

Every time we are aware, as above, of a flash of recognition, we may be engaged in reading the dialectical image. Russell's discussion of this in relation to archive-based films makes sense, and even though her theorization of archiveology is based on cameraless experimental films, her argument applies equally well to documentaries that rely on archival fragments. This is because the dialectical image, in the Benjaminian sense, comes into being in relation to reality; it is an understanding of truth, which aligns well with our assumptions and readings of the documentary mode. *Les Femmes Palestiniennes* and *Atfal Walakin* achieve this through a literal look back at the camera and through causal links between the past to a profilmic "now." The appearance of this "look back" summons the past, and the appearance of the past visually underscores the foundations of the present, demonstrating an awareness and belief in what the camera and film might change.

In Chahal's *Khotwa Khotwa*, however, the looking back is not a process triggered by one specific image or sequence but rather the film as a whole. Relying heavily on footage obtained from the PCI, as well as the Iraqi Film Archive and French news outlets, the film unleashes a rigorous analysis of the conditions in Lebanon that precipitated its descent into war. Chahal presents these conditions as predating the Palestinian resistance and refugee situation and as rooted in class warfare and anti-Arab sentiment. Her film argues that this ethno-nationalist foundation became enabled by a US-brokered step-by-step policy (her words) initiated by the first Egypt-Israeli disengagement agreement of 1974, signed by Henry Kissinger and Anwar Sadat. We are guided by her curation of news footage and film clips obtained from different archives, combined with original footage and her voice-over commentary, which sketches pivotal moments in Lebanon's modern political history, from the French Mandate of 1920 to the late 1970s. Made from the vantage of 1979, looking back at the events of the past and particularly at events anchored to the aftermath of the 1973 war, *Khotwa Khotwa*'s analysis relies on temporal and intentional disparities, or archive effects, produced through a system, perhaps even a language, of images and sounds.

The sound design in *Khotwa Khotwa* is punctuated by a musical passage from *La Pie Voleuse*, adapted by composer Igor Markevitch, which comes in at specific junctures to emphasize significant benchmarks in the step-by-step plan. The narrative within the musical piece is fitting as it centers on the accusation of a poor servant girl for stealing from her masters a spoon and fork, which were actually stolen by a pet magpie, all revealed at the end by the girl's lover, which saves her from hanging. The composition's message of blaming the subaltern subject resonates with the tendency in Lebanon to blame Palestinians for the civil war, when the ingredients for war, as shown in the film, preceded the refugees. The dissonance between things (by which I mean images and image/sound combinations) and the intentional disparity of the archive effect produced by Chahal's selection of material and editing choices reveal her signature preoccupation with the dissimulation of actors, whether as figures featured in the political analysis conducted in her documentaries or written into the characters in her narrative films.

Although *Khotwa Khotwa* deserves a much more thorough study, I refer here to two sequences from this film to demonstrate the dynamic between archive effects and archive affects specifically in delivering the "punch line" of the film. The first sequence demonstrates the economic conditions in the wake of the 1973 Arab–Israeli War that fed social divisions and set the stage for opening of the civil war. This begins with black-and-white documentary footage of leftist demonstrations from the early 1970s showing people with hands upright in peace signals and holding portraits of Gamal Abdel Nasser, the deceased Egyptian leader and icon of Arab nationalism. The motif of *La Pie Voleuse* resumes, accompanying more demonstration footage with men holding guns in the air. This is contrasted with color footage of a marching band and people from the right-wing Kataeb party carrying the Lebanese flag. Politicians drive by in a black, tinted-window car, from which an arm reaches out to wave to the demonstrators.

Chahal's voice-over accompanies this footage, explaining the conditions that led to the unrest in relation to the 1973 war. Her statement, "High prices, repeated Israeli attacks, calls for a new defense policy, reasons that led to the division of the country," is followed by different shots of driving in the countryside and archival black-and-white footage of people fleeing an Israeli attack on the village of Kafarchouba in southern Lebanon. Chahal's voice explains,

> Kafarchouba, seven days of continuous bombardment. A village that was completely destroyed. Dozens of dead and wounded, the living were forced to flee. Alone in the field, the Palestinian Resistance remained, seeking to prevent the advancement of Israeli troops.

A sequence follows, most likely from the PCI archive, presenting shots of *fedaeen* fighting, followed by shots of an Israeli helicopter, Israeli soldiers loading into the helicopter and Israeli tanks. We are then shown villagers fleeing, an explosion in a field, a woman carrying a large bag on her head running near a line of cars and trucks trying to leave (Figure 11.3), a man and woman helping another man lift a heavy bag onto his shoulder, an old woman crying, a child crying, an injured baby crying in a hospital bed, and an empty house with a graffitied message: "Let's make Kafarchouba the symbol of *sumoud* [steadfastness] in the south. No to forced migration." These shots are repeated later in the film's denouement. Similar to the emphasis of the repeated scene discussed in *Atfal Walakin*, these repetitions underscore the film's argument: the step-by-step plan has led to this scene of people fleeing Kafarchouba and its imminent destruction and displacement by the Israeli Occupation Forces.

Just as the sequence above frames a causality between imperial interests, Israeli aggression, and economic hardship that enables the conditions for conflict, the second sequence I'm referring to presents the consequences of war on civilians caught in the cross fires of said conflict. It similarly uses archival images to narrate the official trigger for the civil war in 1975 in the Ain Al-Roummaneh neighborhood of Beirut, where a group of Christian phalangists fired on a school bus transporting Palestinians, mostly women and children, to the Tal Al-Zaatar refugee camp. Chahal relies on photos from news stories, accompanied by her voice-over explanation. We are shown an image of the school bus. The camera pans left and zooms into the bullet-riddled window of

Figure 11.3 A resident of Kafarchouba fleeing, among other residents, from imminent Israeli attack. *Khotwa Khotwa* (Step by Step), directed by Randa Chahal © Hubley Studios 1978.

the driver. A photo of another window with cracked glass from bullets is followed by an image of militiamen with guns peering into the bus after the shooting. The pace of editing alternates rapidly between these two shots, between the bullet-riddled bus window and the militiamen, to simulate the violence of the event.

A photograph of dead children shows what Chahal's voice explains: "Twenty-seven dead. All bus passengers." We see a newspaper page covering the event, with her comment, "Lebanon is hitting one of the darkest phases in its history, a civil war that will divide the country and make it fight itself." The return to the shot of dead children that we just saw is followed by another return, of the aural motif of *La Pie Voleuse*, which forms a transition to color footage of children turning a corner around a barricade of sandbags, as a woman walks the other way. It is the time of war. In the following sequence of the armed militias, Chahal explains, "Very quickly, directions are determined. On one side there are the Progressive Forces and the Palestinian Resistance, being the real target of the maneuver. On the other side there is the right. It was a step in the American step-by-step plan." By contrast, we see two shots of relatively quiet, empty streets where the color film stock is less saturated. *La Pie Voleuse* remains, bridging to a group of middle-class people spilling into a street outside a bakery and a shot of militia trucks in the street. The camera zooms out to show the limited area in which the children ride their bikes, between the trucks and barbed wire, a typical scene of how people "adapt to war."

An Uncanny Resemblance: Between Then and Now

In these three films the selection and use of archival footage makes it impossible to ignore the resemblance of the past to the present political status quo, particularly in view of the live-streamed Israeli genocide in Gaza that began in 2023 and Israel's attempts to obliviate resistance groups like Hezbollah in Lebanon. That is one way the dialectical image might arise from these films, positioning viewers to reflect on the history of wars and aggressions in the region.

As these films affirm, Lebanon is still a contested place, not just between the Israeli Occupation Forces and Lebanese resistance, but also between Lebanese and Syrian forces, Lebanese and Palestinian refugees, external powers competing for influence, and among the Lebanese themselves. For such an afflicted place and history, the past continues to surface and determine social and political realities. This gives value to practices of research, to archives, archivists, and archive-based films such as the ones discussed in this chapter. The significance of archival material in these films casts a series of past temporalities of which only the visible/audible trace remains, and thus they cannot be severed from the historical referent (both to media objects and to technologies of reproduction), no matter the context of repurposing and the intentions of the filmmaker. This is another way one might experience the archive affect, not as an experience of nostalgia but as an awareness of both a dissonance and an uncanny resemblance between these images from the past and the current, albeit different, ideological time.

In the select scenes detailed above, we are responding to disparities of time and space through the suturing of shots and sequences of images and sounds. This does more than produce the layered meanings of the films; it causes our awakenings as active viewers. This is the magical function of editing in archive-based films, and it points to the paradox of archives in their composition of absence and excess, signified by their many gaps, despite the abundance of indexical traces of the past that cannot possibly fit into one film.[30] But the films I have discussed express this paradox somewhat differently: made to respond to a perceived absence of representation or political argument, they produce an excess of meaning that after an initial screening life, and subsequent decades of invisibility (from roughly the early 1980s to the uprisings of the 2010s), has recently undergone recirculation by concerned curators and filmmakers.

These films use archival material to comment on a past that haunts the present and must, as the filmmakers insist, be remembered. But they are also archival documents that inhabit larger archival bodies and interventions, whether as interrupted institutions and subsequent recovery projects (Habashneh's ongoing work to retrieve and restore the scattered film prints of the Palestinian Cinema Institution) or recent preservation projects dedicated to memorializing the work of a single filmmaker (Mathilde Rouxel's recent restoration and programming of the films of Jocelyne Saab; Nour Sabbag's similar project with the films of her late mother, Randa Chahal Sabbag). Indeed, Rouxel, Habashneh, and Sabbag, as archons gathering and preserving remnants from a ruined archive (referring here to archive as a place), echo Lippit's idea that archives are inherently composed of ruins, "preserved by the traces of destruction."[31] Their efforts go a long way to combating the hegemony of information and entertainment industries with their promise of endless addiction to consumption and commodity fetishism. Speaking from shadow archives, by using the limited access that streaming media currently avails, the sharing and streaming of these films allows us to imagine a world where technologies of streaming might emancipate suppressed cinematic histories, making them free to all, so we could more fully engage with the present.

Bibliography

Alkassim, Samirah. "Amiralay and Sabbagh in the Post-Cinematic Age." In *Cinema of the Arab World: Contemporary Directions in Theory and Practice*, edited by Terri Ginsberg and Chris Lippard, 208–30. Cham, Switzerland: Palgrave Macmillan, 2020.

Baron, Jaimie. *The Archive Effect: Found Footage and the Audiovisual Experience of History*. New York: Routledge, 2014.

[30] Baron, *Archive Effect*, 108.

[31] Lippit, *Atomic Light*, 9.

Bruzzi, Stella. *New Documentary.* 2nd ed. New York: Routledge, 2006.

Enterprise. "The Age of Narrowcast Is upon Us: How Streaming Is Changing the Middle East's Media Landscape." February 20, 2022. https://enterprise.press/stories/2022/02/10/the-age-of-narrowcast-is-upon-us-how-streaming-is-changing-the-middle-easts-media-landscape-64729/.

Friedlander, Eli. *Walter Benjamin*. Cambridge, MA: Harvard University Press, 2012.

Habashneh, Khadijeh. *Knights of Cinema: The Story of the Palestine Film Unit*. Cham, Switzerland: Palgrave Macmillan, 2023.

Haugbolle, Sune. "Archival Activists and the Hybrid Archives of the Arab Left." In *The Arab Archive: Mediated Memories and Digital Flows*, edited by Donatella Della Ratta, Kay Dickinson, and Sune Haugbolle, 7–19. Amsterdam: Institute of New Culture Works, 2020.

Hedges, Inez. *World Cinema and Cultural Memory*. New York: Palgrave, 2015.

Lippit, Akira Mizuta. *Atomic Light (Shadow Optics)*. Minneapolis: University of Minnesota Press, 2005.

Marks, Laura U. *Hanan al-Cinema: Affections for the Moving Image*. Cambridge, MA: MIT Press, 2015.

Pad.ma. "10 Theses on the Archive." In *Dissonant Archives: Contemporary Visual Culture and Contested Narratives in the Middle East*, edited by Anthony Downey, 352–63. London: I. B. Taurus, 2015.

Prelinger, Rick. "The Appearance of Archives." In *The YouTube Reader*, edited by Pelle Snickars and Patrick Vonderau, 268–74. Stockholm: National Library of Sweden, 2010.

Russell, Catherine. *Archiveology: Walter Benjamin and Archival Film Practices*. Durham, NC: Duke University Press, 2018.

Sheehan, Edward R. F. "How Kissinger Did It: Step by Step in the Middle East." *Foreign Policy* 22 (1976): 3–70. https://doi.org/10.2307/1148069.

Contributors

Samirah Alkassim writes about Arab cinema and is an experimental documentary filmmaker. She is the co-editor of the Palgrave Studies in Arab Cinema and *Global Horror: Hybridity and Alterity in Transnational Horror Film* (2022). Her publications include the co-authored book *The Cinema of Muhammad Malas* (2018), contributions to *Cinema of the Arab World: Contemporary Directions in Theory and Practice* (2020), the *Historical Dictionary of Middle Eastern Cinema* (2020), *Refocus: The Films of Jocelyne Saab* (2021), *Gaza on Screen* (2023). Her latest documentary, *The Painter of Rust,* is about Jordanian artist Hani Hourani and she is editing *The Palgrave Handbook of Arab Film and Media* (forthcoming 2026).

Alexander Fisher is Director of Graduate Studies and Senior Lecturer in Film Studies in the School of Arts, English and Languages at Queen's University Belfast. His research examines the aesthetic, cultural, and political aspects of international film cultures, with a particular focus on African film. His most recent research concerns digital transformations in international film distribution and culture, and he is co-editor of *World Cinema on Demand: Global Film Cultures in the Era of Online Distribution* (Bloomsbury 2022).

Hadi Gharabaghi is Visiting Assistant Professor of Film, Television, and Media Arts at Fairfield University. Hadi's essays have been published in the edited volume *Cinema of the Arab World: Contemporary Directions in Theory and Practice* (2020), the *Journal of Cinema and Media Studies* (2021), and *The Oxford Handbook of American Documentary* (co-author, 2025). Hadi's co-editing credits include a special issue of the *Journal of e-Media Studies* on the topic of the United States Information Agency (USIA) declassified film and paper-trail collections (2022) and *Governing Genealogies of International Film Education* (2025).

Shakti Jaising is Professor of English and Director of Film Studies at Drew University. Her essays on literary and cinematic narratives pertaining to colonialism, human rights, and the War on Terror appear within publications like *Modern Fiction Studies, Interventions: International Journal of Postcolonial*

Studies, and *Jump Cut: A Review of Contemporary Media*. Her monograph, *Beyond Alterity: Contemporary Indian Fiction and the Neoliberal Script* (2023) was published as part of Liverpool University Press's "Postcolonialism across the Disciplines" series.

Hiju Kim is a PhD Candidate of Italian Studies at the University of California, Los Angeles. She has researched on a wide range of art, literature, film, and media topics, including the Watts Towers and the works of Federico Fellini and Pier Paolo Pasolini. She has worked for curatorial and education departments at cultural institutions, such as the Peggy Guggenheim Collection, the Museum of Contemporary Art in Los Angeles, and the Wende Museum of the Cold War. She is also a writer, editor, and translator for online Korean pop culture media outlets.

Egor Korneev is a PhD candidate in Middle East Studies at the University of Michigan, specializing in mass media and Islam in the Middle East and North Africa (MENA) region. His research centers on how Islamic history is represented on Arab television and through other forms of popular culture, including film, radio, theater, and storytelling. His current project examines the genre of Egyptian historical television drama and its significant role in shaping the image of the premodern Islamic past in modern Arab popular culture.

Daniela Mazur is a post-doctoral researcher at the Postgraduate Program in Communication at Universidade Federal Fluminense (UFF), Brazil, with a fellowship from Fundação de Amparo à Pesquisa do Estado do Rio de Janeiro (FAPERJ). She has a doctorate and master's degree in Communication and a bachelor's in Media Studies from Universidade Federal Fluminense (UFF). She is currently the Director of Culture at the Brazil-Korea Culture Institute. Her current research interests encompass East Asian media studies, hallyu, media counterflows, and television and streaming industries from the Majority World.

Melina Meimaridis is a post-doctoral researcher at the Postgraduate Program in Communication at Universidade Federal Fluminense (UFF), Brazil. Presently, she is developing research on the fictionalization of social institutions in television scripted fiction with a fellowship from Fundação de Amparo à Pesquisa do Estado do Rio de Janeiro (FAPERJ). A member of the Global Internet TV Consortium, her interests also include video streaming services' transnational expansion into Majority World markets, media industries, and television studies.

Wissam Mouawad is Assistant Professor of Media Production and Chair of the Department of Mass Communication at Abu Dhabi University, UAE. A graduate of Paris 1 Panthéon–Sorbonne University, he is also an associate researcher at the ACTE Institute, Sorbonne School of the Arts. His research centers on transnational film production, with a particular focus on Lebanese cinema, exploring co-production dynamics and cultural representation. He has contributed to several academic publications, including edited journal issues, book chapters, and journal articles.

Pamela Nassour holds a PhD in cinema from Saint-Joseph University of Beirut. She has lectured and held leadership positions at multiple universities in Lebanon. An award-winning filmmaker, her short films have received accolades at film festivals worldwide. She has also been the artistic director of the Lebanese Independent Film Festival and regularly writes for the BIMFF online magazine. Pamela Nassour conducts filmmaking workshops between Lebanon and Canada.

Vinicius Navarro is Associate Professor of Film and Media Studies at Emerson College. He is the co-author of *Crafting Truth: Documentary Form and Meaning* (2011) and co-editor of *New Documentaries in Latin America* (2014). His work has appeared in the *Journal of Cinema and Media Studies*, *Film Quarterly*, *Studies in Spanish and Latin American Cinemas*, as well as several edited collections.

Daniel Rios is a post-doctoral researcher at the Postgraduate Program in Communication at Universidade Federal Fluminense (UFF), Brazil, with a fellowship from Fundação de Amparo à Pesquisa do Estado do Rio de Janeiro (FAPERJ). He has a doctorate and master's degree in Communication and a bachelor's in Media Studies from Universidade Federal Fluminense (UFF). His work explores how video streaming services' affordances, business models, and content strategies displace established labor practices in Brazil's video industry. Other areas of interest include media industries, television studies, cultural production, serialized fiction, and audience practices.

Daniel Rudin is an artist-scholar focused on the "Left" and "state," including twentieth-century documentary film and the New Philippine Cinema. His art practice explores dividing lines between documentary and experimental video, while his journalistic work has focused on labor issues. He is currently developing an article on Philippine alternative film, a study of Lino Brocka and Rainer Werner Fassbinder, and a short film about labor in the Philippines. Daniel is a professor at Le Moyne College.

Paul Julian Smith is Distinguished Professor in the Comparative Literature Program at the Graduate Center in City University of New York. He was previously Professor of Spanish at the University of Cambridge, UK. He is the author of twenty-five books and over 100 academic articles. His most recent books are *Multiplatform Media in Mexico: Growth and Change since 2010* (2019); *Mexican Gender, Mexican Genres: Cinema, Television, and Streaming since 2010* (2021); *Reimagining History in Contemporary Spanish Media: Theater, Cinema, Television, Streaming* (2021); and *Y tu mamá también: Film Classic* (Bloomsbury/BFI, 2022).

Valentina Vitali is Professor in Digital Arts at Birmingham City University. Her research explores the intersection of economics, history, and film aesthetics from a comparative perspective. She is a founding member of the *South Asian Cinema and VOD Research Network* and has curated *Contemporary South Asian Films by Women* and *Alia Syed* (Whitechapel Gallery). She has edited a special issue of *BioScope: South Asian Screen Studies* and co-edited *Capital and Popular Cinema*, *Hindi Action Cinema*, and *Theorising National Cinema*. Valentina is currently working on a AHRC-funded project on women's access to creative roles in the UK and Indian film industries.

Bret Vukoder, PhD, is Adjunct Assistant Professor in St. Olaf College's Film and Media Studies program. With Dr. Hadi Gharabaghi, he recently co-edited a special issue, "Motion Picture Legacies of the USIA" (2022), in the *Journal of e-Media Studies*. He is the co-head of the USIA and NARA-WWI projects within Dartmouth College's Media Ecology Project and was a senior consultant on a recent NEH grant, "USIA Moving Images through International Lenses" (2023). His work generally engages USIA and First World War film archives to address how systems of policy, culture, and bureaucracy inform motion picture aesthetics, circulation, and reception.

Index

www.ingramcontent.com/pod-product-compliance
Ingram Content Group UK Ltd.
Pitfield, Milton Keynes, MK11 3LW, UK
UKHW020218290926
15470UKWH00001B/268

9 798765 135860